Scaling
Physiological
Processes

Physiological Ecology
A Series of Monographs, Texts, and Treatises

Series Editor
Harold A. Mooney
Stanford University, Stanford, California

Editorial Board
Fakhri Bazzaz   F. Stuart Chapin   James R. Ehleringer
Robert W. Pearcy   Martyn M. Caldwell   E.-D. Schulze

T. T. KOZLOWSKI (Ed.). Growth and Development of Trees, Volumes I and II, 1971

D. HILLEL (Ed.). Soil and Water: Physical Principles and Processes, 1971

V. B. YOUNGER and C. M. McKELL (Eds.). The Biology and Utilization of Grasses, 1972

J. B. MUDD and T. T. KOZLOWSKI (Eds.). Responses of Plants to Air Pollution, 1975

R. DAUBENMIRE (Ed.). Plant Geography, 1978

J. LEVITT (Ed.). Responses of Plants to Environmental Stresses, 2nd Edition. Volume I: Chilling, Freezing, and High Temperature Stresses, 1980 Volume II: Water, Radiation, Salt, and Other Stresses, 1980

J. A. LARSEN (Ed.). The Boreal Ecosystem, 1980

S. A. GAUTHREAUX, JR. (Ed.). Animal Migration, Orientation, and Navigation, 1981

F. J.VERNBERG and W. B. VERNBERG (Eds.). Functional Adaptations of Marine Organisms, 1981

R. D. DURBIN (Ed.). Toxins in Plant Disease, 1981

C. P.LYMAN, J. S. WILLIS, A. MALAN, and L. C. H. WANG (Eds.). Hibernation and Torpor in Mammals and Birds, 1982

T. T. KOZLOWSKI (Ed.). Flooding and Plant Growth, 1984

E. L. RICE (Ed.). Allelopathy, Second Edition, 1984

M. L. CODY (Ed.). Habitat Selection in Birds, 1985

R. J. HAYNES, K. C. CAMERON, K. M. GOH, and R. R. SHERLOCK (Eds.). Mineral Nitrogen in Plant-Soil System, 1986

T. T. KOZLOWSKI, P. J. KRAMER and S. G. PALLARDY (Eds.). The Physiological Ecology of Woody Plants, 1991

H. A. MOONEY, W. E. WINNER, and E. J. PELL (Eds.). Response of Plants to Multiple Stresses, 1991

List continues at the end of this volume

Scaling Physiological Processes
Leaf to Globe

Edited by

James R. Ehleringer

Department of Biology
University of Utah
Salt Lake City, Utah

Christopher B. Field

Department of Plant Biology
Carnegie Institution of Washington
Stanford, California

Academic Press, Inc.
Harcourt Brace Jovanovich, Publishers
San Diego   New York   Boston
London   Sydney   Tokyo   Toronto

Copyright © 1993 by ACADEMIC PRESS, INC.

All Rights Reserved.
No part of this publication may be reproduced or transmitted in any form or by any
means, electronic or mechanical, including photocopy, recording, or any information
storage and retrieval system, without permission in writing from the publisher.

Academic Press, Inc.
1250 Sixth Avenue, San Diego, California 92101-4311

United Kingdom Edition published by
Academic Press Limited
24–28 Oval Road, London NW1 7DX

Library of Congress Cataloging-in-Publication Data

Scaling physiological processes: leaf to globe / edited by James R. Ehleringer,
Christopher B. Field.
      p. cm. — (Physiological ecology)
    Includes bibliographical references and index.
    ISBN 0-12-233440-X
    1. Plant ecophysiology. I. Ehleringer, J. R. II. Field,
Christopher B. III. Series.
QK905.S33 1992 *1993*
581. 5'01'5118—dc20
                                   92-29452
                                       CIP

PRINTED IN THE UNITED STATES OF AMERICA
92 93 94 95 96 97   QW   9 8 7 6 5 4 3 2 1

Contents

v

Part II
Leaf to Ecosystem Level Integration

4. Scaling Processes between Leaf and Canopy Levels
John M. Norman

5. Scaling Water Vapor and Carbon Dioxide Exchange from Leaves to a Canopy: Rules and Tools
Dennis D. Baldocchi

9. How Ecophysiologists Can Help Scale from Leaves to Landscapes
Richard H. Waring

Part III
Global Constraints and Regional Processes

10. Global Dynamics and Ecosystem Processes: Scaling Up or Scaling Down?
Peter M. Vitousek

11. Observational Strategy for Assessing the Role of Terrestrial Ecosystems in the Global Carbon Cycle: Scaling Down to Regional Levels
Pieter P. Tans

12. Forests in the Global Carbon Balance: From Stand to Region
Paul G. Jarvis and Roddy C. Dewar

13. Prospects for Scaling
Martyn M. Caldwell, Pamela A. Matson, Carol Wessman, and John Gamon

Part IV
Functional Units in Ecology

14. Scaling in Biological Systems: Population and Community Perspectives
Fakhri A. Bazzaz

15. Scaling at the Population Level: Effects of Species Composition and Population Structure
James S. Clark

Part V
Integrating Technologies for Scaling

19. Remote Sensing of Ecological Processes: A Strategy for Developing and Testing Ecological Models Using Spectral Mixture Analysis

Susan L. Ustin, Milton O. Smith, and John B. Adams

20. New Technologies for Physiological Ecology

David S. Schimel

Contributors

Numbers in parentheses indicate the pages on which the authors' contributions begin.

John B. Adams (339), Department of Geological Sciences, University of Washington, Seattle, Washington 98195

Dennis D. Baldocchi (77), Atmospheric Turbulence and Diffusion Division, Air Resources Laboratory, National Oceanic and Atmospheric Administration, Oak Ridge, Tennessee 37831-2456

Fakhri A. Bazzaz (233), Department of Organismic and Evolutionary Biology, Harvard University, Cambridge, Massachusetts 02138

Joseph A. Berry (323), Department of Plant Biology, Carnegie Institution of Washington, Stanford, California 94305

Martyn M. Caldwell (223), Department of Range Science and the Ecology Center, Utah State University, Logan, Utah 84322-5230

F. Stuart Chapin III (287, 313), Department of Integrative Biology, University of California, Berkeley, Berkeley, California 94720

James S. Clark[1] (255), Department of Botany, University of Georgia, Athens, Georgia 30602

Frank Davis (21), Department of Geography, University of California, Santa Barbara, Santa Barbara, California 93106

Todd E. Dawson (313), Department of Ecology and Systematics, Cornell University, Ithaca, New York 14853-2701

Roddy C. Dewar (191), Institute of Terrestrial Ecology, Edinburgh Research Station, Penicuik, Midlothian EH260QB, United Kingdom

James R. Ehleringer (1), Department of Biology, University of Utah, Salt Lake City, Utah 84112

Christopher B. Field (1), Department of Biology, Carnegie Institution of Washington, Stanford, California 94305

John A. Gamon (223), Department of Biology, California State University, Los Angeles, Los Angeles, California 90032

[1] Current address: Department of Botany, Duke University, Durham, North Carolina 27706.

Larry J. Giles (323), Department of Botany, Duke University, Durham, North Carolina 27706

David W. Hilbert (127), Département des Sciences Biologiques, Université du Québec, Montréal, Quebec, Canada H3C 4R1

E. Raymond Hunt, Jr. (141), School of Forestry, University of Montana, Missoula, Montana 59812

Paul G. Jarvis (117, 191), Institute of Ecology and Resources Management, University of Edinburgh, Edinburgh EH9 3JU, United Kingdom

Paul R. Kemp (127), Department of Botany, Duke University, Durham, North Carolina 27708-0340.

Timothy G. F. Kittel[2] (21), Natural Resources Ecology Laboratory, Cooperative Institute for Research in the Atmosphere, Colorado State University, Fort Collins, Colorado 80523

Simon A. Levin (7), Department of Ecology and Evolutionary Biology, Princeton University, Princeton, New Jersey 08544-1003

Pamela A. Matson (223), National Aeronautics and Space Administration, Ames Research Center, Moffett Field, California 94035

John M. Norman (41), Department of Soil Science, University of Washington, Madison, Wisconsin 53706

C. Barry Osmond (323), Research School of Biological Sciences, Australian National University, Canberra 2601, Australia

James F. Reynolds (127), Department of Botany, Duke University, Durham, North Carolina 27706

Stephen W. Running (141), School of Forestry, University of Montana, Missoula, Montana 59812

David S. Schimel[3] (21, 359), Department of Forest and Wood Sciences, Natural Resources Ecology Laboratory, Colorado State University, Fort Collins, Colorado 80523

Milton O. Smith (339), Department of Geological Sciences, University of Washington, Seattle, Washington 98195

Pieter P. Tans (179), Climate Monitoring and Diagnostics Laboratory, National Oceanic and Atmospheric Administration, Boulder, Colorado 80303

Richard B. Thomas (323), Department of Botany, Duke University, Durham, North Carolina 27706

[2] Current address: Climate System Modeling Program, University Corporation for Atmospheric Research, Boulder, Colorado 80307-3000.
[3] Current address: National Center for Atmospheric Research, Boulder, Colorado 80307-3000.

Susan L. Ustin (339), Department of Land, Air, and Water Resources, University of California, Davis, Davis, California 95616

Peter M. Vitousek (169), Department of Biological Sciences, Stanford University, Stanford, California 94305

Richard H. Waring (159), Department of Forest Science, College of Forestry, Oregon State University, Corvallis, Oregon 97331

Carol Wessman (223), Cooperative Institute for Research in Environmental Sciences, University of Colorado, Boulder, Colorado 80309

Dan Yakir (323), Department of Environmental Science and Energy Research, Weizmann Institute of Science, Rehovet 76100, Israel

1

Introduction: Questions of Scale

Christopher B. Field and James R. Ehleringer

I. Scaling from Ecophysiology

Predicting and analyzing the structure and function of ecological systems on large spatial and long temporal scales are research challenges of rare potential but daunting difficulty. The potential derives from both practical need and scientific opportunity. The difficulty reflects the diversity and nonlinearity of ecological responses. This book explores aspects of both the potential and the difficulties, using paradigms and approaches from plant ecophysiology as starting points for capitalizing on opportunities and managing problems.

The traditional focus of plant ecophysiology, understanding how plants cope with often stressful habitats, is organism centered (Mooney *et al.*, 1987a; Mooney, 1991). The questions and approaches focus on diversity in the levels of environmental factors, implications of plant functional diversity for mass and energy exchange, and influences of mass and energy exhange on plant persistence, growth, and reproduction. This organism-centered approach provides a useful framework for predicting the characteristics of organisms likely to be successful in any given habitat and for assessing ecological consequences of physiological mechanisms and morphological characteristics.

In the past, few ecophysiologists emphasized extending these capabilities to problems involving many individuals. However, many of the same individual-level characteristics that determine persistence, growth, and reproduction are primary components of ecosystem-level fluxes of matter and energy, which are, in turn, critical determinants of the biogeochemical cycles of carbon, water, and nutrients. Ecophysiology is, in a sense,

1

preadapted for large-scale problems. This preadaptation is, however, far from complete. Ecophysiology traditionally lacks many of the technical tools for large-scale analyses, and the evolutionary perspective that is so useful at the organism level does not necessarily extend to higher scales.

The clear role of the terrestrial biosphere in global change, including feedbacks on climate (Shukla and Mintz, 1982; Dickenson, 1991), the composition of the atmosphere (Mooney *et al.*, 1987b), and the fate of anthropogenic CO_2 (Tans *et al.*, 1990), generates a critical need for large-scale assessments that are both accurate and generalizable outside the envelope of existing conditions. Because of its focus on the responses of underlying mechanisms to variation in environmental factors, ecophysiology offers the promise of generalization. The accuracy will depend on the effectiveness with which ecophysiological concepts can be integrated with large-scale measurement techniques, global databases, and models from atmospheric sciences, hydrology, biogeochemistry, and population dynamics.

As much as ecophysiology hopefully will contribute new perspectives to large-scale analyses, contributions in the reverse direction are also likely. Global and regional patterns traditionally have provided important stimuli for new hypotheses in ecophysiology. Convergent evolution (Cody and Mooney, 1978) and plant life zones (Woodward, 1987) are clear examples of concepts developed from a geographic perspective. Increasingly quantitative assessments of large-scale patterns are likely to stimulate other advances in ecophysiology. Evidence for the striking generality of the efficiency with which light is used in growth (Goward *et al.*, 1985) already is leading to new research in ecophysiology. The localization of terrestrial sources and sinks of carbon, using global analysis (Tans *et al.*, 1990; Enting and Mansbridge, 1991), almost certainly will lead to intensive ecophysiological studies in the putative source and sink areas.

II. The Art of Scaling

Combining quantitative mechanisms understood precisely at small scales into synthetic assessments appropriate over larger scales of space and time can be a grand expression of scientific confidence, or it can be a sobering warning that information is still missing. Scaling is perhaps most useful between these extremes, when applied as a tool for testing hypotheses and identifying missing components of interpretations.

The need for synthetic assessments based on quantitative mechanisms integrated across scales extends across the sciences. In fields related to ecophysiology, issues of scale are very explicit; the treatment of scale is very sophisticated in landscape ecology (Dale *et al.*, 1989; Turner, 1989), hydrology (McNaughton and Jarvis, 1991), and global change (Rosswall *et al.,* 1988). The chapters in this book take no single approach to scaling. Because they start with the mechanisms underlying the biological processes, the chapters emphasize different aspects of the scaling problem. As a group, they may not present a definitive answer to the general problem of scaling, but they clearly demonstrate that ecophysiology can make major contributions to analysis of ecosystems on large spatial and long temporal scales.

III. Some New Dimensions

This book is a collection of chapters based on presentations and discussions at a meeting in Snowbird, Utah, in December 1990. Some chapters are based on presentations at the workshop that were discussed extensively, and were modified to incorporate concepts and syntheses that emerged from the discussions. For selected topics that are recognized broadly as representing new frontiers, but in which progress will be critically dependent on input from a range of perspectives, the chapters started from discussions at the workshop. The final form of each discussion chapter reflects the enthusiasm of a number of participants and the dedication of one or a few discussion leaders who not only kept the discussions focused, but also built chapters around the concepts covered in the discussions.

The book begins with two chapters that consider conceptual and formal tools for spatial integration. The next two sections address scaling from the two ends of the spatial spectrum: from the bottom up and from the top down. The "bottom-up" chapters develop conceptual frameworks for complex mechanistic models but also assess the quantitative impacts of a number of simplifications. The "top-down" discussions develop general approaches to using global-scale information to constrain smaller scale interpretations.

The fourth section of the book addresses the interface between physiological processes and biological diversity. Two chapters consider scaling of population and community phenomena and two others assess prospects for managing complications of biodiversity by collecting species into functional groups. The three chapters in Part V consider technologies for scaling—stable isotopes, remote sensing, and canopy-flux measurements.

Acknowledgments

The Snowbird meeting was made possible by the support of the Department of Energy, the Electric Power Research Institute, the National Aeronautics and Space Administration, and the National Science Foundation. The staff of the Snowbird resort provided outstanding support, and the Wasatch Mountains provided excellent snow. All the participants in the meeting dove into difficult issues and challenged established disciplinary boundaries with infectious enthusiasm.

References

Cody, M. L., and Mooney, H. A. (1978). Convergence versus nonconvergence in mediterranean-climate ecosystems. *Annu. Rev. Eco. Systemat.* **9**, 265–321.

Dale, V. H., Gardner, R. H., and Turner, M. G. (1989). Predicting across scales: Comments of the guest editors of Landscape Ecology. *Landscape Ecol.* **3**, 147–151.

Dickenson, R. E. (1991). Global change and terrestrial hydrology: A review. *Tellus* **43AB**, 176–181.

Enting, I. G., and Mansbridge, J. V. (1991). Latitudinal distribution of sources and sinks of CO_2: Results of an inversion study. *Tellus* **43B**, 156–170.

Goward, S. N., Tucker, C. J., and Dye, D. G. (1985). North American vegetation patterns observed with the NOAA-7 advanced very high resolution radiometer. *Vegetatio* **64**, 3–14.

McNaughton, K. G., and Jarvis, P. G. (1991). Effects of spatial scale on stomatal control of transpiration. *Agric. For. Meteorol.* **54**, 279–302.

Mooney, H. A. (1991). Plant physiological ecology: Determinants of progress. *Funct. Ecol.* **5**, 127–135.

Mooney, H. A., Pearcy, R. W., and Ehleringer, J. (1987a). Plant physiological ecology today. *BioScience* **37**, 18–20.

Mooney, H. A., Vitousek, P. M., and Matson, P. A. (1987b). Exchange of materials between terrestrial ecosystems and the atmosphere. *Science* **238**, 926–932.

Rosswall, T., Woodmansee, R. G., Risser, P. G. (eds.) (1988). "Scales and Global Change." Wiley, New York.

Shukla, J., and Mintz, Y. (1982). Influence of land-surface evapotranspiration of the earth's climate. *Science* **215**, 1498–1501.

Tans, P. P., Fung, I. Y., and Takahashi, T. (1990). Observational constraints on the global CO_2 budget. *Science* **247**, 1431–1438.

Turner, M. G. (1989). Landscape ecology: The effect of pattern on process. *Annu. Rev. Ecol. Systemat.* **20**, 171–198.

Woodward, F. I. (1987). "Climate and Plant Distribution." Cambridge University Press, Cambridge.

I

Integrating Spatial Patterns

Questions of spatial and temporal scale are unavoidable in biological systems, particularly when one is interested in understanding processes and the implications of interactions among processes. This first section begins with a theoretical consideration of pattern and scaling issues by Levin. In this chapter, he points out that, although there is no correct choice of scale, there may be paradigms or laws that can be used to address the phenomena of interest at higher levels of organization. He provides us with the relevance of such approaches through an examination of spatiotemporal mosaics. Although part of his presentation on patchiness and patch dynamics is for a marine system, he argues that the same principles will apply to terrestrial studies.

Subsequently, Schimel, Davis, and Kittel present an examination of FIFE, a large scale study of ecological processes that spanned leaf-level to landscape-level components. FIFE, First ISLSCP Field Experiment (ISLSCP is International Satellite Land Surface Climatology Project) was an effort to understand ecological and physical processes that regulate gas exchange between the surface and the atmosphere. The project represented a combined effort of different disciplines and approaches (e.g., modeling, remote sensing, geographical information systems), many of which are discussed in later chapters of this volume.

2

Concepts of Scale at the Local Level

Simon A. Levin

I. Introduction

I accepted the writing of this chapter with some uncertainty about what to discuss since I find it hard to separate concepts of scale at the local level from those at any other level. Whereas the importance of such concepts may be manifest differently at different scales, the basic concepts apply across all scales. Thus, I interpret the task as one of relating individual-based mechanisms to patterns that are observed at higher scales.

II. The Ecosystem as an Abstraction

The problem of interrelating processes operating at different scales is a fundamental one in biology and, indeed, in all the sciences. It is the central problem of theoretical biology. The biologist must understand how to relate cells to tissues, tissues to organs, and organs to organisms, being faced in each case with the challenge of relating the behavior of aggregates to the operation of much smaller units. Such problems are equally fundamental in ecology and evolutionary biology, in which individuals are organized into populations, populations into communities, communities into landscapes, and so on. However, two additional complications make these approaches particularly problematic at these higher levels of organization. As one moves up the organization network, the integrity of individual units decreases, and the variability among them increases. These two phenomena are interlinked and are the inevitable consequences of the fact that tight organization is more difficult to maintain as the size of a unit increases: consider the problem of social groupings and the hydrodynamic instability that attends large size. Also, in general, as size increases, so does the interface with the external environment (although not proportionately); hence, so does the exchange with that environment. With miscegenation comes an increase in variability among units, and weaker evolutionary control, thereby compounding the potential for divergence. Indeed, competition and other ecological interactions at lower levels can lead to selection for divergence and to the patterns of diversity that are characteristic of the natural environment.

A consequence of these patterns is that every population and every ecosystem is unique; the use of statistics is much more prevalent in ecology and evolutionary biology than in molecular biology because of an explicit recognition of that variability. As a result the determination of basic laws is much more difficult in ecology than in other fields of biology, and typically must be cast in statistical terms. Indeed, even the definition of the basic unit of study, for example, an ecosystem, involves an arbitrary truncation of the global landscape; arbitrariness similarly arises in the choice of the level of spatial, temporal, or hierarchical detail of interest.

The arbitrariness implicit in the definition of the ecosytem was exposed most clearly by the gradient analyses of Robert Whittaker and his followers (Fig. 2.1), who made clear that the Gleasonian notion of independence in the spatial distributions of species was far more accurate than the Clementsian view of the community or ecosystem as a superorganism, comprising coevolved species whose fates were intertwined ineluctably. The more we understand about the biology of individuals, the more we understand that, even within the genomes of those organisms, there is competition among subunits. Among prokaryotes, the situation is most dramatic: the plasmids that constitute large portions of the (extrachromosomal) genomes of bacteria can be exchanged freely among disparate species; even chromosomal DNA can join that itinerant group. The parasite assemblages of higher organisms similarly are exchanged broadly. As we progress in organizational complexity, we find it more and more difficult to maintain the integrity of the basic unit. Ecosystems are, in general, simply operationally defined; boundaries are chosen for the convenience of the investigator, or according to other externally imposed criteria.

The issue of perceptual scale is, perhaps, even more problematic. The elegant analyses (e.g., Cohen, 1990) of regularities in the organization of trophic webs expose statistical regularities that seem impervious to the level of detail chosen in the description of those webs. According to the

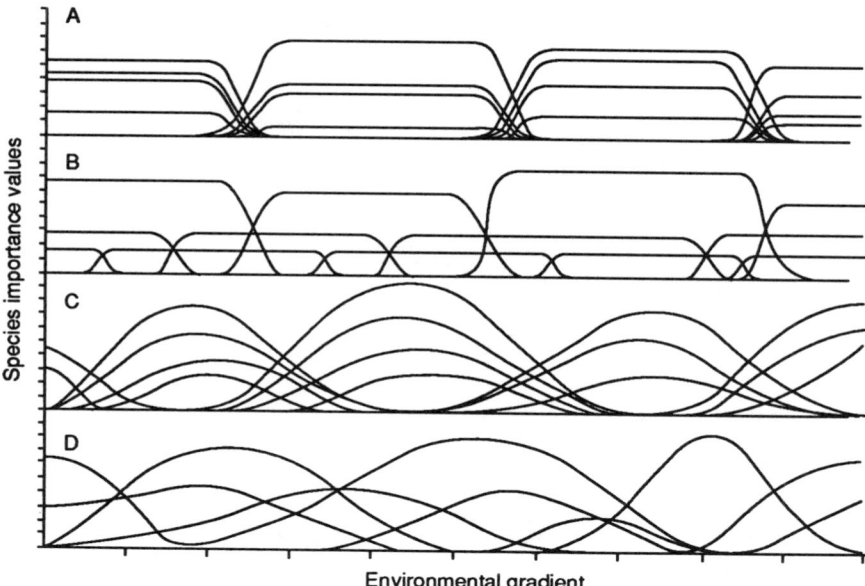

Figure 2.1 Four hypotheses on how species populations might relate to one another along an environmental gradient. Each curve in each part of the future (A–D) represents one species population and the way it might be distributed along the environmental gradient. Figure reprinted, with permission, from Whittaker (1970).

whims of the investigator, a particular bird species, for example, might be given its own category, equal in status to the entire insect world; other taxa might be divided by species, by genus, by feeding habit, or by age class. We have yet to develop the techniques to deal adequately with the interrelationships among such complementary views of the biota, despite several notable efforts (e.g., O'Neill *et al.*, 1986; Cohen, 1990).

In oceanography, perceptual bias has been well recognized. Steele (1978a) has emphasized the limitations placed on the description of any system by the choice of the window through which the investigator views the system (Fig. 2.2). At any scale or range of scales on which one chooses to view a system, a unique view arises: variability, the embodiment of pattern in nature, is a concept that makes sense only with respect to particular scales of space and time, as well as organizational complexity; such relationships may be captured in graphs that relate variability to the spatial and temporal window of choice, as in the Stommel diagrams for oceanography (Fig. 2.3).

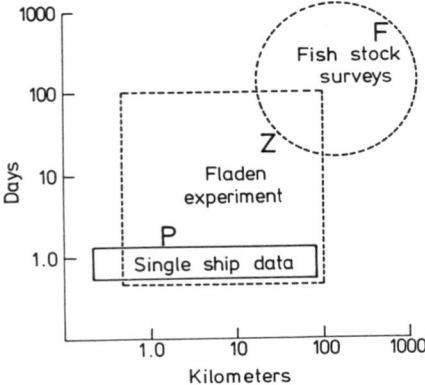

Figure 2.2 An indication of the space and time scales covered by various types of sampling program. Figure reprinted, with permission, from Steele (1978a).

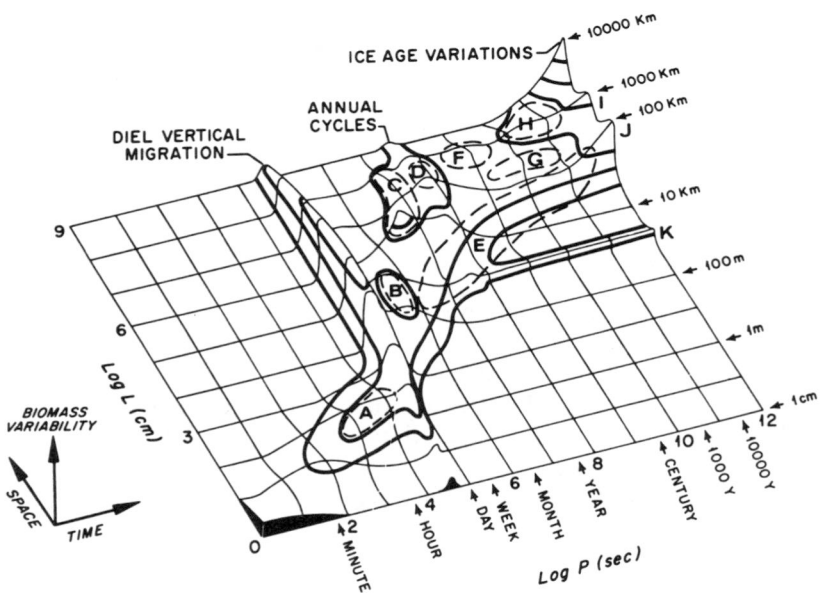

Figure 2.3 The Stommel Diagram, a conceptual model of the time–space scales of zooplankton biomass variability and the factors contributing to these scales. I, J, and K are bands centered about 1000s, 100s, and 10s of kilometers in space scales, with time variations between weeks and geological time scales. A, "Micro" patches; B, swarms; C, upwelling; D, eddies and rings; E, island effects; F, "El Niño" type events; G, small ocean basins; H, biogeographic provinces; I, currents and oceanic fronts (length); J, currents (width) and K, oceanic fronts (width). Figure reprinted, with permission, from Haury et al. (1978).

III. There Is No Correct Scale, but There May Be Scaling Laws

The realization that the choice of scale affects description is not the intellectual property of ecologists. Indeed, it relates to one of the most fundamental paradoxes in physics: increased precision regarding the spatial localization of a measurement carries with it increased uncertainty regarding the measurement (Heisenberg, 1932). The dependence of any description on the scale of measurement is one of the cornerstones of the theory of fractals (Mandelbrot, 1983), which has had a pervasive influence on all the sciences. The most familiar and striking example of this principle is the dependence of the measurement of coastline or frontier on the scale of measurement (Richardson, 1961); the continuous change in these measurements as the length of the measuring stick is altered is a dramatic illustration of the fact that even measurements that we might be tempted to take for granted, for example, the perimeter of a country, hold no meaning at all without reference to a scale of measurement. Further, there is no correct scale of measurement; rather, there is as much essential information in how measurements change with scale (the slope of the graph of border length versus scale, for example) as there is in the absolute length on any particular scale.

The other major cornerstone of the theory of fractals, and perhaps the more surprising one, emerges in the elucidation of how such measurements do change with scale. Remarkably, in a wide variety of cases, change is approximately linear over very broad scales, providing scaling laws that can be used to relate the descriptions of the system on disparate scales. As the physicist Kenneth Wilson noted in his Nobel Prize acceptance speech (1983), a relationship exists with the self-similarity seen in critical phenomena in physics, for which his renormalization group methods proved so powerful. Unfortunately, the relationship remains, as Wilson noted, "murky."

What are the implications of such observations for ecology? Clearly, the reliance of description on scale is a problem as fundamental to ecological phenomena as it is to geomorphic features. Are there similar laws for scaling ecological processes, and can we discover them? What are the limits of those scaling laws? Attention to self-similarity over broad ranges of scales, as expressed in linear relationships such as those just discussed, should not obscure the fact that those relationships cannot, in general, hold over all scales. In critical phenomena in phase transitions, for example, such self-similarity will hold, roughly, at scales less than the correlation length of the system, beyond which a different sort of scaling law will hold. Similar conclusions will apply to ecological phenomena; we simply must discover what they are.

IV. Relevance to Ecological Problems

The most striking parallel to these phenomena in ecology is, perhaps, in the spatiotemporal mosaics that characterize most ecological systems. The relationship of variability to scale, in particular, the tendency for variability and uncertainty to increase, in otherwise homogeneous systems, with the spatial localization of the measurement, is similar in terms of importance with the identical dilemma in physics. In forests, grasslands, intertidal zones, and elsewhere, spatially localized and essentially random disturbances interrupt orderly processes that would otherwise drive the system uniformly toward relatively monotonous end-states. The result is that ecological systems are patchy on virtually every level of space and time (see Steele, 1978b); the elucidation of that patchiness (i.e., variability) and its determinants is one of the fundamental challenges of ecosystem theory (Levin, 1989).

Patchiness has fundamental biological implications. The structure of communities, indeed, the survival of species, is determined by the patterns of spatiotemporal variability (especially fragmentation) in resources, be they food or space. Biogeochemical cycles depend critically on these patterns of internal heterogeneity and on the mosaic structure of ecosystems (Bormann and Likens, 1979). Not surprisingly, therefore, such variability is one of the strongest selective pressures shaping the life histories of species that inhabit these ecosystems. Dispersal, dormancy, and foraging strategies are only a few among the essential modes of evolutionary response to such variability. By averaging over space and time, a genome buffers its bearers against environmental fluctuations, effectively changing the perceptual scale and the actual variability experienced; such redistribution mechanisms also will alter the realized densities of individuals across environments, thereby modifying (to the extent that interspecific or even intraspecific mechanisms are important) the variability experienced by other organisms. Thus, we have the additional complication that environmental variability is not an absolute, even on a particular scale; rather, it is a property of the interaction of the biota and the environment, in terms of both a real effect and a perceptual one. Every organism, and every aggregate of organisms (e.g., species), reads the spatiotemporal fluctuations of the environment uniquely and affects it uniquely.

One of the most important manifestations of intraspecific variation in the way the environment is perceived is in the differing perspectives of ecologists and general circulation modelers, although no genetic basis for these differences has yet been suggested. General circulation models operate on grids whose smallest elements are hundreds of kilometers on a side (Fig. 2.4), whereas ecological investigations usually operate on a

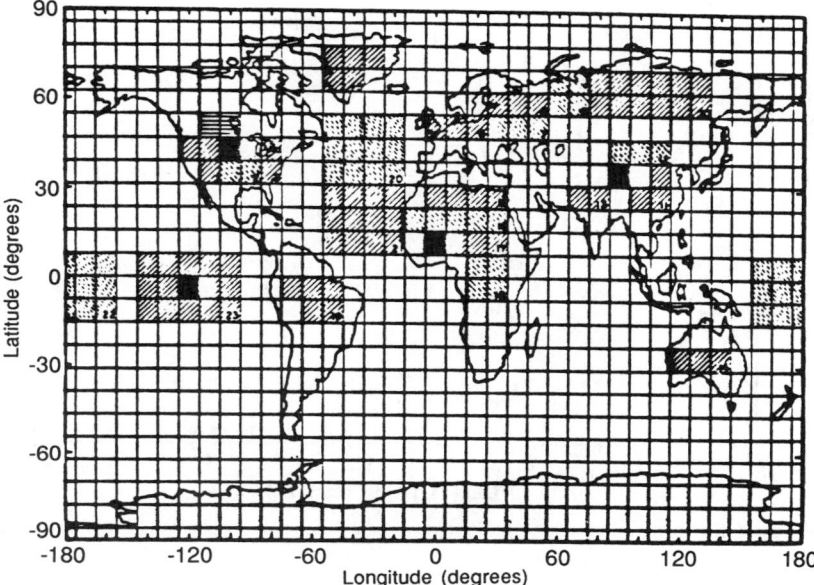

Figure 2.4 Global grid for climate model. Figure reprinted, with permission, from Hansen *et al.* (1987).

scale only a few meters on a side (Fig. 2.5). Finding ways to translate information among these scales and intermediate ones is one of the fundamental challenges in applied ecology.

V. Theories and Bases for Scaling

A variety of tools are available for scaling, involving a combination of correlation, extrapolation, and modeling, all designed to relate patterns across wide ranges of scale. For short-term or small-scale prediction, direct extrapolation of observed trends may be the best technique, but application of such methods can give no hint about when the method will break down or about how patterns will change beyond already observed ranges or in response to novel environmental changes. This limitation has been ignored and models have been extended beyond their range of validity inappropriately in a plethora of examples from applied ecology (see, e.g., Levin, 1979). Thus, the firmest basis for scaling involves the development of an understanding of the mechanisms determining and

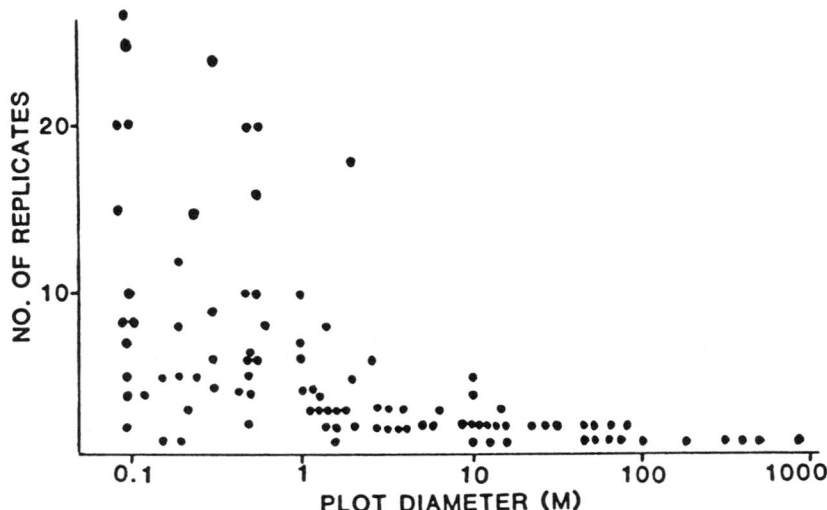

Figure 2.5 Size and replication in experimental community ecology. Each data point is from a different published paper in *Ecology* between Janaury 1980 and August 1986. Figure reprinted, with permission, from Kareiva and Anderson (1988).

governing patterns and processes. This can be achieved only through an integrated theoretical and empirical approach.

Although it is well understood that correlations are no substitute for mechanistic understanding of relationships (e.g., Lehman, 1986), correlations can play an invaluable role in suggesting candidate mechanisms for investigation. The first approach to the study of any system should involve an examination of the scales of variation of key variables and a separation of those variables into ones that change across scales similar enough that there is some potential for interaction. Consider, for example, the spectral relationships exhibited in Fig. 2.6 for the spatial variation in temperature, fluorescence, and krill in the Southern Ocean. The concordance of distributions over broad scales suggests that physical factors cannot be rejected as adequate for determining the broad-scale distributions of both phytoplankton and zooplankton, but that, on finer scales, alternative explanations are needed for the distribution of krill. Biological mechanisms involving the swimming and aggregation behavior of krill are the most likely explanations for the fine-scale patchiness of krill, although that conclusion cannot be derived from Fig. 2.6. Such correlations therefore have stimulated us (S. Levin, T. Powell, A. Okubo, D. Grünbaum, and E. Hofmann) to propose and initiate a two-level modeling effort in which the fluid dynamics of the ocean determine the move-

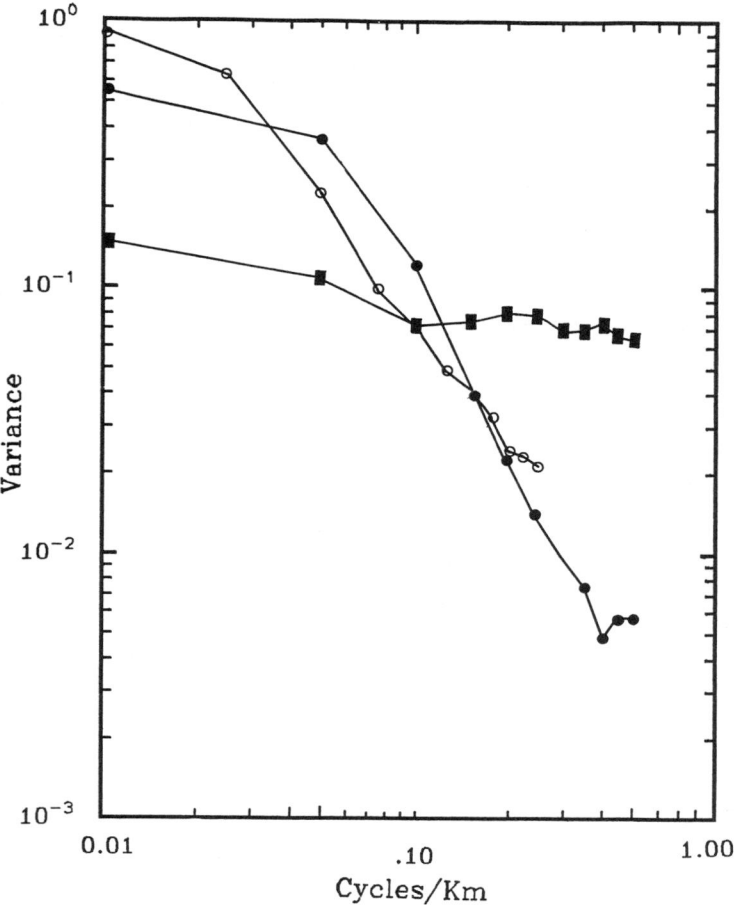

Figure 2.6   Mean spectral plots for krill (■), *in vivo* fluorescence (●), and temperature (○). Figure reprinted, with permission, from Weber *et al.* (1986).

ment of large patches of krill, within each of which an individual-based model of krill swimming behavior must be implemented.

Modeling efforts also face similar limitations. Relating pattern to process is, as stated earlier, the fundamental challenge of theoretical biology; understanding spatial pattern formation, in areas ranging from developmental biology to ecology, has been one of the most active and productive areas of research. However, although a number of instructive generalities, for example, involving the interplay between short-range activation and long-range inhibition, have emerged, these are not specific enough to discriminate among a wide variety of candidate mechanisms. The general lesson is that, for any set of patterns, there almost certainly will be a number of feasible mechanisms that could give rise to those patterns. Investigation of models can help reject mechanisms, produce a slate of candidates for further investigation, and guide the empirical investigations that are needed to distinguish among candidates; but they do not suffice by themselves. Many efforts in theoretical biology have failed because this fact was forgotten.

There are several approaches to building mechanistic models. One of the most productive, and most satisfying, is the individual-based approach, in which one begins with the factors impinging on an individual, develops a model for the dynamics of that individual, and uses that as a basis for understanding the behavior of aggregates of such individuals. For example, for the spatial dynamics of krill, Dan Grünbaum has developed a model that begins by considering the forces and factors impinging on an individual animal, including the influences of the positions of other animals. A stochastic model is developed and analyzed. From this Lagrangian approach, in which one can account for the movement of each animal in response to others, one can proceed to models for the statistical behavior of aggregates, and ultimately to Eulerian models, in which the locations of individuals are replaced by density functions for the distribution of animals within particular volume elements. Approaches of this sort are extensions of the more familiar and highly successful application of diffusional models to the spread of propagules and populations (see, e.g., Levin, 1976; Okubo, 1980). The difference is that, in the simple diffusion approach, individuals are assumed to move independently of one another; thus, the aggregate behavior is simply the sum of individual behaviors. Aggregation (swarming or schooling) models are much more complicated because the movements of individuals are correlated.

In landscape models, two approaches are possible, again paralleling the Lagrangian and Eulerian descriptions of fluid dynamics. In the vector-based approach (e.g., Pacala, 1986), the basic units are individuals, and one takes account of distances to other individuals, for example, to determine competitive effects. In raster-based models, in contrast, the basic unit is a piece of the landscape; these units interact with one another through the exchange of individuals or materials, by shading, and so on. Such models can be used to investigate a wide variety of problems, ranging from ones in conservation biology to ones in community and ecosystem dynamics, including basic theoretical questions as well as applied issues such as the effects of ozone or climate change on tree species abundance, or the spread of disturbances.

In the basic raster approach, each cell is treated as a homogeneous unit, with a specified mix of species. A mosaic of such cells is arrayed on a grid, providing a model of the landscape (Fig. 2.7) (Moloney *et al.*, 1992). A local vegetation simulator imitates local growth and competition, using functional relationships that may be derived and parameterized from laboratory or field measurements. Interactions among cells occur through dispersal and nearest-neighbor interactions; additional spatial and temporal correlations are introduced by localized disturbances, underlying environmental gradients, or other external forcing.

Modeling programs of this sort provide an ideal tool for the investigation of scaling relationships. Statistical tools such as semivariograms, spectral plots, or nested evaluations of variance can be used to quantify details of spatial and temporal pattern and to study how variance and pattern change from one scale to another. Such studies can be carried out, similarly, albeit with more difficulty, on real data; in the latter case, however, the problem is to sort out which of the manifold influences in nature is responsible for determining pattern. In model systems, factors can be altered individually or in concert; use of the model as an experimental tool in this way provides invaluable information concerning scaling relationships and the ways in which information is transferred up

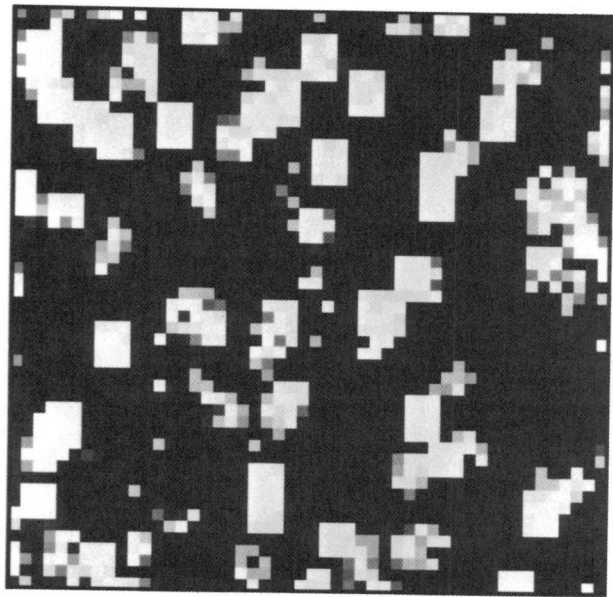

Figure 2.7 A portion of the grassland landscape, from a stimulation of the dynamics of the annual plant *Plantago*. Figure reprinted, with permission, from Moloney *et al.* (1992).

and down the spatial hierarchy. Of course, comparison of model output with data from particular systems also provides a powerful tool for refining the model or for achieving effective experimental design.

VI. Program for Research on Scaling in Terrestrial Systems

The success of coupled correlative and modeling efforts, of the type just described, in marine systems provides strong support for the application of similar methods to studying the terrestrial landscape. The starting point for such an approach must be embedded in the data, both remotely sensed and collected from field studies. Spatial pattern analyses of various types, coupled with knowledge of the life history characteristics of component species, allow an association of elements that are changing on similar time and space scales. Such examination of scales must be the first step in any modeling approach (Denman and Powell, 1984). The development of individual plant models, starting from leaf-level responses to environmental factors, and details of dispersal, establishment, growth, and competition provide the basis for either vector- or raster-based models of interplant interaction. These models can be used to generate patterns that can be compared with actual patterns in nature, in a coupled empirical and theoretical program that can provide guidance for modification in model structure and experimental design. More powerfully, the model can be used as an experimental tool to identify critical parameters and experiments, to develop approaches to simplification and aggregation, and to study how information is transferred across scales. The potential for such models in landscape studies is enormous.

Acknowledgments

It is a pleasure to acknowledge the support of the National Science Foundation under Grant BSR-8806202, the Department of Energy under Grant DE-FG02-90ER60933, and the National Aeronautics and Space Administration under Grant NAGW-2088. Additional support was provided by the U.S. Environmental Protection Agency, through Cooperative Agreement CR-812685-03 with Cornell University, and by Hatch Award NYC-183430 and McIntire-Stennis Award NYC-183550. The views expressed herein are the responsibility of the author alone, and not of the sponsoring agencies or Cornell University.

References

Bormann, F. H., and Likens, G. E. (1979). "Pattern and Process in a Forested Ecosystem." Springer-Verlag, New York.
Cohen, J. E., Briand, F., and Newman, C. M. (1990). Community Food Webs: Biomathematics, Vol. 20. Springer-Verlag, Berlin.

Denman, K. L., and Powell, T. M. (1984). Effects of physical processes on planktonic ecosystems in the coastal ocean. *Oceanogr. Mar. Biol. Annu. Rev.* **22**, 125–168.

Hansen, J., Fung, I., Lacis, A., Lebedeff, S., Rind, D., Ruedy, R., and Russell, G. (1987). Prediction of near-term climate evolution: What can we tell decision-makers now? *In* "Preparing for Climate Change," pp. 35–47. Proc. of the First North American Conference on Preparing for Climate Changes, October 27–29. Government Institutes, Washington, D.C.

Haury, L. R., McGowan, J. A., and Wiebe, P. H. (1978). Patterns and processes in the time space scales of plakton distributions. *In* "Spatial Pattern in Plankton Communities" (J. H. Steele, ed.), pp. 277–327. NATO Conference Series, Series IV, Marine Sciences, Vol. 3. Plenum Press, New York.

Heisenberg, W. (1932). Nobel Prize in Physics Award Address. Nobel Foundation and Elsevier.

Kareiva, P. M., and Andersen, M. (1988). Spatial aspects of species interactions: The wedding of models and experiments. *In* "Lecture Notes in Biomathematics" (A. Hastings, ed.), Vol. 77, pp. 35–50. Springer-Verlag, Berlin.

Lehman, J. T. (1986). The goal of understanding in limnology. *Limnolog. Oceanogr.* **31**, 1160–1166.

Levin, S. A. (1976). Population dynamic models in heterogeneous environments. *Annu. Rev. Ecol. Systematics* **7**, 287–311.

Levin, S. A. (1979). "The Concept of Compensatory Mortality in Relation to Impacts of Power Plants on Fish Populations." Written testimony prepared for the U.S. Environmental Protection Agency.

Levin, S. A. (1989). Challenges in the development of a theory of community and ecosystem structure and function. *In* "Perspectives in Ecological Theory," (J. Roughgarden, R. M. May, and S. A. Levin, eds.), pp. 242–255. Princeton University Press, Princeton, New Jersey.

Mandelbrot, B. B. (1983). "The Fractal Geometry of Nature." Freeman, San Francisco.

Moloney, K. A., Levin, S. A., Chiariello, N. R., and Buttel, L. (1992). Pattern and scale in a serpentine grassland. *Theoret. Popul. Biol.* (in press).

O'Neill, R. V., DeAngelis, D. L., Waide, J. B., and Allen, T. F. H. (1986). "A Hierarchical Concept of Ecosystems:" Monographs in Population Biology, Vol. 23. Princeton University Press, Princeton, New Jersey.

Okubo, A. (1980). "Diffusion and Ecological Problems: Mathematical Models." Biomathematics, Vol. 10. Springer-Verlag, New York.

Pacala, S. W. (1986). Neighborhood models of plant population dynamics. 4. Single-species and multispecies models of annuals with dormant seeds. *Am. Nat.* **128**, 859–878.

Richardson, L. F. (1961). The problem of contiguity: An appendix of statistics of deadly quarrels. *Gen. Syst. Yearbook* **6**, 139–187.

Steele, J. H. (1978a). Some comments on plankton patches. *In* "Spatial Pattern in Plankton Communities," (J. H. Steele, ed.), pp. 1–20. NATO Conference Series, Series IV, Marine Sciences, Vol. 3. Plenum Press, New York.

Steele, J. H. (ed.) (1978b). "Spatial Pattern in Plankton Communities." NATO Conference Series, Series IV, Marine Sciences, Vol. 3. Plenum Press, New York.

Weber, L. H., El-Sayed, S. Z., and Hampton, I. (1986). The variance spectra of phytoplankton, krill and water temperature in the Antarctic Ocean south of Africa. *Deep-Sea Res.* **33**,(10), 1327–1343.

Whittaker, R. H. (1970). "Communities and Ecosystems." Macmillan, New York.

Wilson, K. (1983). The renormalization group and critical phenomena. Nobel Lecture, December 8, 1982. *Rev. Mod. Phys.* **55**, 583–600.

3

Spatial Information for Extrapolation of Canopy Processes: Examples from FIFE

David S. Schimel, Frank W. Davis, and Timothy G. F. Kittel

I. Introduction

That biosphere–atmosphere exchanges of water and energy are significant determinants of regional climates is increasingly clear (Avissar and Pielke, 1989; Shukla *et al.*, 1990). In addition, the carbon balance of ecosystems is coupled closely to climate through atmospheric CO_2 and trace gas concentrations. These interactions present a series of difficult methodological challenges to ecology because, although the processes governing matter and energy exchange on a local basis are often known, exchanges aggregated over kilometers (for water and energy) or globally (for long-lived trace species) are needed to investigate coupled atmosphere–ecosystem dynamics. Because the rates and magnitudes of atmosphere–ecosystem exchanges vary dramatically in time and space, an understanding of the principles governing exchange is insufficient for prediction. Rather, this understanding of principles must be coupled to a knowledge of the geography of controls and constraints to answer scientific questions about atmosphere–ecosystem interaction (Matson *et al.*, 1989; Burke *et al.*, 1990; Costanza *et al.*, 1990; Esser, 1990; Schimel *et al.*, 1990). This coupling is being achieved through rapid advances in computerized process models, satellite observations for characterizing earth surfaces, and geographic information systems (GIS) for integrated spatial data analysis.

From 1987 to 1989, a large interdisciplinary project studied techniques for extrapolation of ecological and physical processes regulating exchange with the atmosphere. This project—the First ISLSCP (International Satellite Land Surface Climatology Project) Field Experiment (FIFE)—set out to develop systematically a landscape-based approach to extrapolation that would use advanced measurement techniques, remote sensing, and GISs to measure important atmosphere–biosphere exchanges of matter and energy (Sellers *et al.*, 1988; Davis *et al.*, 1992). The experiment studied a 256-km^2 area in the Flint Hills of eastern Kansas, a hilly and heterogeneous area of mixed rangeland, agricultural lands, and intermittent woodlands (Schimel *et al.*, 1991).

In this chapter, we review the use of geographic information in designing the FIFE experiment, the implementation of a GIS for analysis and extrapolation of ground and image data, and the implications of FIFE results for modeling exchange between the atmosphere and terrestrial ecosystems. Additionally, we discuss the intellectual contributions to biology resulting from the study of the atmosphere and ecosystems as parts of a coupled system.

II. Experiment Overview

The FIFE experiment included field measurements on a number of scales. Airborne sensors measured carbon dioxide and sensible and latent heat flux at low altitudes. Flight patterns were laid out to determine average flux from the entire site and spatial variability in that flux (at a scale of about 1 km) (Desjardins *et al.*, 1990). Ground-based measurements were of three types. First, measurements of energy balance were made using eddy correlation or Bowen ratio techniques at a number of sites (Fowler and Duyzer, 1989; Smith *et al.*, 1992). Some eddy correlation measurements of CO_2 flux were also made (Verma *et al.*, 1990). These flux sites served additionally for measurements of phytomass, soil moisture, spectral reflectance, and other properties (Sellers *et al.*, 1990; Verma *et al.*, 1990; Walter-Shea *et al.*, 1990). Next, systematic sampling of topographic features was used to measure finer variability in vegetation, trace gas fluxes, soil moisture, and rainfall (Wood, 1990; Schimel *et al.*, 1991; Turner *et al.*, 1992). A number of measurements of leaf gas exchange (Schimel *et al.*, 1991) and soil respiration (Norman *et al.*, 1990) were made using field-portable gas exchange systems to aid in quantifying the source and sink contributions to net CO_2 exchange as measured using eddy correlation. Finally, detailed experiments using small plot techniques addressed specific controls over surface exchange and radiometric properties (Turner *et al.* 1992).

These studies of matter and energy exchange were designed to develop methodology for the systematic calculation of fluxes for the entire FIFE site. Thus, the location of the ground measurements and their distribution within the site was a critical aspect of the experiment. Because of the need for extrapolation, geographic aspects of the experimental design were crucial. We dealt with spatial variation using a two-step approach. First, we classified the site based on our a priori understanding of the sources of landscape variability in atmosphere–ecosystem exchange. This, a priori stratification was carried out at the outset of the project using preliminary topographic data (referred to as the "initial implementation") and later reimplemented with more complete digital geographic data ("DEM-based implementation"). Subsequently, we developed an alternate, less subjective stratification based on regression-tree analysis of remotely sensed data.

III. A Priori Stratification

A. Rationale

The FIFE experiment in the Flint Hills is a region of dissected uplands with dendritic drainages. The geology consists of alternating layers of limestone and shale. The limestone layers produce steep slopes with shallow soils, whereas the shale layers weather to shallower slopes and deeper soils. The interbedding of these materials produces a stairstep-like topography in which total relief ranges from 50 to 75 m. The soils range from moderately to very deep silt loams in lowlands and on lower hillslopes. Shallow stony soils form on the steepest slopes. Upland soils are shallow and rocky, except for some areas with a deep loess cap. A false color image of the experimental area produced using the airborne multispectral AVIRIS sensor (Plate 1) shows the spatial variability due to topography and management.

The vegetation is dominated by the C_4 grasses *Andropogon gerardii*, *A. scoparius*, and *Sorghastrum nutans*. Woody invasion by red cedar (*Juniperus virginiana*), smooth sumac (*Rhus glabra*), and American elm (*Ulmus americana*) occurs in unburned areas. Oak forests (*Quercus* spp.) occupy the lower portions of some drainages and steep north-facing slopes. Roughly a fourth of the FIFE site is occupied by the Konza Prairie Long-Term Ecological Research (LTER) site, which is subject to a range of controlled burning practices. The remainder of the FIFE site is occupied by rangeland, most of which is burned every year or two. Some inclusions of cereal cultivation occur in the larger valleys.

At the outset of the FIFE experiment, we recognized that diversity in topography, range management, or vegetation type would cause spatial variation in rates of atmosphere–ecosystem exchange. Topography and fire management were the principle controls over biophysics that could be mapped. Topography exerts influence over biophysical fluxes through several mechanisms. These include slope and aspect controls over incoming and net solar radiation (Dubayah *et al.*, 1990) and downslope movement of water (Wood, 1990). In addition, soil depth and texture vary with slope position as a consequence of variation in both bedrock geology and sediment transport (Schimel *et al.*, 1991). Grazing intensity is also an important control, but its spatial distribution was less readily documented. Detailed rationale for the a priori stratification and its application are described in Davis *et al.* (1992).

B. Initial Implementation

To account for the hypothesized controls over surface climate and biophysical exchange, we identified 14 strata, including burned and unburned prairie on different topographic positions ("hillslope zones"), cropland, and wooded areas (Fig. 3.1A). In the initial implementation, these strata were mapped using preliminary digital elevation data and aerial photographs; photographic information was required because only crude land-use data were available outside of the Konza Prairie LTER site. Hillslope zones (e.g., uplands, moderate and steep slopes, and lowlands) were chosen to capture the variance in plant productivity and soil moisture along topographic gradients that was observed during reconnaissance and later verified with field sampling (Schimel *et al.* 1991).

The initial implementation of the a priori stratification was used to allocate primary flux measurement sites over the landscape. We recognized at the outset that we had insufficient resources to sample all the strata adequately; therefore we attemped to allocate at least two stations to each of the most extensive surface types. In the end, a disproportionate number of stations was allocated to upland sites because of their micrometeorological attributes with respect to flux station sitting criteria. Moderate and steep slopes were undersampled because of the difficulty of making flux measurements on slopes. In addition, systematic sampling of biomass, light interception, and leaf gas exchange was designed to capture effects of topography and management treatments. These measurements allowed inferences from the small number of core measurement sites to be tested using a larger number of less subjectively chosen sites (Schimel *et al.*, 1991).

The site-stratification approach chosen was based on our best a priori understanding of the controls over biophysical exchange and their distribution within the landscape. Random location of ground samples was rejected for three reasons. First, we believed that we had considerable insight into landscape variation in exchange rates from prior studies on

A

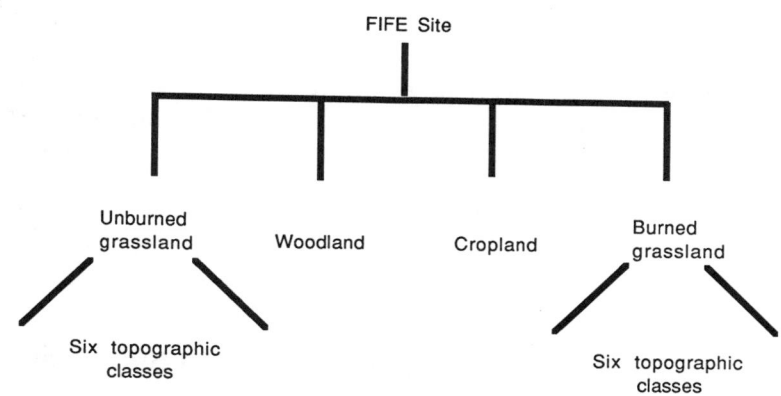

B

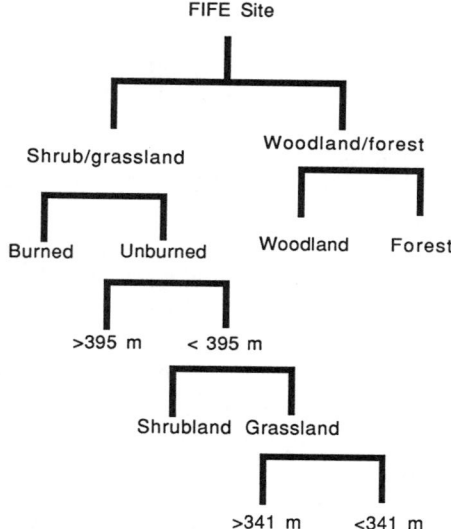

Figure 3.1 (A) Schematic of the initial stratification. Topographic classes, not shown, are uplands, valley bottoms, moderate slopes, steep north slopes, steep south slopes, and steep east or west slopes (six in all). (B) The regression-tree stratification.

variation in productivity, standing biomass, and species composition due to topography, burning, and grazing (Knapp, 1984; Ojima, 1987; Hobbs *et al.*, 1991). Second, we felt that random allocation of measurement sites would result in high uncertainty due to the low number of samples relative to variability. Also, it was crucial to have in place a strategy to ensure coverage of the rugged and dissected portions of the site where flux measurements were difficult because site criteria for micrometeorological measurements favored collection in uncomprised situations, that is, sites level with a long fetch.

IV. Digital Elevation Model-Based a Priori Stratification

A. Methods

The initial implementation of the a priori stratification was not based on a quantitative landscape analysis. Subsequently, we used a digital elevation model (DEM) and other geographic data compiled from remotely sensed images and traditional maps to enhance the objectivity of the stratification based on topography and burning. Topographic variables were computed from a 30-m horizontal resolution DEM obtained by digitizing contours of U.S. Geological Survey 7.5-minute topographic quadrangles and interpolating elevations to a regular grid. We specified six topographic classes: lowlands, moderate slopes (3–7°), steep slopes with north (N), south (S), or east/west (E/W) facing aspects, and uplands (Fig. 3.1A). To map these strata, we combined three parameters from the DEM to tag each 30-m pixel with its topographic class (Plate 2A). The parameters were (1) slope angle and aspect computed from local derivative operations, (2) absolute elevation, and (3) drainage area above the pixel computed using a water-routing algorithm. For example, uplands were defined as having slopes less than 3° and elevations above 420 m or having drainage basins of less than 1800 m^2. Slopes over 7° were classified as steep and were further identified as being N, S, or E/W facing.

The stratification also identified burned and unburned areas. These were mapped by interpretation of aerial photographs obtained on May 31, 1987; subsequently, the results were digitized. Croplands were also identified by photointerpretation. Woody vegetation was mapped with multitemporal SPOT images using a classification procedure and subsequent manual interpretation by co-registration of image classes to aerial photography. In the final analysis, woody areas covered 7% of the site and cropland 3%. In 1987, 43% of the site was burned; this value varies considerably from year to year. The final classification included 14 classes in a hierarchical stratification (Fig. 3.1A, Plate 2A).

B. Analysis

The DEM stratification was intended as a predictor of spatial variability in controls of atmosphere–surface exchange across the tallgrass landscape. We tested the stratification using a one-way analysis of variance of biological and micrometeorological data collected at the flux measurement sites and of satellite-derived vegetation variables (Davis *et al.*, 1992). Because little or no ground data were collected for crop and woodland cover types, these areas were not fully included in the analysis. For satellite estimates of plant-related parameters, we used the LANDSAT Thematic Mapper (TM) greenness vegetation index (GVI), a principal component transform of the original spectral data. This index is a powerful observation tool because it is highly correlated with grassland biophysical attributes, such as biomass and LAI, and because it is relatively insensitive to illumination geometry and background color (Davis *et al.* 1990). The analysis of variance showed that while the stratification significantly reduced the variance of estimated plant biomass, GVI, and physical parameters (e.g., Bowen ratio, soil moisture) over the site, significant unexplained variance remained (Davis *et al.* 1992).

With completion of this more quantitative stratification, several problems became apparent. First, for 8 of the 28 measurement stations, the initial-implementation stratum assignments based on field observations were inconsistent with the DEM-based stratification. Most disagreements were in assignment of stations to one of the three hillslope classes, reflecting the different criteria used by the two implementations for class membership. Observers in the field assigned stations to classes based on subjective definition of lower, middle, and upper slopes, while digital classes were defined by the objective criteria described previously. Based on the analysis of variance results, the DEM appeared to produce better topographic classes than did subjective observations. Second, there was evidence that the relatively small sample of flux measurement sites was not representative of the site as a whole. GVI—a correlate of biomass—was highest on slopes and lowest in bottomlands when analyzed for measurement stations only. This is in conflict with both systematic field sampling of biomass (Schimel *et al.*, 1991) and a joint random sampling of 300 DEM and GVI 30-m pixels. This latter analysis showed that bottomlands had the highest GVI values. The disagreement between the flux station analysis and the more spatially intensive field results suggest either misassignment of flux stations to strata or a biased sampling in the location of stations within each stratum.

A final source of variance not captured by the stratification was grazing. Grazing produces considerable heterogeneity in plant cover within our strata (Hobbs *et al.*, 1991); however, we had no spatial data on the occurrence or intensity of grazing. The influence of grazing as a source of error is suggested by the observation that the stratification performed better when restricted to the ungrazed Konza Prairie portion of the FIFE site (Davis *et al.*, 1992).

V. Regression-Tree Stratification

A. Methods

We developed the a priori spatial stratification and implemented it as a map and table of weights for area averaging to support the original strategy for spatial extrapolation. That strategy was based on locating some number of measurement stations in each stratum proportional to the area of that stratum. Although it was essential to the rigor of the experiment that this strategy be carried out, we also explored alternative stratifications. We hypothesized that multitemporal satellite data could be analyzed in conjunction with digital terrain data to identify appropriate terrain variables for a biophysical classification of the site.

Specifically, we used regression-tree analysis in which a continuous dependent variable is partitioned hierarchically based on a set of categorical or continuous variables (Breiman *et al.*, 1984; Davis *et al.*, 1990). In the case of continuous independent variables, the splits are made at threshold values. For categorical variables, splits are made based on membership in a category, for example, burned or unburned. Whether a split is based on categorical or continuous data, the split chosen is that which produces the greatest reduction in mean squared error of the dependent variable. An excessively large tree is created and then pruned back by removing branches with low predictive value. The tree is developed with a random selection of half of the data and then tested with the other half; this procedure is then reversed (cross-validation; Mosteller and Tukey, 1977).

The goal of the regression-tree analysis was to provide an objective stratification that would select strata based on measured variability over the entire site, in contrast with the informed but subjective choice of strata in the a priori strategy. GVI provided a continuous variable mapped over the entire site for use in assigning categories. We used a GVI image for August 15, 1987. The biophysical justification for choosing the satellite index was the strong relationship between GVI and biomass observed in predominantly grassland parts of the site ($R^2 = 0.5$). Biomass is a reason-

---

Plate 1   The experimental area, in false color from the NASA airborne high resolution sensor AVIRIS. Each plane in the colored data cube represents 1 of 224 AVIRIS spectral channels.

AVIRIS
KONZA PRAIRIE, KANSAS
31 AUGUST 1990
Red = 2209nm   Green = 815nm   Blue = 557nm

Plate 1

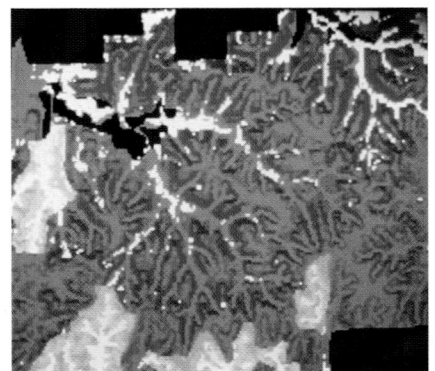

Plate 2A

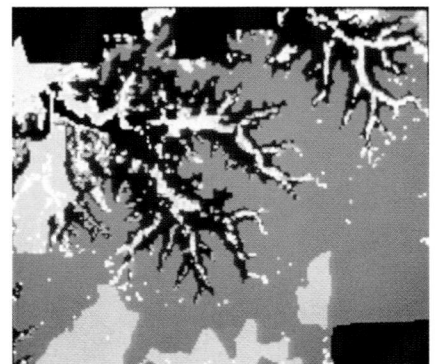

Plate 2B

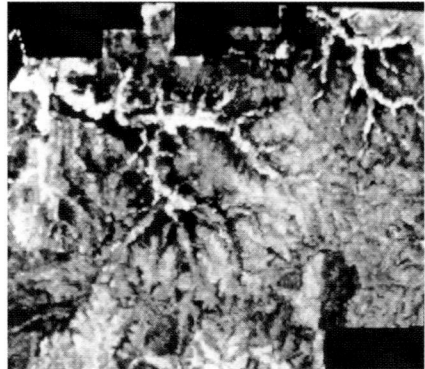

Plate 2C

able surrogate variable for variation in matter and energy exchange because biomass accumulation is both an integration of gas exchange processes over time and a control over instantaneous rates of exchange through LAI and surface roughness (Schimel *et al.*, 1991). Smith *et al.* (1992) made a similar argument for using NDVI (Normalized Difference Vegetation Index) computed from SPOT imagery as a surrogate in an analysis of sample-size adequacy of the flux measurement network.

B. Results

We tested the regression-tree method for a joint sample of GVI and mapped variables used to stratify the site. To increase the interpretability of the results, we restricted the analysis to the Konza LTER site, for which complete land management information was available. The regression-tree analysis produced a stratification (Fig. 3.1B) that was very similar to the a priori stratification and that supported the original ecological hypotheses concerning the sources of variation (fire, topography). There were several significant differences. First, the GVI-based stratification was sensitive to variations in vegetation that we did not consider in the a priori classification specifically, unburned vegetation was partitioned into shrub- and grass-dominated phases. Second, whereas the a priori stratification used hillslope zones, the regression tree showed elevation to be a more powerful predictor. A major split within unburned vegetation occurred at 395 m, an elevation that corresponds to the boundary between the Benfield–Florence soils and the Clime–Sogn soils. Despite this clear soil boundary, we did not include soils as an explicit data layer because digital soil maps were not available over the entire site. A second elevation split within the grassland phase occurred at 341 m, which corresponds closely to the boundary of deep alluvial soils in the region. Finally, the regression tree did not separate out croplands, although anomalously high GVI values in the low-elevation unburned grassland stratum may be due to crops. The tree identified two woodland phases: forest and open woodland. Overall, the 7-strata regression tree reduced the GVI error variance by 30% over that of the 14-strata a priori stratification for the Konza LTER site.

The regression-tree stratification did not identify topographic strata within burned vegetation (Fig. 3.1B). This failure highlights several features of the use of GVI in this type of stratification. First, field measurements show strong effects of topography on biomass. However, these

---

Plate 2 (A) Map of the DEM-based a priori stratification. The map was produced by overlaying digitial maps of burning treatment, hillslope position, slope angle, slope aspect, wooded areas, and cropped areas onto semivariograms of the GVI. (B) Map of the regression-tree stratification. (C) August 15, 1987 GVI image.

effects were more evident earlier in the growing season than in mid-August (Schimel *et al.*, 1991) when the image used in the regression-tree analysis was obtained. Second, the most productive of the lowland areas occur in valley bottoms and are long sinuous features with typical widths of 50–100 m. In 30-m resolution satellite data, these features do not appear as well-defined discrete features but are blurred (Plate 2C). Thus, the use of satellite data suppresses identification of these extensive but irregular valley-bottom areas.

Third, the vegetation index used appears to be sensitive to abundance of dead plant material and shrub density. In unburned areas, the effects of topography are emphasized by the accumulation of dead plant material and shrub biomass in lowland areas. The range of green biomass is similar in burned and unburned areas. In contrast, dead plant material is at trace levels in burned areas, compared to 1000 kg/ha or more in unburned areas. Similarly, whereas burning almost completely suppresses shrub growth, unburned areas may have high shrub densities (Launchbaugh and Owensby, 1978). Finally, many burned areas are grazed, whereas unburned areas are often not managed for cattle. The removal of biomass by grazing reduces the apparent effect of topography on biomass (Kittel *et al.*, 1990; Hobbs *et al.*, 1991; Turner *et al.*, 1992). Thus, in GVI data, topographic effects are more evident in unburned than in burned areas.

In summary, the remote sensing-based stratification has problems as well as advantages. These problems arise from (1) interactions between burning, grazing, and topography that obscure their individual control over landscape variability and (2) the low resolution of GVI relative to important linear features. Topographic effects can be assumed to be similar in burned and unburned areas, such that topographic substrata can be added to the burned strata based on field-derived understanding of processes (Hobbs *et al.*, 1991; Schimel *et al.*, 1991). However, this is a move toward a subjective stratification. No easy solution to the limitations of satellite data in capturing process information or to the limitations of process studies in capturing spatial pattern is evident. This highlights the problems and potential compromises in spatial stratification for extrapolation of canopy processes.

VI. Scale Dependence in GVI and Terrain Variables

To explore the association of surface biophysical properties and terrain variables such as land management and soils, we performed geostatistical analyses of GVI imagery acquired over the Konza LTER site (Davis *et al.*, 1990). Isotropic semivariograms (Oliver and Webster, 1986) were calculated to describe the spatial autocorrelation of GVI data. We examined the covariance of GVI and terrain variables by stratifying a GVI image by each terrain variable, subtracting the mean GVI for each stratum, and recalculating the image variogram using the residuals.

In 1987, spatial variation in GVI was greatest in June and declined over the rest of the growing season (Fig. 3.2A). As illustrated by the variogram for June 12, GVI exhibited positive spatial dependence up to 210 m, with most of the variation at 30- to 60-m length scales (Fig. 3.2B). Variation at scales finer than 30 m cannot be assessed with this image data. When the effect of forest vs. grassland was removed ('Forest' curve, Fig. 3.2B), the image variance was reduced by 30.4%. This is reflected in the variogram by a lower sill (quasi-asymptotic variance level). The selective removal of some high frequency variation is suggested by the reduced slope of the variogram at shorter distances. Further stratification of the image into burning and soil classes reduced image variance by an additional 13.3% (Fig. 3.2B). Similar effects of forest, burning, and soil were found in imagery from late August (Davis *et al.,* 1989). Important results from this analysis were the large contribution of forests to image variance (despite their limited extent, covering only 8.9% of the study area) and the very localized nature of variance in GVI (<210-m length scale).

VII. Spatial Analysis of Flux Measurements

A careful analysis of sources of variability was done by the FIFE surface-flux group (Smith *et al.,* 1992). They investigated the relative effects of stratification variables (using the initial implementation of the a priori stratification) on latent (QE), sensible (QH), and soil (QG) heat fluxes and derived variables, such as evaporative fraction [QE/(net radiation + QG)] and Bowen ratio (QH/QE). The reader is referred to Smith *et al.* (1992) for a detailed discussion of their results which highlights a difficult problem in flux-measurement extrapolation.

An important source of variation in the averaging of fluxes over the site was cloudiness, influencing evaporation and photosynthesis via the surface energy budget and stomatal resistance (Knapp and Smith, 1989). Spatial autocorrelation analysis showed distance dependence to be a function of the overlying cloud field rather than of landscape structure. Thus, interpretation and modeling of instantaneous fluxes and their time series are very sensitive to the cloud regime. The significance of clouds is supported by physiological studies; for example, Knapp and Smith (1989) showed that carbon gain and water-use efficiency are influenced strongly by cloud-mediated changes in light through the speed at which

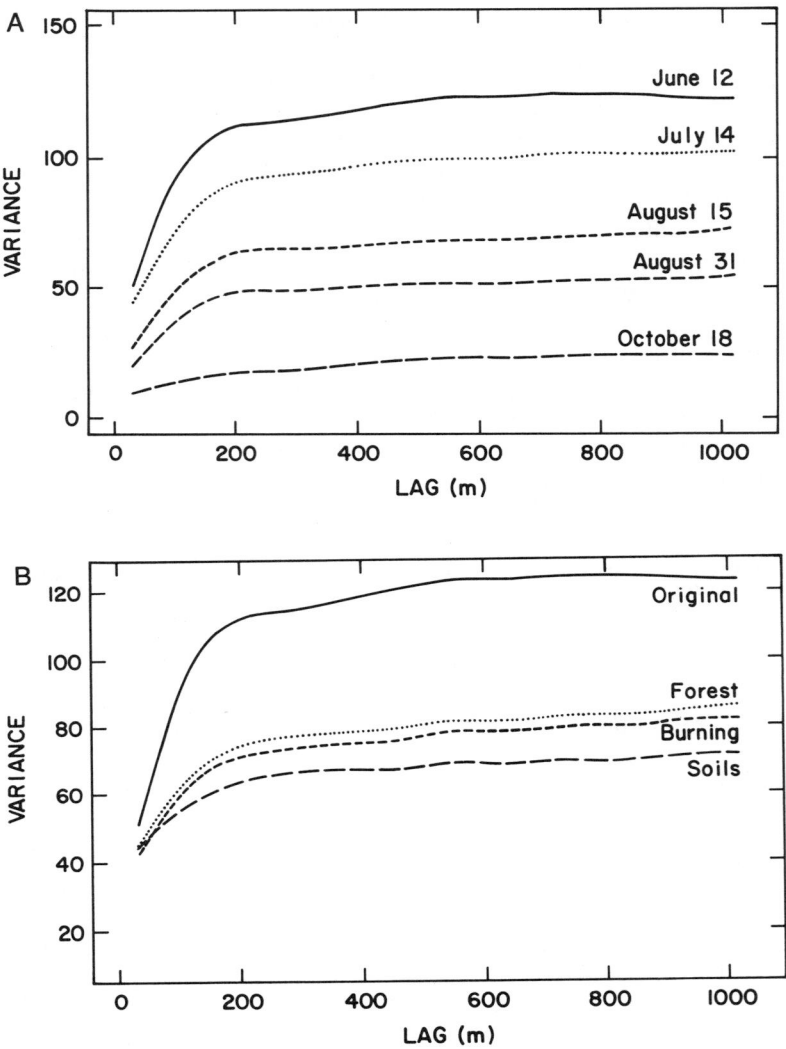

Figure 3.2 (A) Semivariograms of the GVI by date. (B) Sources of spatial variance in the June 12 GVI. Stratification by major vegetation class (forest vs. grassland; Fig. 3.1B) reduced the image variance by 30.4%, burning, 3.5%, and soils, 9.8%, for a total variance reduction of 43.7%. Data are for 1987.

plants acclimate to transient changes in radiation. Different plant functional groups may have different strategies for responding to cloud-induced radiation changes through lags in stomatal response to such changes. Despite the significance of clouds in determining instantaneous fluxes, analysis of all integral parameters (such as light response of photosynthesis, biomass, LAI, GVI, root properties) shows strong dependence on landscape structure (Schimel *et al.*, 1991), suggesting that incoming radiation is uniform (before topographic effects; Dubayah *et al.*, 1990) given a sufficient averaging time. Cloudiness undoubtedly influences (1) the time integral of fluxes at a point and (2) the spatial structure of fluxes at an instant. However, the effect of cloudiness on the spatial structure of the time integral of fluxes may be minimal because of a lack of spatial bias given sufficient time. The spatial structure of time-integrated fluxes may not have been well resolved with the temporal sampling strategy used by Smith *et al.* (1992).

Short simulations of atmosphere–biosphere coupling, as have been conducted by Avissar and Pielke (1989), clearly will require information on or simulation of the autocorrelation function of the overlying cloud field (Lovejoy *et al.*, 1988; Kiehl and Ramanathan, 1990). Longer simulations may be less sensitive to the spatial structure of the cloud field, assuming that the cloud bias is eliminated given sufficient averaging time. The problems posed by the spatial structure of the cloud field were not considered in the FIFE experimental design; cloud statistics should be evaluated in the design of future experiments addressing spatial and temporal integration of fluxes. Obviously, any experiment studying fluxes along a gradient or at a boundary where the spatial autocorrelation functions of the cloud field change within the experimental domain should include cloud regime as a variable in the study. The role of clouds illustrates the difficult and general issues raised in data analysis and modeling of coupled systems, particularly when widely disparate rates of change (e.g., biomass accumulation compared with incident radiation) are involved. Although the atmospheric sciences have focused much attention on the problems of cloud physics, it is clear from FIFE (Smith *et al.*, 1992) and other work (Knapp and Smith, 1989) that ecologists should be interested in how cloud fields are simulated in atmospheric models for future modeling studies of the coupled atmosphere–biosphere system.

VIII. Lessons for Physiological Ecology

1. The problem of extrapolation requires rethinking of traditional experimental design for ecological field studies (Harriss, 1989).

Sampling strategies adequate for understanding leaf and plant physiological processes may be inadequate or misleading when used for studies of whole landscapes. The a priori stratification was based on understanding gained by subjectively analyzing perceived extremes (end-members) of the topography (e.g., ridgetops and valley bottoms; Knapp, 1985; Schimel *et al.*, 1985) and fire management (Knapp, 1984; Knapp and Seastedt, 1986; Hobbs *et al.*, 1991) and extrapolating those results to the entire landscape. Our FIFE analyses suggested that (1) certain landscape strata captured very little variance (e.g., topography within burned grazed areas), (2) factors we had not considered were important (e.g., shrub abundance related to soil type), and (3), even at 30-m resolution, significant variation was inadequately captured (e.g., the most productive lowland areas were poorly represented). Because atmosphere–ecosystem interactions occur over inherently large areas relative to the spatial scale of leaf or plant physiology, experimental designs must address all the significant variation in the domain of study. This contrasts with studies of process, for which experiments in end-member situations or along one axis of variation may suffice. The need to integrate state variables and fluxes over entire areas presents conceptual and logistical challenges to experimentalists and modelers (Schimel *et al.*, 1990).

2. Remote sensing and geographic data, such as digital topography, aid in the design of large-area experiments (Matson *et al.*, 1990) but are not a panacea. This is illustrated by our inability to deal adequately with the stratification and analysis of grazed landscapes using remotely sensed information. Remote sensing and GIS are of most value when easily mapped proxy variables can be related directly to the fluxes of interest and when the mapped scale of these variables matches or exceeds the predominant scale of variation in the fluxes.

3. In the analysis of atmosphere–ecosystem interactions, consideration must be given to vertical structure of the atmosphere (e.g., height of the planetary boundary layer, cloud cover) that, although transient, may have as large an influence as edaphic structure or landscape hydrology. This point was made for gas exchange in relation to atmospheric moisture fields by Jarvis and McNaughton (1986), who argued that, over sufficiently large areas, evapotranspiration becomes a control upon itself as relative humidity increases. Cloudiness also exerts a control over gas exchange, such that the spatial structure of cloudiness must be considered when analyzing interactions occurring at certain combinations of time and space scales (Knapp and Smith, 1989). Further, cloudiness in midcontinental areas is not under purely physical control because many processes influencing cloud formation—convection, atmospheric moisture, and boundary layer

dynamics—are under a degree of biotic control through surface energy and water exchange (Segal *et al.*, 1988; Avissar and Pielke, 1989; Sato *et al.*, 1989; Lakhtakia and Warner, 1990; Pielke *et al.* 1991).

IX. Conclusion

It is well accepted that the biota influence soil and geomorphic processes and that biological and geochemical or geomorphic changes do not proceed independently (Jenny, 1941; Schimel *et al.*, 1985; Pastor and Post, 1988). It is becoming increasingly clear that coupling between ecosystems and the atmosphere is also significant. In the short term, ecosystems influence the atmosphere through patterns of energy exchange at the surface and, in the long term, through the composition of the atmosphere (Avissar and Pielke, 1989; Melillo *et al.*, 1989; Berner, 1990; Shukla *et al.*, 1990). The significance of these interactions is being studied intensively as part of the global change research agenda, but relatively little thought has been given to their significance in the context of our basic assumptions about plant biology and evolution (Jarvis and McNaughton, 1986). Future collaborations among ecologists and atmospheric scientists should give these biological questions additional weight relative to the physical issues that motivated FIFE.

X. Summary

An interdisciplinary project—the First ISLSCP (International Satellite Land Surface Climatology Project) Field Experiment (FIFE)—studied techniques for extrapolation of ecological and physical processes regulating exchange with the atmosphere. We reviewed the use of geographic information in FIFE, the implementation of a geographic information system in analyzing and extrapolating the ground and image data, and the implications of FIFE results for modeling exchange between the atmosphere and terrestrial ecosystems. FIFE set out to develop systematically approaches to extrapolation using advanced measurement techniques, remote sensing, and geographic information systems to measure important atmosphere–biosphere exchanges of matter and energy (Sellers *et al.*, 1988; Davis *et al.*, 1992). In addition to studying methods for quantifying flues of matter and energy at multiple spatial scales, FIFE made significant contributions to physiological ecology on large scales and to the study of the atmosphere and ecosystems as parts of a coupled system.

Acknowledgments

This research was supported by NSF and NASA FIFE grants to Dave Schimel and Frank Davis. We gratefully acknowledge support from Forrest Hall, Piers Sellers, Ghassam Asrar, Don Strebel, Ralph Dubayah, and other investigators in the FIFE Integrative Science Group. Thanks also go to Carol Wessman, who provided the AVIRIS image in Plate 1, Susan Chavez and Gaylynn Potemkin, for assistance in manuscript preparation, and to chapter reviewers for invaluable comments.

References

Avissar, R., and Pielke, R. A. (1989). A parameterization of heterogeneous land surfaces for atmospheric numerical models and its impact on regional meteorology. *Monthly Weather Rev.* **117,** 2113–2136.

Berner, R. A. (1990). Atmospheric carbon dioxide levels over phanerozoic time. *Science* **249,** 1382–1386.

Breiman, L., Friedman, J., Olshen, R., and Stone, C. (1984). "Classification and Regression Trees." Wadsworth, Belmont, California.

Burke, I. C., Schimel, D. S., Yonker, C. M., Parton, W. J., Joyce, L. A., and Lauenroth, W. K. (1990). Regional modeling of grassland biogeochemistry using GIS. *Landscape Ecol.* **4,** 45–54.

Costanza, R., Sklar, F. H., and White, M. L. (1990). Modeling coastal landscape dynamics. *BioScience* **40,** 91–107.

Davis, F., Dubayah, R., Dozier, J., and Hall, F. (1989). Covariance of greenness and terrain variables over the Konza Prairie. *In* "Proceedings IGARSS '89," pp. 1322–1325.

Davis, F. W., Schimel, D. S., Friedl, M. A., Michaelsen, J. C., Kittel, T. G. F., Dubayah, R., and Dozier, J. (1992). Correspondence of surface climate variables and digital terrain data over a tallgrass prairie. *J. Geophys. Res.* (in press).

Davis, F. W., Michaelson, J., Dubayah, R., and Dozier, J. (1990). Optimal terrain stratification for integrating ground data from FIFE. *In* "Symposium on the First ISLSCP Field Experiment (FIFE)" (F. G. Hall and P. J. Sellers, conveners), pp. 6–12. American Meteorological Society, Boston.

Desjardins, R. L., Schuepp, P. H., and MacPherson, J. I. (1990). Spatial and temporal variations of CO_2 sensible and latent heat fluxes over the FIFE site. *In* "Symposium on the First ISLSCP Field Experiment (FIFE)" (F. G. Hall and P. J. Sellers, conveners), pp. 46–50. American Meteorological Society, Boston.

Dubayah, R., Dozier, J., and Davis, F. W. (1990). Topographic distribution of clear-sky radiation over the Konza Prairie, Kansas. *Water Resources Res.* **267,** 679–690.

Esser, G. (1990). Modeling global terrestrial sources and sinks of CO_2 with special reference to soil organic matter. *In* (A. F. Bowman, ed.), "Soils and the Greenhouse Effect" pp. 247–262. Wiley, Chichester.

Fowler, D., and Duyzer, J. H. (1989). Micrometerological techniques for the measurement of trace gas exchange. *In* (M. O. Andreae and D. S. Schimel, eds.) "Exchange of Trace Gases between Terrestrial Ecosystems and the Atmosphere" pp. 189–208. Wiley, Berlin.

Hall, F. G., Markhamm, B. J., Wang, J. R., and Huemmerich, F. (1990). FIFE: Results overview. *In* "Symposium on the First ISLSCP Field Experiment (FIFE)" (F. G. Hall and P. J. Sellers, conveners), pp. 17–24. American Meteorological Society, Boston.

Harriss, R. C. (1989). Experimental design for studying atmosphere-biosphere interactions.

In "Exchange of Trace Gases between Terrestrail Ecosystems and the Atmosphere" (M. O. Andreae and D. S. Schimel, eds.), pp. 291–301. Wiley, Berlin.

Hobbs, N. T., Schimel, D. S., and Owensby, C. E., and Ojima, D.S. (1991). Fire and grazing in the tallgrass prairie: Contingent effects on nitrogen budgets. *Ecology* **72,** 1374–1382.

Jarvis, P. G., and McNaughton, K. G. (1986). Stomatal control of transpiration: Scaling up from leaf to region. *Adv. Ecol. Res.* **15,** 1–49.

Jenny, H. (1941). "Factors of Soil Formation." McGraw-Hill, New York.

Kiehl, J. T., and Ramanathan, V. (1990). Comparison of cloud forcing derived from the Earth Radiation Budget Experiment with that simulated by the NCAR Community Climate Model. *J. Geophys. Res.* **95,** 11679–698.

Kittel, T. G. F., Knapp, A. K., Seastedt, T., and Schimel, D. S. (1990). A landscape view of biomass, LAI, and photosynthetic capacity for FIFE. *In* "Symposium on the First ISLSCP Field Experiment (FIFE)" (F. G. Hall and P. J. Sellers, conveners), pp. 66–69. American Meteorological Society, Boston.

Knapp, A. K. (1984). Post-burn differences in solar radiation, leaf temperature and water stress influencing production in a lowland tallgrass prairie. *Am. J. Bot.* **71,** 220–227.

Knapp, A. K. (1985). Early season production and microclimate associated with topography in a C$_4$ dominated grassland. *Acta Oecologia* **6,** 337–346.

Knapp, A. K., and Seastedt, T. R. (1986). Detritus accumulation limits productivity of tallgrass prairie. *Bioscience* **36,** 662–668.

Knapp, A. K., and Smith, W. K. (1989). Influence of growth on ecophysiological responses to variable sunlight in subalpine plants. *Ecology* **70,** 1069–1082.

Lakhtakia, M. N., and Warner, T. T. 1990. Sensitivity of simulated surface fluxes to the use of time-dependent surface parameters. *In* "Proceedings, Eighth Conference on Hydrometeorology." Kananaskis Park, Alta, Canada.

Launchbaugh, J. L., and Owensby, C. E. (1978). "Kansas Rangelands: Their Management Based on a Half Century of Research." Kansas Agricultural Experiment Station Bulletin 622. Kansas State Univ., Manhattan, Kansas.

Lovejoy, S., Gabriel, P., Austin, G. L., and Schertser, D. (1988). Modeling the scale dependence of visible satellite images by radiative transfer in fractal clouds. *In* "Third Conference on Satellite Meteorology and Oceanography." American Meteorological Society, Anaheim, California.

Matson, P. A., Vitousek, P. M., and Schimel, D. S. (1989). Regional extrapolation of trace gas flux based on soils and ecosystems. *In* "Exchange of Trace Gases between Terrestrial Ecosystems and the Atmosphere" (M. O. Andreae and D. S. Schimel, eds.), Wiley, Berlin.

Matson, P. A., Vitousek, P. M., Livingston, G. P., and Swanberg, N. A. (1990). Sources of variations in nitrous oxide flux from Amazonian ecosystems. *J. Geophys. Res.* **95,** 16,789–16,798.

Melillo, J. M., Steudler, P. A., Aber, J. D., and Bowden, R. D. (1989). Atmospheric deposition and nutrient cycling. *In* "Exchange of Trace Gases between Terrestrial Ecosystems and the Atmosphere" (M. O. Andreae and D. S. Schimel, eds.), pp. 263–280. Wiley, Berlin.

Mosteller, F., and Tukey, J. W. (1977). "Data Analysis and Regression." Addison-Wesley, Reading, Massachusetts.

Norman, J. M., Garcia, R., and Verma, S. B. (1990). Soil surface CO$_2$ fluxes on the Konza prairie. *In* "Symposium on the First ISLSCP Field Experiment (FIFE)" (F. G. Hall and P. J. Sellers, conveners), pp. 29–31. American Meteorological Society, Boston.

Ojima, D. S. (1987). "The Short-term and Long-term Effects of Burning on Tallgrass Ecosystem Properties and Dynamics." Ph.D. Dissertation. Colorado State University, Fort Collins, Colorado.

Oliver, A. O., and Webster, R. (1986). Semi-variograms for modelling the spatial pattern of landform and soil properties. *Earth Surf. Processes Landforms* **11,** 491–504.

Pastor, J., and Post, W. M. (1988). Response of northern forests to CO_2-induced climate change. *Nature* **334,** 55–58.

Pielke, R. A., Dalu, G., Snook, J. S., Lee, T. J., and Kittel, T. G. F. (1991). Nonlinear influence of mesoscale landuse on weather and climate. *J. Climate* **4,** 1053–1069.

Sato, N., Sellers, P. J., Randall, D. A., Schneider, E. K., Shukla, J., Kinter, J. L., III, Hou, Y.-T., and Albertazzi, E. (1989). Effects of implementing the Simple Biosphere Model in a general circulation model. *J. Atmos. Sci.* **46,** 2757–2782.

Schimel, D. S., Stillwell, M. A., and Woodmansee, R. G. (1985). Biogeochemistry of C, N and P in a soil catena of the shortgrass steppe. *Ecology* **66,** 276–282.

Schimel, D. S., Kittel, T. G. F., Knapp, A. K., Seastedt, T. R., Parton, W. J., and Brown, V. B. (1991). Physiological interactions along resource gradients in a tallgrass prairie. *Ecology* **72,** 672–684.

Schimel, D. S., Parton, W. J., Kittel, T. G. F., Ojima, D. S., and Cole, C. V. (1990). Grassland biogeochemistry: Links to atmospheric processes. *Climatic Change* **17,** 13–25.

Segal, M., Avissar, R., McCumber, M. C., and Pielke, R. A. (1988). Evaluation of vegetation effects on the generation and modification of mesoscale circulation. *J. Atmos. Sci.* **45,** 2268–2392.

Sellers, P. J., Hall, F. G., Markham, B. J., Wang, J. R., Strebel, D. E., Kanemasu, E. T., Kelly, R. D., and Blad, B. L. (1990). Experiment design and operations. *In* "Symposium on the First ISLSCP Field Experiment (FIFE)" (F. G. Hall and P. J. Sellers, conveners), pp. 1–5. American Meteorological Society, Boston.

Sellers, P. J., Hall, F. G., Asrar, G., Strebel, D. E., and Murphy, R. E. (1988). The First ISLSCP Field Experiment (FIFE). *Bull. Am. Meteorol. Soc.* **69,** 22–27.

Shukla, J., Nobre, C., and Sellers, P. (1990). Amazon deforestation and climate change. *Science* **247,** 1322–1325.

Smith, E. A., Hsu, A. Y., Crosson, W. L., Field, R. T., Fritschen, L. J., Gurney, R. J., Kanemasu, E. T., Kustas, W., Nie, D., Shuttleworth, W. J., Stewart, J. B., Verma, S. B., Weaver, H., Wesley, M., (1992). Area averaged surface fluxes and their time-space variability over the FIFE experimental domain. *J. Geophys. Res.* (in press).

Turner, C. L., Seastedt, T. R., Dyer, M. I., Kittel, T. G. F., and Schimel, D. S. (1992). Effects of management practices and topography on the radiometric response of tallgrass prairie. *J. Geophys. Res.* (in press).

Verma, S. B., Kim, J., and Clement, R. J. (1990). Fluxes of momentum, sensible heat, water vapor and CO_2 over a tallgrass prairie. *In* "Symposium on the First ISLSCP Field Experiment (FIFE)" (F. G. Hall and P. J. Sellers, conveners), p. 28. American Meteorological Society, Boston.

Walter-Shea, E. A., Blad, B. L., Starks, P. J., Hays, C. J., and Mesarch, M. A. (1990). Bidirectional reflectance, leaf optical and physiological properties of prairie vegetation. *In* "Symposium on the First ISLSCP Field Experiment (FIFE)" (F. G. Hall and P. J. Sellers, conveners), p. 70. American Meteorological Society, Boston.

Wood, E. F., (1990). Water balance model for Kings Creek. *In* "Symposium on the First ISLSCP Field Experiment (FIFE)" (F. G. Hall and P. J. Sellers, conveners), pp. 163–167. American Meteorological Society, Boston.

II

Leaf to Ecosystem
Level Integration

Much of the past work in physiological ecology has focused on processes at the whole-leaf level. Scaling these gas exchange and water relations processes to the canopy level began in the 1950s with pioneers such as Monsi and Saeki in Japan, who modeled light attentuation through canopies and its impact for photosynthesis. Similar efforts were underway in Europe by de Wit and colleagues who developed models to assess the impact of changes in environmental conditions on crop productivity. Since that time, such models have been expanded and new approaches developed to relate the impact of physiological processes at landscape and regional levels. In this section, we explore the development and utility of process-based models for addressing scaling issues from whole-leaf to canopy levels and from stand to ecosystem levels. Norman begins with a conceptual framework for developing process-based models. He then proceeds to describe how such models can be constructed from first principles and how we can add components to the models later as our understanding of those processes increases. His perspective is that of a physiologist, linking leaf-level physiological models with canopy-level phenomena. In the next chapter, Baldocchi examines how photosynthesis and transpiration can be scaled from leaf to canopy levels from the perspective of a micrometeorologist.

Models originate from one of two possible approaches: bottom up or top down. Bottom-up modeling approaches begin with a smaller spatial scale and shorter temporal scale than the output. Bottom-up models are open-ended in their model output. On the other hand, in top-down models the output of the models is constrained totally through an experimentally determined relationship with a crucial driving variable. Top-down models tend to be more empirical. In a chapter developed from group discussions, Jarvis presents the advantages and limitations of each approach.

Reynolds, Hilbert, and Kemp discuss how bottom-up and top-down approaches can be used together to develop better models of both types. Top-down models have less mechanistic insight and thus are limited in their application to new environmental regimes, whereas bottom-up models can be too complicated to be of general use in scaling to higher levels. In their chapter, Reynolds and colleagues discuss how they have melded the two modeling approaches to better address questions of climate change and its impact on ecological processes at leaf and landscape levels.

In the next chapter, Running and Hunt examine the generality of combined bottom-up and top-down models. Their model, developed for coniferous forest ecosystems, is applied to a study of a grassland ecosystem. Running and Hunt suggest that a hierarchy of the variables influences most ecosystem processes and that, as a consequence, models developed for particular ecosystems may find general application across a diversity of landscapes and ecosystems.

In the final chapter of this section, Waring examines ways in which ecophysiologists can contribute more productively to studies directed at higher scales, for example, the landscape and regional levels. He argues that better linkages must be developed between those interested in a detailed mechanistic understanding at the leaf level and those investigators interested in understanding the consequences of environmental perturbations at landscape and global levels.

4

Scaling Processes between Leaf and Canopy Levels

John M. Norman

I. Introduction

Natural phenomena occur over a wide range of spatial and temporal scales. The scales of particular interest in this chapter span the spatial domain from individual organs such as leaves to vegetation canopies and range in time from minutes to months. Because these scales are similar to the natural scales of human functioning (that is, scales with which human beings most easily identify), extensive data from many different kinds of observations are available. A wide range of physiological data, anatomical data, chemical constituent analyses, soil and atmospheric environmental observations, and aspects of the entire vegetation classification system contributes to this range of scales as the anchor of vegetation studies on larger and smaller scales. Microscopic (cellular) and macroscopic (regional) scales involve observations that require more effort for a similar amount of information than do observations on this leaf-to-canopy scale. Clearly, the understanding derived from the wealth of information on this leaf-to-canopy scale is of tremendous benefit to developing an understanding of phenomena on other scales.

The integration, synthesis, and insight required to use knowledge from leaf observations to infer canopy function have occurred in the minds of plant scientists for a long time. Unfortunately, such integrative capabilities are difficult to transfer from person to person, and they die with the scientist. The principles of mathematics, physics, and chemistry provide a means for combining information on many scales to improve our understanding of phenomena on some specific scale; these methodologies reside collectively with a community of scientists. The objective of this chapter is to explore some basic concepts involved in scaling from leaf-level processes to canopy-level processes, and contrast the requirements for doing this with vegetation to the requirements for scaling in a purely physical system. Living systems are more complex than physical systems and analogs to physical scaling techniques may be of limited usefulness. Therefore, a strategy for dealing directly with the complexity is outlined and an example of a hierarchy of approaches for scaling leaf photosynthesis to canopy photosynthesis is developed.

II. What Is Scaling and Why Do It?

In this chapter "scaling" refers to the use of information on one spatial or temporal scale to infer characteristics on another scale. In this case, the word "scale" refers to a characteristic size or time step. Two related terms come to mind as we consider relating phenomena on different spatial and temporal scales: integration and scaling. A useful distinction between integration and scaling is that integration is an orderly process of bringing component parts together to form a whole and scaling implies an intuitive leap that provides a quantitative connection between distant phenomena—a short cut. Both intuitive leaps and methodical consistency are essential parts of scientific exploration. As Mark Twain notes, "There is something fascinating about science. One gets such wholesale returns of conjecture out of such a trifling investment of fact."

Why is this process, which we refer to as "scaling," of interest to us in ecology and agriculture, and what is our objective in engaging in it? As plant scientists, we recognize the importance of the influence of plants on field-, regional- and global-scale phenomena such as weather, climate, and human habitability. However, we have only a limited ability to make representative measurements on larger canopy scales; therefore, we use integration and scaling techniques to try to apply small-scale, short-term measurements to make large-scale, long-term inferences. Although we can measure on larger scales (km) with micrometeorological techniques and aircraft, applying these results to other ecosystems, locations, time, or weather conditions has proven difficult for two reasons: (1) Limited resources constrain measurement periods and conditions, so less-than-representative observations are available. (2) The system behavior is the result of interactions among many factors on small and large scales and, without accommodating the most important of these interactions, generalizations are difficult. Successful scaling from leaf to canopy will depend on whether the important processes are understood. To gain confidence in our ability to scale, we use different methods. We will explore several methods in this chapter.

III. Issues in Scaling from Leaf to Canopy

Scaling some process, such as photosynthesis, from the leaf level to the canopy level might be divided into two components: (1) a biological component that consists of the canopy architecture and plant physiological responses without considering environmental gradients that occur between the vegetation elements and the bulk environment above the canopy, and (2) a fluid-dynamical component that considers the environmental gradients that occur between some reference level above the canopy and the vegetation elements. Using the leaf area index (LAI) to scale the mean leaf photosynthetic rate to a canopy photosynthetic rate is an example of what we might refer to as "biological scaling." Straight fluid-dynamical scaling does not appear to have been used in leaf-to-canopy scaling. However fluid dynamics equations have been used to describe the combined vegetation–environment integration problem by considering the canopy as a fluid source or sink and using biological scaling to derive an equation for the magnitude of this source or sink as a function of vegetation characteristics (Stewart and Lemon, 1969; Goudriaan, 1977; Norman and Campbell, 1983). Because of the level of detail associated with treating the canopy as a source or sink in fluid dynamics equations, the three references just cited would be considered detailed integration approaches, not scaling approaches. The approaches used by Deardorf (1978) or Sellers (1985) may be simplified sufficiently to qualify as combined biological–fluid-dynamical scaling; however, even in these simplified examples the fluid dynamics part of the solution more nearly represents coarse integration rather than scaling.

Biological scaling of the vegetation component alone, or fluid-dynamical scaling of the environmental component alone, seems to be a tractable problem since a single entity is considered, namely, plant leaves in one case and fluid air in the other case. When vegetation and environmental effects are considered simultaneously, the simplified concepts of scaling may not apply, and the complex process of integration in space and time may be required. At the very least, solutions to the complex problem may be required to evaluate simple scaling schemes because a systematic approach to simple scaling seems unlikely.

We can illustrate the complexity of leaf-to-canopy scaling by considering some of the factors that are known to be important: (1) canopy architectural effects—leaf area index, leaf orientation functions, leaf location distributions (random, clumped, regular, or any combination), plant parts (stems, leaves, flowers, and branches), and architectural changes with wind and light; (2) temporal changes in the environment—transient light fluctuations of sunflecks associated with wind and movement of the sun, diurnal environmental changes, and seasonal changes; (3) site characteristics—latitude, longitude, altitude, slope, and aspect; and (4) gradients between environment outside of the canopy and leaves within the canopy—environmental gradients of wind, radiation, temperature, humidity, and variation of physiological characteristics of plant elements with position in the canopy. If we can identify a few factors that are most important, then we may be able to use scaling to good advantage; however, if most of these factors can dominate under one condition or another, then scaling may not work and the arduous process of detailed integration is likely to be necessary.

IV. Can an Investigative Paradigm in Physics Be Applied Directly to Biology?

The possibility exists that a radically new approach to vegetation scaling may be necessary to accommodate the broad array of factors that could be important, including fluid-dynamical factors and biological considerations such as plant physiological characteristics, competition, ontogeny, genetic plasticity, reproductive strategies, and seed, disease, or insect dispersal characteristics.

Expanding the scientific knowledge base necessarily involves issues of scale. Very large spatial and temporal scales can be explored by inference using analogs to processes on small scales that have similar patterns to the large-scale observations; this might be referred to as "pattern scaling." Thus, in a way, we can extrapolate to large scales through the process of reductionism if appropriate patterns are recognizable. An example of this technique in fluid dynamics is the use of principles from molecular diffusion to describe turbulent eddy diffusion. An example of this technique in biology is the Gaia hypothesis (Lovelock, 1979). As the knowledge base expands, direct integration rather than pattern scaling may be possible. For example, in fluid dynamics, cellular automata are being used to predict gas behavior from direct simulation of molecular-scale collisions on a computer with lattice-gas concepts (Frisch, 1990).

The numerous complex equations that describe the various physical processes that affect leaves, plants, or canopies could be integrated, using the standard scientific process as it has been applied to physical systems. Alternatively, could the same process of scientific investigation be applied directly to living systems such as leaves or plants? That is, could fundamental properties, axioms, and invariant quantities (some quantity analogous to the speed of light in a physical system) be derived for living or natural systems apart from such an approach for physical systems? Perhaps chaos or fractals might provide mathematical formalism that could be applied directly to living systems. However, this process of scientific investigation need not be limited to existing mathematical constraints;

new formalism could be created. Are there any fundamental nonphysical properties that living systems possess? Could these properties be related to each other? The approach used by Cowan (1982) to connect photosynthesis and stomatal conductance by hypothesizing an "optimal behavior" borders on a new paradigm. Examples of new paradigms are difficult to find because most of us are too deeply entrenched in past successes to risk much investment in the unknown.

V. Scaling in Fluid Dynamics

Scaling in fluid dynamics has been remarkably successful and has led to a broad range of engineering applications. Since fluid-dynamical concepts are directly relevant to the fluid portion of leaf-to-canopy scaling, is it possible that these concepts might be applied in a more general way to biological scaling?

Two approaches to scaling are used in fluid dynamics: similitude and dimensional analysis (Vennard and Street, 1976). The principles of similitude can be applied to a differential equation, which describes fluid flow in some particular system, to establish the criteria for obtaining geometrical (shape of objects), kinematic (velocities and accelerations), and dynamic (forces) similarity. The application of similitude analysis requires an appropriate differential equation to describe the momentum (Navier–Stokes equation), heat (energy equation), or mass exchange. Although similitude imposes rigid constraints on scaling, it has provided numerous "dimensionless" numbers, such as Reynold's number, the Prandtl number, and the Nusselt number, to name just a few, that have served a critical function in the application of fluid dynamics to practical problems. In fact, the relationship between the Nusselt number, the Prandtl number, and Reynold's number is the basis for estimating the boundary layer conductance of leaves in canopies (Grace, 1981). However, how to use these dimensionless numbers directly to scale from a leaf-level result is not obvious and has not been done. Similitude analysis, however, has been applied to water-flow properties of soils (Miller and Miller, 1956; Warrick *et al.*, 1977).

Dimensional analysis is a field of mathematics, closely related to similitude, that deals with dimensions of quantities (Vennard and Street, 1976). Whereas similitude requires an appropriate differential equation, dimensional analysis can be used when the identity of important variables is known but the functional relationship between the variables is not known. Such dimensional analysis is based on Fourier's Principle of Dimensional Homogeneity (1822), which states that an equation expressing a physical relationship between quantities must have the same dimensions on each side of the equation. An example is probably most useful here. Suppose we know from experiments that power [P (watts)] is related to flow [X (kg·m^{-3})], force per unit volume [Y (Nt·m^{-3})], and length [Z (m)]. If we want to obtain a dimensionally consistent equation, the variables must be multiplied or divided because addition and subtraction can only be done with variables of similar dimensions. Therefore an appropriate equation would take the form

$$P = k\,X^a Y^b Z^c, \tag{1}$$

where k is a constant of proportionality and a, b, and c must be obtained subject to the constraint that dimensions on the left and right are consistent. If L is length, T is time, and M is mass, then the dimensions of X are $L^3 T^{-1}$, of Y are $ML^{-2}T^{-2}$, and of Z are L. With appropriate algebra $a = b = c = 1$, so the appropriate equation is

$$P = k\,XYZ. \tag{2}$$

Dimensional analysis has been used in the characterization of plant–environment relationships. For example, LAI is dimensionless because the variable of leaf area is more useful in this form. Characterizing the height dependence of wind speed in a canopy is made dimensionless by dividing height above the ground by the height of the canopy (Thom, 1971). Dimensional analysis potentially may be quite useful to us in scaling from leaf-level processes to canopy-level processes, and already is being used to characterize atmospheric resistance to momentum, heat, and mass fluxes above canopies. For example, through the log wind profile formulation, wind speed is nondimensionalized with a friction velocity ($u*$) that is related to the momentum flux and apparent height ($z - d$) is scaled by the canopy aerodynamic roughness for momentum (z_0).

$$u/u_* = k^{-1}\ln[z - d)/z_0]. \tag{3}$$

This relationship forms the basis for defining an aerodynamic resistance for momentum (see Brusaert, 1984, for a discussion).

In a fluid-dynamical sense, scaling between two systems is based on dimensionless quantities, for example, Re, being invariant. This constant Re must hold over the full range of spatial scales from small (dx) to large (x) and temporal scales from short (dt) to long (t). The influence of the outside world comes from initial conditions, boundary conditions, and volume sources or sinks; these also must meet a separate set of rigid conditions for scaling to be complete.

Scaling in fluid dynamics is difficult enough because of constraints on the equation of motion and similarity requirements for the boundary conditions, but it obviously is much simpler than scaling the combined fluid–vegetation system. Therefore, analogs between fluid-dynamical scaling and leaf-to-canopy scaling do not seem to be useful, because vegetation systems do not appear to meet similarity constraints on boundary conditions and a single set of equations may not be appropriate for all scales. Of course equations similar in form to fluid flow equations may be applied to vegetation by assigning some new definitions to dependent and independent variables; although such equations incorporate advantages of scaling, they usually clash with known properties and processes in biology by requiring the invention of fictitious quantities that cannot be observed.

Fluid dynamics itself is an important part of scaling from leaves to canopies. After all, leaves are immersed in the fluid air, and the primitive equations (consisting of three component equations of motion, the continuity equation, the ideal gas law, the energy transport equation, and the water transport equation) form the basis for the general circulation models (GCMs) that are used to predict regional and global impacts of vegetation. Thus, fluid dynamics are important to understanding plant–environment relationships, but scaling concepts from fluid dynamics do not appear to be useful directly for leaf-to-canopy scaling.

Clearly, dimensional analysis and similitude are already in use for scaling the fluid-dynamical part of the leaf-to-canopy scaling problem. However, methods for combining fluid-dynamical and biological scaling are not yet available. In summary, fluid systems are simple compared with vegetation systems and analogies between the two are difficult.

VI. Comprehensive Plant–Environment Models

One approach to developing scaling methods is to create detailed plant–environment (PE) models that integrate from leaf levels to canopy levels with equations that are as comprehensive as possible. Then simpler scaling methods can be derived from the comprehensive models. This approach requires appropriate PE models that have withstood rigorous testing against field measurements.

Living communities are too complex to be described by known equations at the present time; no general, quantitative relationships exist to connect properties and processes at various levels in a formal and coherent way. However, directly integrating from the leaf scale to the canopy scale using all the physical and biological information is one possibility. The direct-integration approach at first appears hopelessly complicated. Of course, this tends to be true for anything that is not understood because of the infinite number of possibilities that can be imagined by the human mind when it is not constrained by experience, ignorance, or faith. This direct-integration approach is not so intimidating when viewed as an interlaced hierarchy of models that are based on well-formed equations at all the various levels of interest. Scientists have been using this approach to explore natural sciences for a long time (Thornley, 1976).

The backbone of an interlaced hierarchy of models is the detailed integration of well-formed basic equations that are associated with each scale. Simplified equations can be derived from the more detailed equations and used for application studies or for scaling to the next higher level. In the particular case of scaling from the leaf to the canopy, PE models provide a means of using leaf-level parameterizations to infer canopy-level characteristics.

A. What Are PE Models?

PE models use parameterizations of important processes at the leaf level (cm) and integrate mechanistic equations to the canopy level (10–100 m) with appropriate initial and boundary conditions. These models may attempt to describe the full range of major factors that affect plant growth, development, and environmental interactions. Leaf-level parameterizations may be obtained from results of models that describe processes at the next level down in the hierarchy of models.

Consider photosynthesis and respiration. Diffusion equations may be used to describe the movement of CO_2 throughout the soil and canopy air spaces. Boundary conditions may be specified some distance above the canopy and below the root zone. These equations describe the CO_2 gas fluxes and concentrations throughout the fluid portion of the PE system. Vegetation may be considered a volume source or sink in the fluid, so organ photosynthetic and respiratory rates may be characterized in terms of local environmental conditions. Therefore leaf photosynthetic and respiratory rates must be known as functions of leaf temperature, local humidity, incident light, water status of the leaf, nutrition, and so on, and the dependence of root respiration on soil temperature, oxygen concentration, available water, and other factors must be known. Solving the CO_2 equations is not particularly difficult, but unfortunately the state of the leaf depends on the solution of many other equations, including radiative, convective, and conductive energy equations in the soil and atmosphere; water transport equations for both liquid and vapor phases in the soil and atmosphere; soil nutrient transformation (related to microbial activity) and transport and root uptake equations; equations to describe transport and concentrations of various constituents inside the plant; and equations to describe the growth and development process of various plant organs. For a community of organisms, many of these equations must be specified for various species.

To say that PE models are complicated is an understatement. Committing energy to attacking such a problem could be considered folly unless some reasonable likelihood of success is apparent. Thus, a carefully planned approach is desirable.

B. Early History of PE Models

Simple empirical equations have long been used to represent the functioning of complex organisms. Scientists then spend years or careers trying to find the best combination of equations and parameters in hopes of securing generality and simplicity. The record of the simple approaches is not a sterling example of development in scientific thought. For example, in 1920, Wallace (1920) proposed several simple three-parameter regression equations to predict corn yields in several Midwestern states. Since then, hundreds of regression equations have been studied (Baier, 1973). A review of over 200 such models and a quantitative evaluation of the 20 best ones revealed that the original equations of Wallace produced the best predictions of corn yield in central Pennsylvania (Johnson, 1976). In fact, Wallace's equations were nearer to measurements than two more complex process-oriented models. Even today, most agricultural yield models require "calibration." In this context, "calibration" seems to mean the following: "If you provide a measurement, then the model will provide a prediction of that measurement."

A second example is evapotranspiration modeling over the past 50 years. The early work (between 1910 and 1920) by L. J. Briggs, H. L. Shantz, and T. A. Kiesselbach, summarized by Tanner and Sinclair (1983), recognized the importance of atmospheric humidity in studies of evapotranspiration. However, for decades after their work was published, researchers largely ignored this important factor. In 1948, a paper by Penman (1948) formalized the importance of humidity and is a classic work. How much progress have we made since then? In 1965, Monteith (1965) reworked some of Penman's ideas; these formulations still are the basis for most applications and research in evapotranspiration. Some refinements have been made in considering soil effects, but it is far from clear that this research effort has been more useful than simply using most of the scientific papers as a mulch, analogous to a light-hearted suggestion by Stanhill (1970).

Unless these simple empirical models are based in invariant quantities, they are mainly useful in localized applications rather than for the wider scientific community as fundamental knowledge. Unfortunately, the natural optimism of many scientists and the excessively burdensome literature lead some individuals to think that they can find the elusive, simple general equations; alternatively, others yield to the pressure from a short-sighted granting system and pursue the engineering approach by seeking a limited solution to their specific problem with no intention of contributing to the knowledge base of the wider scientific community. In any case, many scientific person-years can be wasted. Searching for simple solutions to complex problems is the backbone of science, but a quotation attributed to Einstein may be appropriate: "Everything should be made as simple as possible, but not simpler."

C. How Do We Approach PE Modeling?

Numerous attempts have been made to construct PE models for various purposes (Stewart and Lemon, 1969; Goudriaan, 1977; Normal and Campbell, 1983). Each of these models has some portions that are faithful to known mechanisms and some parts that are gross simplifications. However, the complexity is now beginning to overwhelm us because we may not have an adequate means to organize all the essential information. The equations of motion provide a powerful and relatively simple structure for organizing exploration of fluid-flow phenomena. Where would fluid dynamics be without such equations? Perhaps where vegetation scaling is now. The accomplishments of Kepler in organizing the mountains of astronomical data into a few simple equations of planetary motion provide an interesting case that may parallel the vegetation scaling problem. Kepler simply faced the seemingly insurmountable problem directly, instead of looking for some clever and expedient short cut, and his accomplishments are a landmark in science. Perhaps we also need to face this mountain of data directly but, unlike Kepler, who worked essentially alone, we may need to establish a larger community of cooperating scientists to accomplish this goal rather than continue in competition among individual scientists or very small groups.

The scientific method rests on the assurance that any competent and educated person can repeat experiments and verify theories and concepts put forward by another. Although the scientific method is clear in concept, practical limitations may corrupt its use. One such example is the generation of large, complex computer models of biological systems. The details of such models may be buried in thousands of lines of program code in one of many possible languages. Although independent verification is possible, in principle, it is not practical because of the impediments associated with programming. Untold logical atrocities and inconsistencies can be hidden beneath the surface in the name of progress because of gaps in fundamental knowledge that would cause unacceptable delays if they were faced head-on. Announcing such weak links to the reader can cause an overly negative reaction that may obscure many of the positive accomplishments of a particular model. A usual means of dealing with this dilemma is to make the program available and allow the user to find the weak links.

Perhaps we in the biological community could benefit from an approach used in global climate modeling; namely, the establishment of a center where a Community Plant-Environment Model could be developed and maintained much like the Community Climate Model (CMM) is handled at the U.S. National Center for Atmospheric Research (NCAR). This can reduce the redundant effort of duplicating program codes of complex systems and greatly enhance progress by focusing creative energy on scientific rather than programming problems. Further, the code and documentation are maintained by professionals and readily available for execution and evaluation.

Several impediments to accomplishing such a goal of community cooperation exist. (1) An organizational structure is needed with some central staff and location. (2) Analytical and numerical procedures need to be developed and implemented for coping with the large programming code and making it simple for many researchers to link into that code; for example, modular program development (Acock and Reynolds, 1990) or parallel processing techniques. (3) A funding base needs to be set up for such a center. (4) Publication of contributions will have to be credited appropriately, perhaps through an electronic journal (Acock *et al.*, 1990). Setting up such an effort would be challenging, indeed.

VII. Examples of Scaling Leaf Photosynthesis to Canopy Photosynthesis

Leaf photosynthetic rates can be measured directly with portable field chambers (Field *et al.*, 1989), but canopy photosynthetic rates are much more difficult to measure (Goel and Norman, 1990). Therefore, accurate and general methods for scaling from leaf measurements to canopy estimates of photosynthetic rate can be most useful. In this section, I will compare several methods of scaling leaf-to-canopy photosynthetic rate and transpiration rate with predictions from a comprehensive PE model called Cupid (Norman and Campbell, 1983; Norman and Arkebauer, 1991). Cupid accommodates all the generality inherent in a comprehensive PE model whereas the simplified scaling methods are based on only one or two variables. However, the few inputs that are required for the simplified scaling methods are similar to the input Cupid uses. Predictions of radiation interception, canopy photosynthetic rate, and transpiration rate from the Cupid model compare reasonably well with measurements when measured leaf characteristics are used as inputs for native prairie grasses (Norman and Polley, 1989) and corn (Norman, 1988; Norman and Arkebauer, 1991). In the following discussion, predictions from Cupid are used as a reference for comparing various simplified

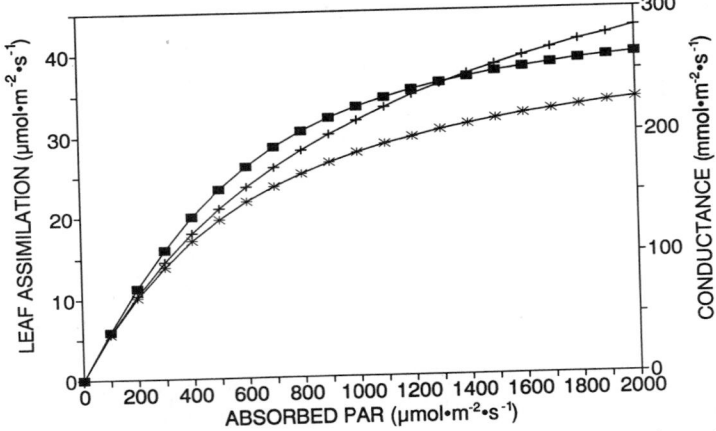

Figure 4.1   Leaf assimilation rate (μmol CO_2 m^{-2} sec^{-1}) and stomatal conductance (mmol H_2O m^{-2} sec^{-1}) as a function of absorbed PAR (μmol quanta m^{-2} sec^{-1}) used in the simplified scaling methods for comparison with the Cupid model. The nonrectangular hyperbola curves for conductance (*) and photosynthetic rate (■) match responses used in Cupid and the rectangular hyperbola (+) curve is used in scaling Method 2b (Eq. 12).

scaling approaches because equations in Cupid accommodate more factors that are known to be important and results from Cupid compare favorably with measurements. The following examples will use light-versus-photosynthesis and light-versus-conductance relationships typical of C_4 prairie grasses (Fig. 4.1). Actual weather data from 2 days (one clear and one partly cloudy) will be used as input data for the Cupid model for comparison with scaling methods (Table 4.1). However the leaf gross assimilation-versus-light relationship (A vs $a_\ell R_{PAR}$) is considered independent of temperature to simplify interpretation of these comparisons. Dark respiration is temperature dependent, with a $Q_{10} = 1.8$ and a rate of 1.55 μmol · m^{-2} · sec^{-1} at 25°C (see Polley *et al.*, 1991).

Three LAI values (4.5, 1.5, 0.5) are used with a spherical leaf angle distribution (LAD) and one LAI (1.5) is used with an erect and horizontal LAD. The spherical LAD has a mean leaf inclination angle of 57°, the horizontal 27° and the erect 70°.

A. Method 1: Scale Using Leaf Photochemical Efficiency and APAR

The simplest method for estimating canopy photosynthesis from leaf photosynthesis is to use the maximum photochemical efficiency (α) at low light, which appears to be constant for C_4 plants, and multiply it by the absorbed photosynthetically active radiation [APAR (μmol q·m^{-2}·sec^{-1})] From Ehleringer and Pearcy (1983), a reasonable value of α is 0.062 mol

Table 4.1 Environmental Conditions That Are Used as Boundary Conditions for the Comparison of the Cupid Model with Simple Scaling Models

Time (hr)	Air temperature (°C)	Vapor pressure (mbar)	Wind speed (m·sec^{-1})	Solar radiation (w·m^{-2})	PAR (μmol quanta m^{-2} sec^{-1})	F beam
0.9	23	23	3	0	0	0
1.9	23.1	22	3.4	0	0	0
2.9	22.5	21	3.2	0	0	0
3.9	22.3	20.5	3.2	0	0	0
4.9	21.6	20.1	2.7	0	0	0
5.9	21.2	20.5	2.2	59	125	0.33
6.9	22.8	21.1	3.7	232	492	0.59
7.9	25.8	22.5	4.3	461	988	0.86
8.9	29.1	23.5	5.3	655	1409	0.9
9.9	31.6	24.5	3.8	819	1764	0.92
10.9	34.6	24.5	3.8	938	2021	0.93
11.9	35.8	25	3.1	984	2121	0.93
12.9	37.5	25	1.7	984	2121	0.9
13.9	37.5	25	1.6	910	1961	0.82
14.9	37	24.5	1.6	804	1732	0.79
15.9	36	24	2	626	1348	0.68
16.9	32.8	24	5	436	935	0.58
17.9	31	24	4.4	243	517	0.46
18.9	28.8	23.5	3.7	69	143	0.21
19.9	25.6	23.5	3	0	0	0
20.9	24.2	23.5	2.9	0	0	0
21.9	22.9	23.5	2.8	0	0	0
22.9	21.7	23.2	2.8	0	0	0
23.9	20.1	23.2	2.9	0	0	0
0.9	20.5	23	2.7	0	0	0
1.9	19.7	22.8	2.8	0	0	0
2.9	19	21.7	3.1	0	0	0
3.9	18.8	21.3	3.1	0	0	0
4.9	18.7	21.4	3.4	0	0	0
5.9	18.4	20.9	3.5	40	83	0.16
6.9	18.9	20.9	2.3	124	263	0.19
7.9	20.5	21.1	1.5	333	714	0.44
8.9	22.6	21.8	2.7	375	807	0.31
9.9	22.8	22.5	3.4	474	1021	0.32
10.9	21.8	21.5	3.4	225	486	0.02
11.9	21.8	22	3.2	196	423	0
12.9	22.6	22	2.5	263	566	0.04
13.9	23.9	22.5	2.2	367	791	0.14
14.9	25.3	23.2	2	538	1159	0.37
15.9	25.8	23.5	2.5	536	1153	0.51
16.9	26.1	23.8	2.5	358	768	0.41
17.9	25.3	23.5	1.7	83	176	0.03
18.9	24.8	23.2	2.3	14	29	0
19.9	24.2	23.2	2.8	0	0	0
20.9	23.9	23	2.6	0	0	0
21.9	22.3	22.9	1.8	0	0	0
22.9	21.6	22.4	1.9	0	0	0
23.9	21.4	22.5	1	0	0	0

$CO_2 \cdot (\text{mol quanta})^{-1}$. Therefore, the scaled canopy assimilation rate is given by

$$A_{la} = 0.062 \text{ APAR} - r_d \text{LAI}, \tag{4}$$

where APAR is given by the difference of net PAR on the horizontal above the canopy and below the canopy and r_d is leaf dark respiration rate. Hourly or daily values of APAR can be used to obtain hourly or daily scaled canopy photosynthetic rates. Although this method is simple, it overestimates canopy photosynthetic rates on clear days by a factor of two and provides modest overestimates on cloudy days.

An alternative method of using APAR is based on using the average slope of the A_ℓ vs R_{PAR} curve between the light compensation point and the PAR flux density on the horizontal above the canopy (R_{PAR}). The curve in Fig. 4.1 is represented by the following nonrectangular hyperbola:

$$A_\ell = [x - (x^2 - 4z A_{max} \alpha_\ell R_{PAR})^{0.5}]/(2z), \tag{5}$$

where

$$x = \alpha_\ell R_{PAR} - A_{max}$$

with $z = 0.7$ and $A_{max} = 45 \ \mu\text{mol m}^{-2}\text{sec}^{-1}$ and a_ℓ is leaf absorptivity to PAR. The slope of Eq. 5 is given by

$$dA/dR = \alpha[1 - (x - 2 z A_{max})/(x^2 - 4z\alpha A_{max} a_\ell R_{PAR})]^{0.5}/(2z) \tag{6}$$

The mean slope is obtained from an average of α and the value from Eq. 6, so

$$P_{1b} = \text{APAR} \ (\alpha + dA/dR)/2 - r_d \text{LAI}. \tag{7}$$

Equation 7 provides remarkably good agreement with estimates of canopy photosynthesis from the Cupid model under clear conditions at a range of LAI values, even for an erect leaf angle distribution (Fig. 4.2); for horizontal leaves it tends to overestimate midday rates. Under cloudy and partly cloudy conditions, Eq. 7 compares well except at the highest LAI, where it underestimates (Fig. 4.3). Values of APAR can be measured with light sensors above and below a canopy or estimated from measurements of incoming PAR above the canopy and LAI using a light model.

B. Method 2: Scale Using Average Illumination and LAI

The average radiation incident on a horizontal plane in the canopy might be represented by the mean of incident radiation above the canopy (R_{PAR}) and radiation that is transmitted through the canopy and incident on the soil surface (R_T). Therefore the amount of radiation absorbed by an average leaf might be

$$R_\ell = (R_{PAR} + R_T) \ a_\ell/(2\text{LAI}). \tag{8}$$

Substituting R_ℓ for $a_\ell R_{PAR}$ in Eq. 5 yields an estimate of photosynthesis for this average leaf, so canopy photosynthetic rate can be estimated from photosynthetic rate can be estimated from

$$P_{2a} = (A_\ell - r_d)\,\text{LAI}. \tag{9}$$

This equation may overestimate or underestimate canopy photosynthesis by 30% or more, and leaves much to be desired as a scaling method. This overestimate occurs because leaves in a canopy experience a wide range of intensities and the average of the assimilation rates of all these leaves cannot be represented by substituting a mean light intensity in the nonlinear light assimilation relationship. The more linear the leaf photosynthesis rate is with light, the smaller will be the canopy overestimate from Eqs. 8 and 9.

The deficiencies in the averaging method represented by Eqs. 8 and 9 can be overcome partially by integrating leaf photosynthetic rate over LAI and accommodating the nonlinear light extinction with LAI and the nonlinear dependence of photosynthesis on absorbed light. For this integration to be useful in simplified scaling approaches, we must be able to integrate the appropriate equations analytically. Instead of assuming that all leaves in the canopy are at some average illumination, as implied by Eqs. 8 and 9, this method assumes that all leaves in a layer are at some average illumination. Although this method does not accommodate the fact that leaves in a layer may be sunlit or shaded and thus experience a range of light intensities, it does account for a general decrease in illumination with depth in the canopy. The canopy assimilation can be obtained from the leaf assimilation with

$$P_{2b} = \int A_\ell \, dF, \tag{10}$$

where dF is the incremental LAI. Since the nonrectangular hyperbola (Eq. 5) is difficult to integrate when combined with exponential light extinction, we will use a rectangular hyperbola (Fig. 4.1). The light transmitted below an LAI of F and absorbed by a leaf is $a_\ell R_{PAR}\exp(-KF)$, where K is the extinction coefficient or the fraction of the leaf area projected onto the horizontal. One of the difficulties in applying Eq. 10 is that direct and diffuse light have different extinction coefficients and are distributed over different amounts of leaf area. Setting $K = 0.7$ is a reasonable compromise for the extinction coefficient, but direct beam effectively illuminates KdF leaf area whereas diffuse light illuminates dF. Therefore, we can multiply the assimilation calculated from the rectangular hyperbola by an adjustment factor $f_b K + 1 - f_b$ to adjust for the different areas illuminated by direct and diffuse radiation, so

$$A_\ell = \frac{[(f_b K + 1 - f_b)A_{max}\alpha]}{\{\alpha + A_{max}/[a_\ell R_{PAR}\exp(-KF)]\}}, \tag{11}$$

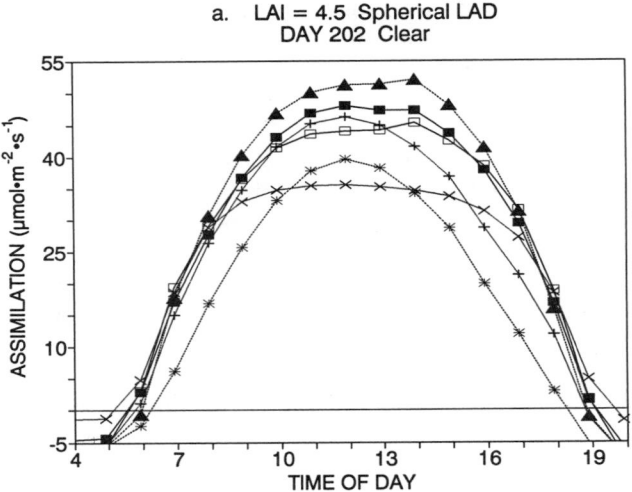

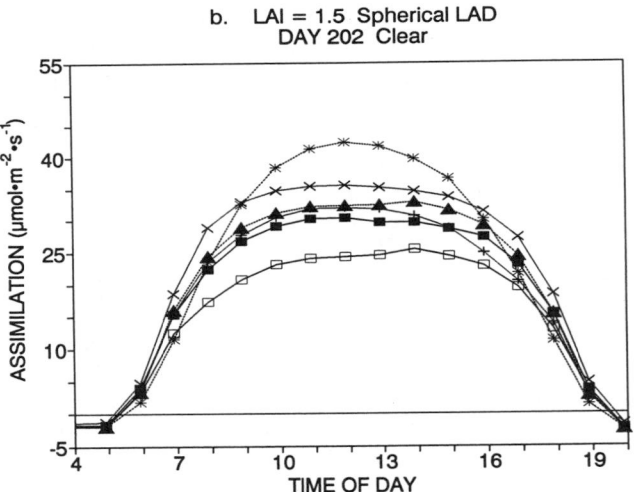

Figure 4.2 Comparison between canopy assimilation predictions from the plant–environment model Cupid (■) and five simplified scaling methods (Eq. 7,+; Eq. 9, *; Eq. 10, □; Eq. 13, x; Eq. 14,▲) for a clear day for five different canopy architectures.

c.   LAI = 0.5 Spherical LAD
DAY 202 Clear

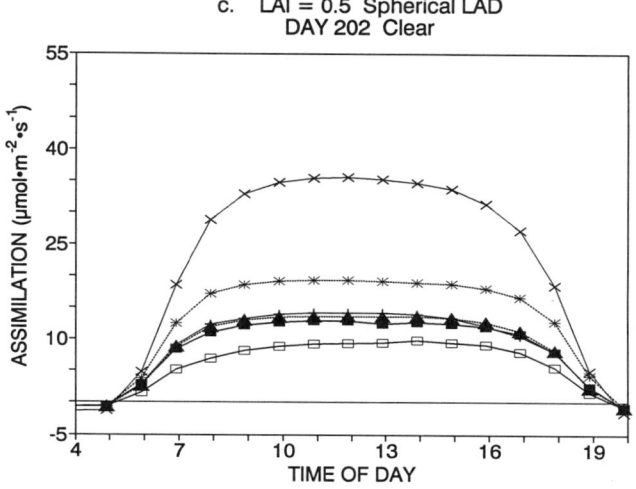

d.   LAI = 1.5 Erect LAD
DAY 202 Clear

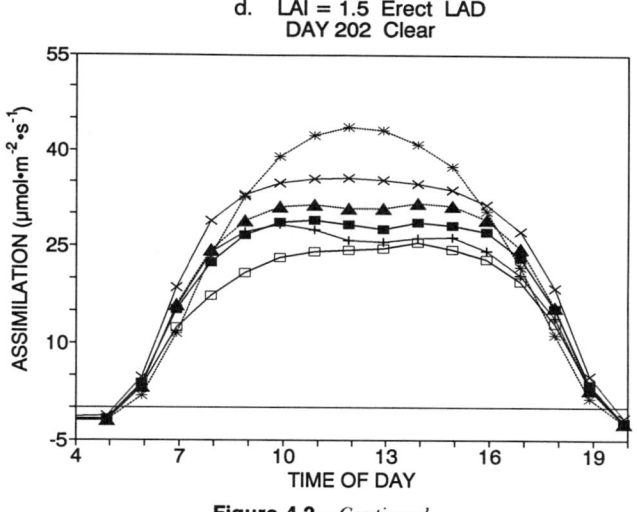

Figure 4.2—*Continued*

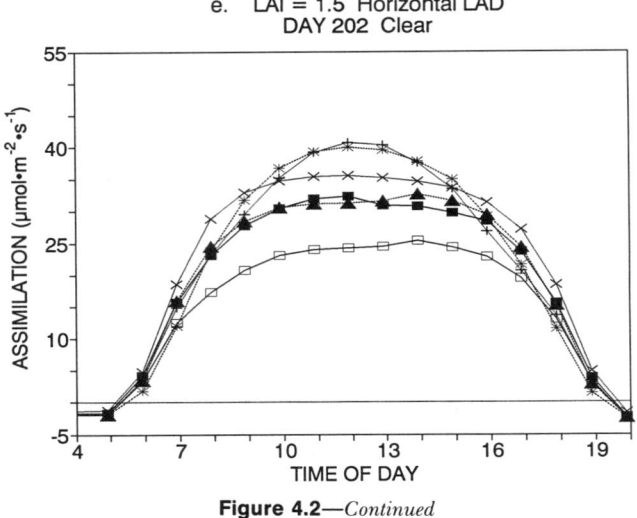

Figure 4.2—*Continued*

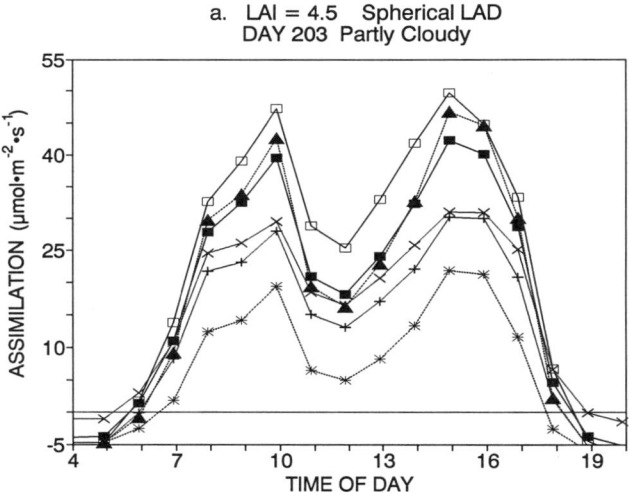

Figure 4.3 Comparison between canopy assimilation predictions from the plant–environment model Cupid (■) and five simplified scaling methods (Eq. 7,+; Eq. 9,*; Eq. 10,□; Eq. 13, x; Eq. 14, ▲) for a partly cloudy day for five different canopy architectures. The incident radiation is 100% diffuse at noon.

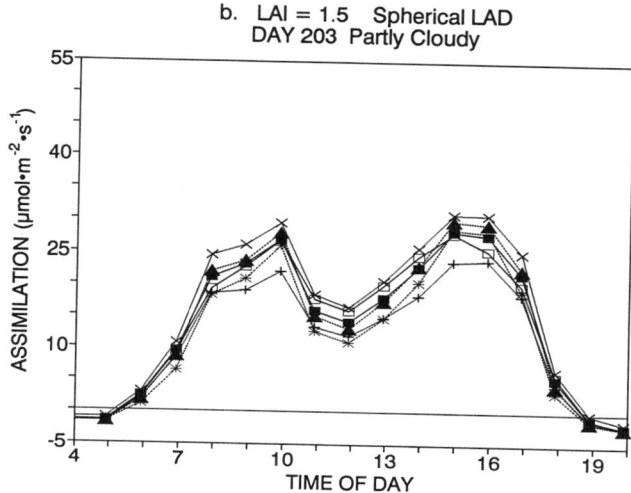

b.   LAI = 1.5   Spherical LAD
DAY 203 Partly Cloudy

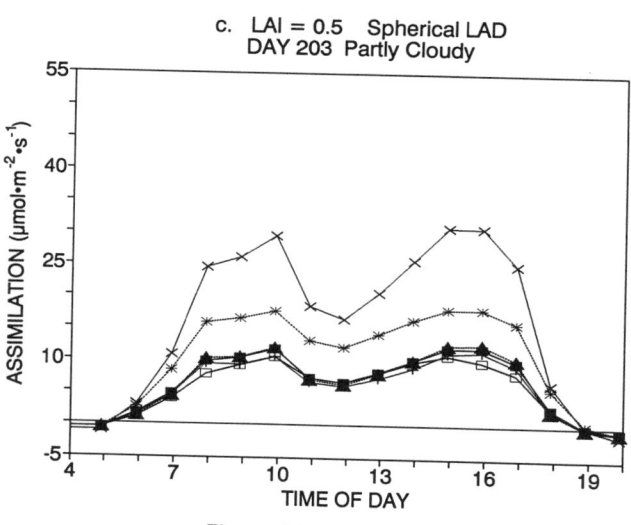

c.   LAI = 0.5   Spherical LAD
DAY 203 Partly Cloudy

Figure 4.3—*Continued*

60 *John M. Norman*

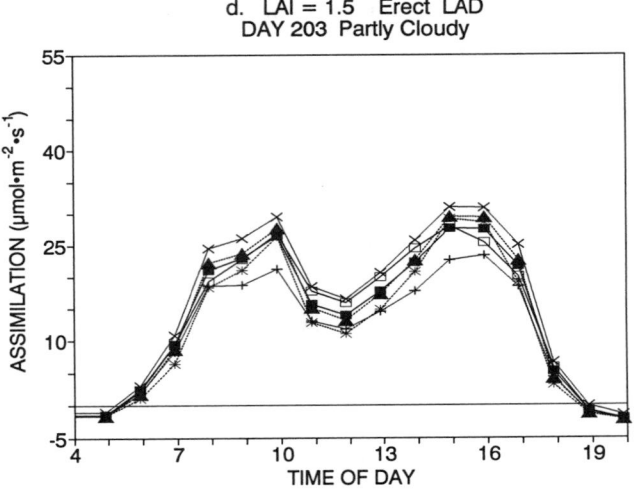

d. LAI = 1.5   Erect LAD
DAY 203 Partly Cloudy

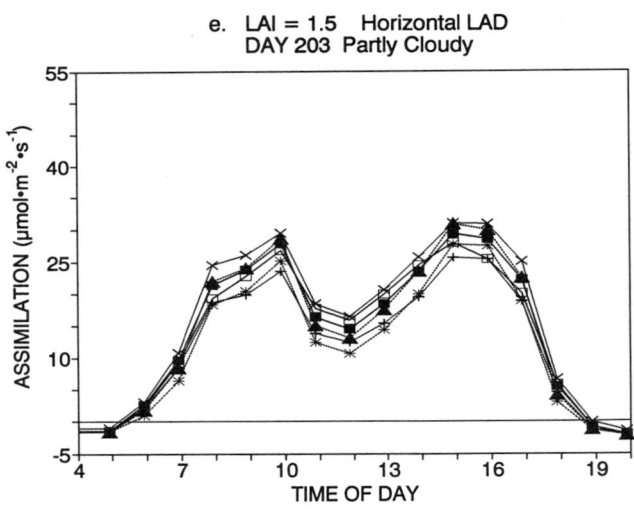

e. LAI = 1.5   Horizontal LAD
DAY 203 Partly Cloudy

Figure 4.3—*Continued*

where f_b is the fraction of incident PAR above the canopy on the horizontal that is direct beam. Substituting Eq. 11 into Eq. 10 and integrating yields

$$P_{2b} = (f_b K + 1 - f_b) \, A_{max}(\text{LAI} - C_1 + C_2) - r_d \text{LAI}, \qquad (12)$$

where

$$C_1 = (1/K) \, \ln\{1/(A_{max}) + 1/[\alpha_\ell R_{PAR} \exp(- K \text{ LAI})]\}$$

and

$$C_2 = (1/K) \, \ln[1/(A_{max}) + 1/(\alpha_\ell R_{PAR})].$$

This equation for scaling from leaf to canopy photosynthetic rates gives reasonable agreement with the Cupid model under cloudy conditions (Fig. 4.3) and tends to underestimate canopy rates when skys are clear (Fig. 4.2), if f_b is known. Approximate value of f_b can be obtained from solar radiation using equations in Weiss and Norman (1985). However, if f_b is set to unity because it is not known, then Eq. 12 provides reasonable estimates of canopy assimilation under clear skys but underestimates canopy photosynthetic rate when $f_b = 0$ by 30%. Since K is known to vary with f_b and sun zenith angle, this method can be refined by making adjustments for K. The leaf assimilation curve for upper sunlit leaves was used to obtain the results from Eq. 12 by fitting the best rectangular hyperbola to the curve in Fig. 4.1. However, leaves are known to adapt to lower light levels by reducing their maximum photosynthetic and dark respiration rates; this effect is included in the Cupid model, so overestimation by Eq. 12 is expected. Use of some average leaf assimilation curve would improve agreement of the results.

Sellers (1985) has used an integration procedure similar to but more complex than that represented by Eq. 10. The result is an analytical solution for canopy photosynthesis as a function of incident light, leaf physiological properties, and canopy architecture considering the interception and scattering of diffuse and direct beam radiation separately. However, he did not partition leaves into sunlit and shaded fractions; in fact, he states that using a PAR flux density that is averaged over leaves in each layer is more appropriate than considering sunlit and shaded leaves separately when applying the nonlinear light-assimilation relationship (Sellers, 1985, p. 1350). Although, at present, I tend to disagree with Sellers (1985) and believe that applying the nonlinear light-assimilation relationship to sunlit and shaded leaf fractions separately is preferable, a definitive resolution of this question awaits a complete solution of the time-dependent behavior of photosynthesis and stomatal conductance in a fluctuating light environment. The analytical solutions of Sellers (1985) are considerably more complicated than Eq. 12 and presumably more accurate.

C. Method 3: Consider the Canopy as One Large Horizontal Leaf

When considering scaling of leaf assimilation to canopy assimilation, an obvious question that might be asked is, "How well is a canopy represented by a large horizontal leaf?" In other words, how well does Eq. 5 represent canopy photosynthesis? In this case, the canopy photosynthesis rate is estimated from

$$P_3 = A_\ell - r_\mathrm{d}. \tag{13}$$

Based on the results in Figs. 4.2 and 4.3, the answer is "suprisingly well for canopies of LAI near 2." However, at higher LAI values it underestimates and at lower LAI values it overestimates canopy photosynthetic rate.

D. Method 4: Stratify Canopy into Sunlit and Shaded Leaves

Leaves in a canopy are illuminated by light from three sources: diffuse sky light, light scattered from the soil and other leaves in the canopy, and direct light from the beam of the sun. For a clear day, illumination of a leaf by the direct beam of the sun depends mainly on the angle between the leaf and the sun and not on depth in the canopy. Therefore, if we assume that a single curve represents the light-assimilation relationship for all leaves in the canopy, leaves can be divided into two classes: sunlit and shaded. If we know the LAI that is sunlit ($\mathrm{LAI}_\mathrm{sun}$), then contributions of sunlit and shaded leaves can be calculated separately and summed, so

$$P_4 = A_\mathrm{sun}\, \mathrm{LAI}_\mathrm{sun} + A_\mathrm{shad}\, (\mathrm{LAI} - \mathrm{LAI}_\mathrm{sun}), \tag{14}$$

where A_sun is the leaf net photosynthetic rate appropriate for the illumination received by sunlit leaves, A_shad is for shaded leaves, and $\mathrm{LAI} - \mathrm{LAI}_\mathrm{sun}$ is the shaded LAI. This procedure has been used in the past (Norman, 1981; Forseth and Norman, 1992). Separating the canopy into sunlit and shaded leaves has an advantage because sunlit leaves are the only leaves likely to be light saturated and shaded leaves are normally operating in the near-linear portion of the light-assimilation relationship. Therefore the equations derived by Norman (1982) can be used for predicting the mean illumination of shaded leaves (R_shad); that illumination level can be substituted into Eq. 5 to obtain A_shad. The flux density incident on sunlit leaves is

$$R_\mathrm{sun} = f_\mathrm{b} R_\mathrm{PAR} K + R_\mathrm{shad}, \tag{15}$$

where f_b is the fraction of the incident PAR above the canopy that is in direct beam and K is the extinction coefficient that can be represented by the two-parameter β distribution (Goel and Strebel, 1984) or the single parameter ellipsoidal distribution (Campbell, 1986). This extinction coefficient, K, is also the cosine of the mean leaf–sun angle divided by the cosine of the sun zenith angle. The sunlit LAI is calculated from

$$\text{LAI}_{\text{sun}} = [1 - \exp(-K\,\text{LAI})]/K. \qquad (16)$$

This method provides good estimates of canopy photosynthesis from leaf photosynthesis, as has been shown in the past (Norman, 1981); the main source of differences between P_4 and canopy photosynthesis rate from Cupid is the change in leaf photosynthetic characteristics with depth in the canopy. Equation 14 can be used to accommodate the change in leaf photosynthetic characteristics with depth by applying it layer by layer. For example, if the canopy were divided into three layers, then Eq. 14 would be applied three times, once for each layer, and a different light assimilation curve used for each layer. From agreement of Eq. 14 with results from the Cupid model in Figs. 4.2 and 4.3, this normally may not be necessary.

E. Method 5: Consider Leaf Energy Balance and Environmental Gradients

Leaf and canopy photosynthetic rates do not depend only on light, but also on temperature, air humidity, wind speed (through leaf boundary layer conductance and aerodynamic conductance), and CO_2 concentration. Gradients of all these quantities exist between the bulk air and the leaves within the canopy. In addition, leaf photosynthesis is coupled closely to stomatal conductance, so the relationship between light and leaf photosynthetic rate depicted in Fig. 4.1 is simplistic. Because of this close coupling between leaf photosynthesis and stomatal conductance, environmental gradients and the leaf energy balance can have a strong influence on scaling from the leaf to the canopy.

A useful leaf model that accommodates both photosynthesis and stomatal conductance has been described by Ball *et al.* (1986). This model works well for describing the leaf behavior of the native prairie grasses used in this simulation (Polley *et al.*, 1991). Leaf photosynthetic rate is dependent on leaf temperature (T_ℓ), light, and internal CO_2 concentration (C_i). Stomatal conductance depends linearly on leaf photosynthetic rate, CO_2 concentration at the surface of the leaf (C_s), and relative humidity at the surface of the leaf (RH_s). The leaf energy balance must be solved along with the photosynthesis equations to obtain estimates of leaf temperature, stomatal conductance, and photosynthetic rate.

To illustrate scaling in the presence of environmental gradients, we will consider estimating canopy transpiration from knowledge of leaf characteristics. This can be done by two methods: (1) using the somewhat nebulous concept of "canopy conductance" and (2) solving the energy balance of a "representative" leaf and scaling this result to the canopy level.

The concept of canopy conductance that currently is in wide use was formalized by Monteith (1965). The canopy transpiration (Tr_1) in units of W·m^{-2} can be represented by

$$Tr_1 = [s(R_n - G) + q]/[s + \gamma(1 + r_c/r_a)], \tag{17}$$

where s is the slope of the saturation vapor pressure relation with temperature (Pa·°C^{-1}), R_n is the net allwave radiation above the canopy (W·m^{-2}), G is the soil heat flux (W·m^{-2}), γ is referred to as the psychrometer constant (Pa·°C^{-1}), r_c is the canopy resistance (m^2·sec·mol^{-1}), r_a is the aerodynamic resistance to heat or water vapor arising from the atmosphere above the canopy (m^2·sec·mol^{-1}), and

$$q = \rho\, C_p\, [e_s(T_{air}) - e_{air}]/r_a,$$

where ρ is the molar density of air (mol·m^{-3}), C_p is the molar specific heat of air (J·mol^{-1}·°C^{-1}), $e_s(T_{air})$ is the saturation vapor pressure at the temperature of the air (Pa), and e_{air} is the vapor pressure of the air (Pa). The canopy conductance, $k_c = 1/r_c$, can be calculated from Eq. 14 by replacing A_{sun} by k_{sun} and A_{shad} by k_{shade} using a stomatal conductance versus light curve as in Fig. 4.1 (Forseth and Norman, 1992). The aerodynamic resistance is calculated from the diabatically corrected log wind profile and is discussed later (Eq. 35). The soil heat flux can be evaluated empirically (Clothier *et al.*, 1990), set to 10% of the net radiation above the canopy, or neglected. The canopy temperature can be obtained from solving the following equation for T_c (°C).

$$R_n - G - Tr_1 = Q_1 = \rho\, C_p\, (T_c - T_{air})/r_a, \tag{18}$$

where Q_1 is the sensible heat flux from the canopy and the soil surface evaporation and sensible heat fluxes are assumed negligible. Equations 17 and 18 provide reasonable estimates of canopy transpiration for dense canopies. For sparse canopies of low LAI, the soil surface can become a dominant factor and it is not considered adequately in Eqs. 17 and 18. Exchanges with the soil can be accommodated in scaling methods like those we are considering here, but are beyond the scope of this brief discussion. Therefore, we will estimate the net radiation divergence in the canopy alone and replace G in Eqs. 17 and 18 by $R_{n,s}$, this net radiation just above the soil surface, which can be approximated by

$$R_{n,s} = R_n \exp(-0.4\, \text{LAI}), \tag{19}$$

if we assume that net radiation has an extinction coefficient of 0.4. Actually, net radiation can have extinction coefficients that vary from 0.3 (for low sun zenith angles) to 0.6; for sun zenith angles greater than 70°, the extinction coefficient can exceed even 0.6. Diffuse net radiation typically has an extinction coefficient near 0.5.

Equations 17–19 provide good estimates of canopy transpiration at high LAI under clear and cloudy conditions (Figs. 4.4a and 4.5a) because warming of the canopy air by heat that is absorbed at the soil surface is small. When LAI is 1.5 or less, the heating of the soil surface from absorbed net radiation causes elevated air temperatures in the canopy and increased transpiration. Cupid predicts these effects but Eq. 17 does not, so Eq. 17 underestimates canopy transpiration. In the case of horizontal leaves (Figs. 4.4e and 4.5e), the net radiation divergence of the canopy is larger than predicted by Eq. 19; this causes a further underestimate by Eq. 17. If the soil surface were wet, than Eq. 17 probably would overestimate canopy transpiration.

A second method for estimating canopy transpiration considers the leaf energy budget explicitly on a "representative" leaf, then scales this leaf value to the canopy. I refer to this method as a "scaled-leaf" method. The energy budget equation for a leaf can be solved for leaf temperature by linearizing the equation in terms of the temperature difference between the leaf (T_ℓ) and the canopy air (T_h) immediately surrounding that leaf ($\Delta T = T_\ell - T_h$), resulting in (Norman, 1979):

$$\Delta T = \{R^*_{n,\ell} - \lambda g_\ell \, [e_s(T_h) - e_h]/P\}/Y, \tag{20}$$

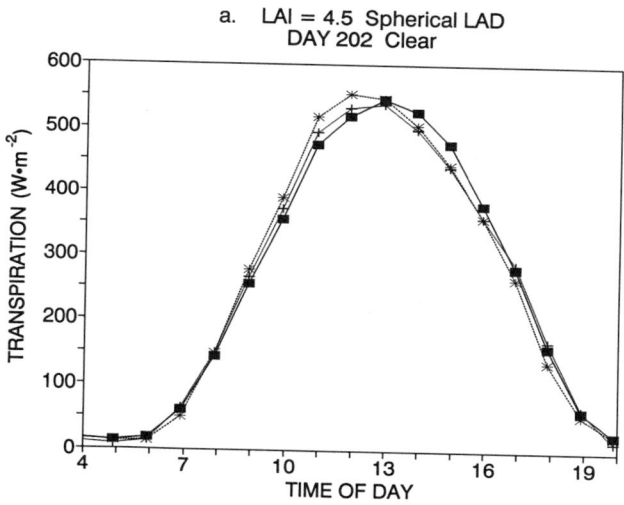

Figure 4.4 Comparison between canopy transpiration predictions from the plant–environment model Cupid (■) and two simplified scaling methods (Eq. 17, +; Eq. 26, *) for a clear day for five different canopy architectures.

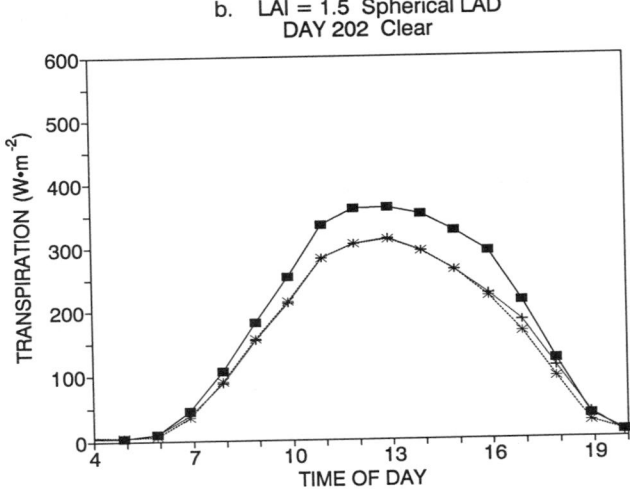

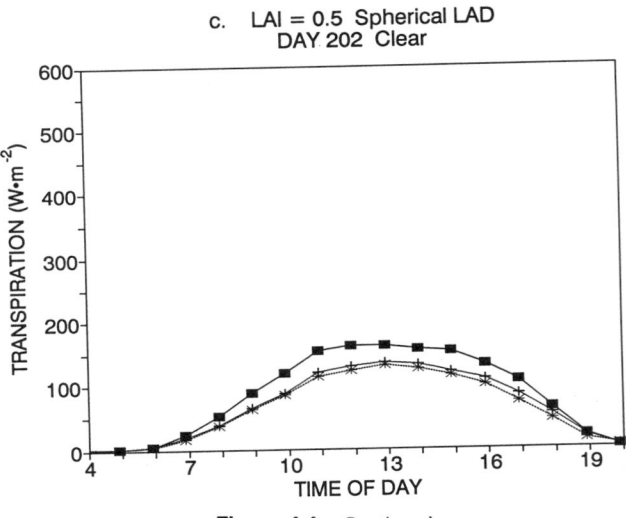

Figure 4.4—*Continued*

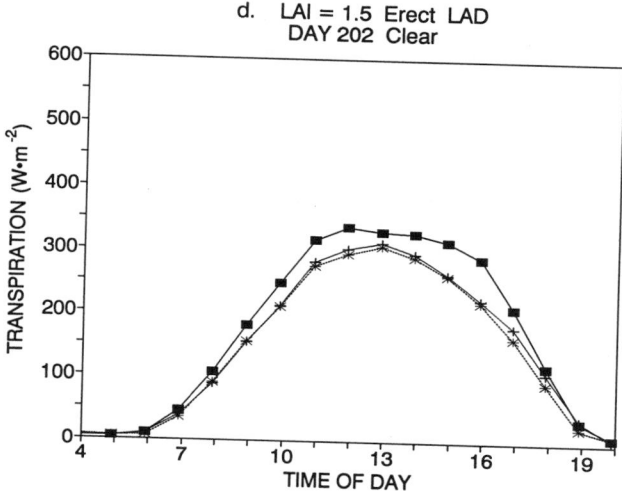

d.   LAI = 1.5 Erect LAD
DAY 202 Clear

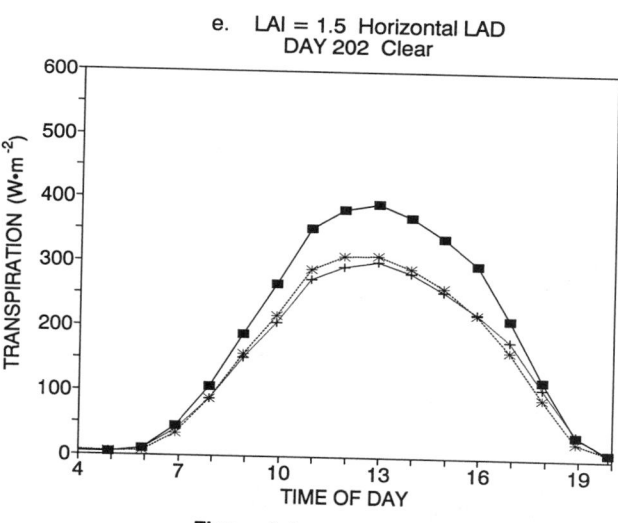

e.   LAI = 1.5 Horizontal LAD
DAY 202 Clear

Figure 4.4—*Continued*

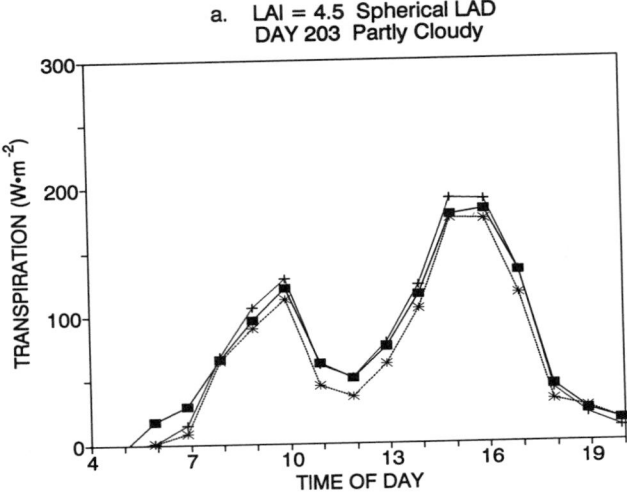

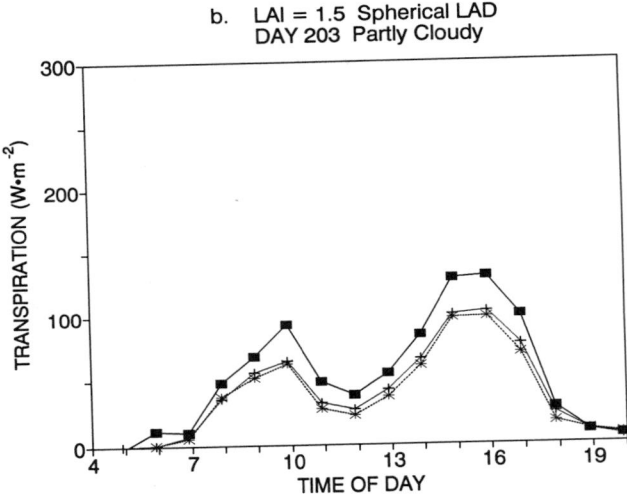

Figure 4.5 Comparison between canopy transpiration predictions from the plant–environment model Cupid (■) and two simplified scaling methods (Eq. 17,+; Eq. 26, *) for a partly cloudy day for five different canopy architectures. The incident radiation is 100% diffuse at noon.

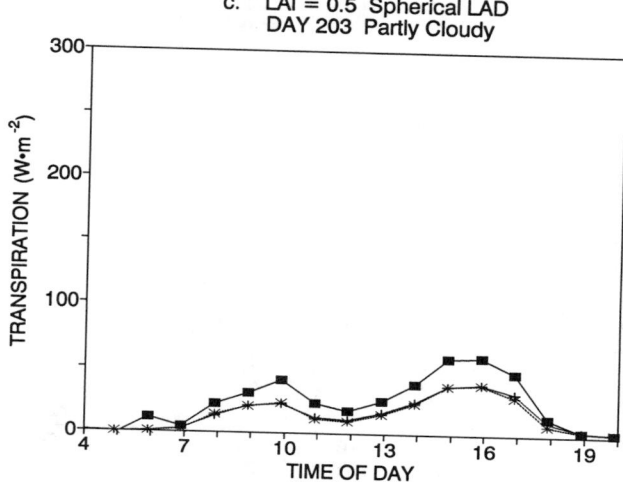

c. LAI = 0.5 Spherical LAD
DAY 203 Partly Cloudy

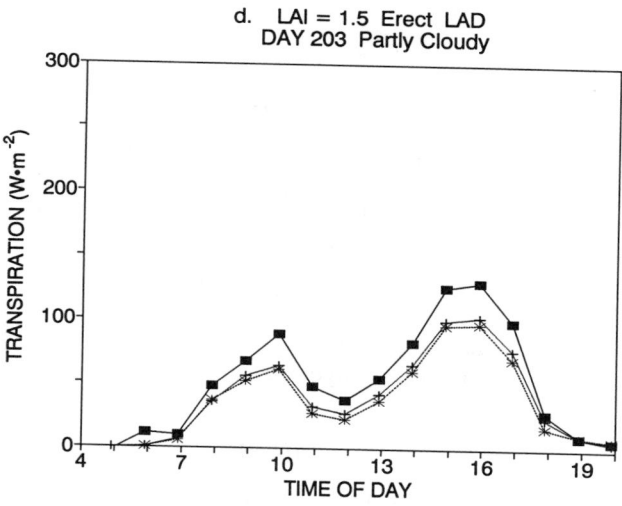

d. LAI = 1.5 Erect LAD
DAY 203 Partly Cloudy

Figure 4.5—*Continued*

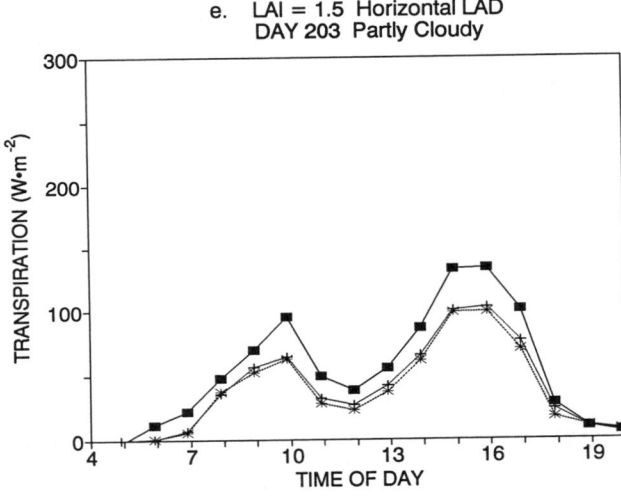

e.   LAI = 1.5 Horizontal LAD
     DAY 203 Partly Cloudy

Figure 4.5—*Continued*

where

$$Y = \lambda g_\ell/P + C_p/r_h + 4\sigma\varepsilon(T + 273)^3. \tag{21}$$

e_h is the vapor pressure of the canopy air, $e_s(T_h)$ is the saturation vapor pressure at the temperature of the canopy air, and $R^*_{n,\ell}$ is the leaf net radiation assuming the leaf to be at the temperature of the canopy air. The leaf conductance to water vapor, g_ℓ, is given by

$$g_\ell = 1/(r_h + r_s), \tag{22}$$

where $r_s = 1/g_s$ is the leaf stomatal resistance in units of $m^2 \cdot sec \cdot mol^{-1}$ (see Fig. 4.1) and the boundary layer resistance, including both sides of a leaf, is

$$r_h = 2.2 \ (D/u)^{0.5} \tag{23}$$

in units of $m^2 \cdot sec \cdot mol^{-1}$, where D is the appropriate leaf dimension (m) and u is the wind speed (m $\cdot$ sec^{-1}). In addition, P is the atmospheric pressure (Pa), σ is the Stefan–Boltzman constant (J $\cdot$ mol$^{-1} \cdot$ K^{-1}), and ε is the leaf thermal emissivity. Results from this treatment of the leaf energy balance are most accurate if s in Eqs. 20 and 21 and T in the thermal radiation term in Eq. 21 use the mean of T_ℓ and T_h.

The leaf net radiation, sensible heat, and transpiration are given by the following, respectively:

$$R_{n,\ell} = R^*_{n,\ell} - 4\sigma\varepsilon(T + 273)^3 \Delta T \tag{24}$$

$$Q_\ell = C_p \Delta T / r_h \tag{25}$$

$$Tr_\ell = \lambda g_\ell[e_s(T_h) - e_h + s\Delta T]/P. \tag{26}$$

Using Eqs. 20–26 requires estimates of leaf net radiation, as well as wind speed, air temperature, and air vapor pressure in the canopy surrounding the "representative" leaf. The leaf net radiation can be estimated from the net radiation divergence of the canopy by

$$R^*_{n,\ell} = R_n[1 - \exp(-0.4 \text{ LAI})]/\text{LAI}. \tag{27}$$

If wind speed, U, is available at some height, z, then the wind speed at the top of the canopy (u_h in m · sec^{-1}) can be estimated from the log wind profile equation

$$u_h = u_* \{\ln[(h - d)/z_0] - \Psi_m\}/0.4, \tag{28}$$

where

$$u_* = 0.4 \ U/\{\ln[z - d)/z_0] - \Psi_m\}. \tag{29}$$

The height of the canopy is h (m); z_0 (m) is the canopy momentum roughness ($z_0 = 0.13 \ h$); d (m) is the displacement height ($d = 0.63 \ h$); and Ψ_m is the diabetic profile correction factor for momentum. (See Brussaert, 1984, for a discussion of profile correction factors.) The wind speed at the height $d + z_0$ can be taken as appropriate for our "representative" leaf; however, the wind speed predicted by the log wind profile is zero at this height. Therefore an equation developed by Thom (1971) is used to predict the wind speed at the height $d + z_0$:

$$u_d = u_h/\{1 + m[1 - (d + z_0)/h]\}^2, \tag{30}$$

where m is a profile coefficient that is about 1.5 for grass canopies. Substituting u_d into Eq. 23 provides an estimate of leaf boundary layer resistance for our "representative" leaf.

The temperature (T_h), vapor pressure (e_h), and CO_2 concentration (C_h in μmol·mol^{-1}) of the air surrounding the leaf can be obtained from values at the reference height, z (T_z, and C_z), by evaluating the log profiles for these quantities at the height of the top of the canopy (h). Therefore,

$$T_h = T_z + (Q_\ell \text{LAI W})/ (0.4 \ u_* \ \rho \ C_p) \tag{31}$$

$$e_h = e_z + [Tr_\ell \text{LAI R}(T_h + 273)W]/(0.4u_*\lambda) \tag{32}$$

$$C_h = C_z - [P_i R(T_h + 273)W]/(0.4u_*P), \tag{33}$$

where P_i is the canopy photosynthetic rate by one of the scaling methods discussed earlier, $Q_\ell \mathrm{LAI}$ and Tr_ℓ/LAI are canopy heat and water fluxes, respectively, and

$$W = \{\ln[(z - d)/z_0] - \Psi_h\} - \{\ln[(h - d)/z_0] - \Psi_h\}. \tag{34}$$

Equations 20–34 can be solved by a method of iteration because the diabatic correction factors, Ψ_m and Ψ_h (see Brusaert, 1984, for calculation of diabatic correction factors), in the log profile equations depend on the canopy sensible heat flux ($Q_\ell \mathrm{LAI}$), so the leaf energy balance equation must be solved consistent with the profile equations.

Earlier, Eq. 17 was presented as a means of scaling transpiration from leaf to canopy using a canopy resistance (r_c), which can be estimated from leaf characteristics (Fig. 4.1) and aerodynamic resistance to heat exchange (r_a). This r_a between the canopy and a reference height z above the ground can be estimated from the log wind profile equation and is related to W in Eq. 34:

$$r_a = r_b + \{\ln[(z - d)/z_0] - \Psi_h\}^2/(0.16U), \tag{35}$$

where r_b is an excess resistance, as discussed by Verma (1989) or Monteith and Unsworth (1990, pp. 248–251), or a scalor interfacial resistance from Brusaert (1984, pp. 103–110). Sometimes this excess resistance is accommodated by using a roughness length for heat (z_H) in Eq. 35 instead of z_0 and eliminating the additive term r_b. Numerous methods are available for estimating r_b in the three references just cited, but considerable uncertainty surrounds all the equations. The "scaled-leaf" approach (Eqs. 20–34) appears to accommodate an excess resistance as the leaf boundary layer resistance; one advantage of the "scaled-leaf" approach. Therefore, profile Eqs. 31–34, which usually would contain a roughness length for heat (z_H) instead of momentum (z_0) because of an additional interfacial resistance for heat (Brusaert, 1984; Verma, 1989), use the roughness length for momentum.

A major advantage of the leaf-to-canopy scaling described by Eqs. 20–34 is that chamber measurements made on individual leaves can be used to fit a photosynthesis model such as that of Ball *et al.* (1986) under the conditions appropriate for the leaf chamber; these equations can be used to scale the chamber results to the canopy level using the environment appropriate for the canopy. This is particularly important for boundary layer conductance because it can be high in chambers ($\sim 3 \ \mathrm{mol \cdot m^{-2} \cdot sec^{-1}}$) and low in the canopy ($0.8 \ \mathrm{mol \cdot m^{-2} \cdot sec^{-1}}$) for grass leaves. This difference in boundary layer resistance between leaf chambers and a canopy can affect leaf temperature, partitioning of absorbed radiation between sensible heat, and transpiration and affect the stomatal conductance through changes in the relative humidity at the leaf surface

(Ball *et al.,* 1986). When the leaf photosynthetic rate is known, the canopy rate can be estimated from Method 2b (Eq. 12) or Method 4 (Eq. 14).

The results from Eqs. 20–34 agree well with the results from the scaling Eqs. 17–19 (Figs. 4.4 and 4.5), so departures from predictions of the Cupid model for both scaling methods are similar.

F. Summary

The various examples described in this section illustrate many of the principles that can be used to scale from leaf-level processes to canopy-level processes. The methods vary widely in their complexity; a comparison such as this is possible because of the availabilty of a comprehensive PE model that has been compared with field measurements.

The examples discussed here have considered spatial variations and slow temporal variations (hourly variations in time). More rapid temporal variations do occur in canopies because of light fluctuations from clouds, wind, and changing sun angles. Leaf responses to light fluctuations have been studied (Pearcy, 1988) but little is known about the implications of these leaf responses for canopy-level phenomena.

VIII. Summary

This chapter contains a discussion of scaling leaf-level processes to canopy-level processes. Scaling has been a powerful concept in fluid dynamics and it is useful in dealing with the fluid–dynamical part of scaling from leaves to canopies. However, fluid–dynamical scaling appears to have limited usefulness in systems dominated by biological phenomena where detailed mathematical equations are not available for describing the characteristics. One approach to scaling biophysical systems is to develop and test detailed soil–plant–atmosphere models and evaluate various scaling schemes against the detailed model. Several examples of scaling photosynthesis and transpiration are included. Clearly the major limitation of this approach is availability of a reliable soil–plant–atmosphere model. This can best be accompanied through a community cooperative effort that exploits a broad base of expertise and avoids unnecessary redundancy. Therefore, a challenge ahead of us may be to create an environment conducive to the development of a Community Plant–Environmental Model. This has been done in the meteorological community with the Community Climate Model. Would such a community biophysical modeling effort be desirable, or would it be too difficult? Perhaps it might be unrealistic, or unfundable, or possibly unmanageable, or even against human nature, but it also might be worth trying.

References

Acock, B., Heller, S. R., and Rawlins, S. L. (1990). An electronic journal for sharing data on crop growth. *In* "Scientific and Technical Data in a New Era" (P. H. Glaeser, ed.), pp. 307–311. Hemisphere, New York.

Acock, B., and Reynolds, J. F. (1990). Model structure and data base development. *In* "Process Modeling of Forest Growth Responses to Environmental Stress" (R. K. Dixon, R. S. Meldahl, G. A. Ruark, and W. G. Warren, eds.), Chap. 13, pp. 169–179. Timber Press, Portland.

Baier, W. (1973). Crop-weather analysis model: Review and model development. *J. Appl. Meteorol.* **12**, 937–947.

Ball, J. T., Woodrow, I. E., and Berry, J. A. (1986). A model predicting stomatal conductance and its contribution to the control of photosynthesis under different environmental conditions. *In* "Progress in Photosynthesis Research" (J. Briggins, ed.), Vol. 4, pp. 221–224. Martinus Nijhoff, Dordrecht, The Netherlands.

Brusaert, W. (1984). "Evaporation into the Atmosphere: Theory, History and Applications." Reidel, Dordrecht.

Campbell, G. S. (1986). Extinction coefficients for radiation in plant canopies calculated using an ellipsoidal inclination angle distribution. *Agric. For. Meteorol.* **36**, 317–321.

Clothier, B. E., Clawson, K. L., Pinter, P. J., Moran, M. S., Jr., Reginato, R. J., and Jackson, R. D. (1990). Estimation of soil heat flux from net radiation during the growth of alfalfa. *Agric. For. Meteorol.* **37**, 319–329.

Cowan, I. R. (1982). Regulation of water use in relation to carbon gain in higher plants. *In* "Encyclopedia of Plant Physiology" (O. L. Lange, P. S. Nobel, C. B. Osmond, and H. Ziegler, eds.), pp. 589–613. Springer-Verlag, Berlin.

Deardorf, J. (1978). Efficient prediction of ground temperature and moisture with inclusion of a layer of vegetation. *J. Geophys. Res.* **83**, 1889–1903.

Ehleringer, J., and Pearcy, R. W. (1983). Variation in quantum yield for CO_2 uptake among C_3 and C_4 plants. *Plant Physiol.* **73**, 555–559.

Field, C. B., Ball, J. T., and Berry, J. A. (1989). Photosynthesis: Principles and field techniques. *In* "Plant Physiological Ecology" (R. W. Pearcy, J. Ehleringer, H. A. Mooney, and P. W. Rundel, eds.), pp. 209–253. Chapman and Hall, New York.

Forseth, I. N., and Norman, J. M. (1992). Modelling of solar irradiance, leaf energy budget and canopy photosynthesis. *In* "Photosynthesis and Production in Changing Environment: A Field and Laboratory Manual" (D. O. Hall, J. M. O. Scurlock, H. Bulhar, R. C. Leegood, and S. P. Long, eds.), Chap. 13, pp. 207–219. Chapman and Hall, London.

Frisch, U. (1990). A new strategy for hydrodynamics: Lattice gasses. *In* "Whither Turbulence: Turbulence at Crossroads" (J. L. Lumley, ed.). Springer-Verlag, New York.

Goel, N. S., and Norman, J. M. (eds.) (1990). "Instrumentation for Studying Vegetation Canopies for Remote Sensing in Optical and Thermal Infrared Regions." Harwood, London.

Goel, N. S., and Strebel, D. E. (1984). Simple beta distribution representation of leaf orientation in vegetation canopies. *Argon. J.* **76**, 800–802.

Goudriaan, J. (1977). "Crop Micrometeorology: A Simulation Study." Centre for Agricultural Publication and Documentation, Wageningen, The Netherlands.

Grace, J. (1981). Some effects of wind on plants. *In* "Plants and Their Atmospheric Environment" (J. Grace, E. D. Ford, and P. G. Jarvis, eds.), pp. 31–56. Blackwell Scientific, London.

Johnson, D. B. (1976). "A Survey of Crop-Weather Models With Data on Application of Selected Models to Centre County Pennsylvania." M. S. Thesis, Pennsylvania State University, University Park.

Lovelock, J. E. (1979). "Gaia, A New Look at Life on Earth." Oxford University Press, Oxford.

Miller, E. E., and Miller, R. D. (1956). Physical theory for capillary flow phenomena. *J. Appl. Phys.* **27,** 324–332.

Monteith, J. L. (1965). Evaporation and environment. *In* "Symposium of the Society for Experimental Biology 19," pp. 205–234. Cambridge University Press, Cambridge, England.

Monteith, J. L., and Unsworth, M. H. (1990). "Principles of Environmental Physics." Edward Arnold, London.

Norman, J. M. (1979). Modeling the complete crop canopy. *In* "Modification of the Aerial Environment of Crops" (B. J. Barfield and J. Gerber, eds.), pp. 249–277. American Society of Agricultural Engineers, St. Joseph, Michigan.

Norman, J. M. (1981). Interfacing leaf and canopy light interception models. *In* "Predicting Photosynthesis for Ecosystem Models" (J. D. Hesketh and J. W. Jones, eds.), Vol. II, pp. 49–67. CRC Press, Boca Raton, Florida.

Norman, J. M. (1982). *In* "Biometeorology in Integrated Pest Management" (J. L. Hatfield and I. J. Thomason, eds.), pp. 65–99. Academic Press, New York.

Norman, J. M. (1988). Systesis of Canopy Processes. *In* "Plant Canopies: Their Growth, Form and Function" (G. Russell, B. Marshall, and P. G. Jarvis, eds.), pp. 161–175. Cambridge, Univ. Press, Cambridge.

Norman, J. M., and Arkebauer, T. J. (1991). Predicting canopy light-use efficiency from leaf characteristics. *In* "Modeling Plant and Soil Systems" (J. T. Ritchie and J. Hanks, eds.), American Society of Agronomy, Madison, Wisconsin.

Norman, J. M., and Campbell, G. S. (1983). Application of a plant-environment model to problems in irrigation. *In* "Advances in Irrigation" (D. I. Hillel, ed.), pp. 155–188. Academic Press, New York.

Norman, J. M., and Polley, W. R. (1989). Canopy Photosynthesis. *In* "Photosynthesis" (W. R. Briggs, ed.), pp. 227–241. A. R. Liss, New York.

Pearcy, R. W. (1988). Photosynthetic light utilization of light flecks by understory plants. *Aust. J. Plant Physiol.* **15,** 223–238.

Penman, H. L. (1948). Natural evaporation. *Proc. R. Soc. London Ser. A* **193,** 120–143.

Polley, H. W., Norman, J. M., Arkebauer, T. J., Walter-Shea, E. A., Greegor, D. H., Jr., and Bramer, B. (1991). Leaf gas exchange of *Andropogon gerardii* Vitman, *Panicum virgatum* L., and *Sorghastrum nutans* (L.) Nash in a tallgrass prairie. Submitted for publication.

Sellers, P. J. (1985). Canopy reflectance, photosynthesis and transpiration. *Int. J. Remote Sens.* **6,** 1335–1372.

Stanhill, G. (1970). A new method of reducing evaporation. *J. Irreproducible Results* **18,** 55–57.

Stewart, D. W., and Lemon, E. R. (1969). "The Energy Budget at the Earth's Surface: A Simulation of Net Photosynthesis." Interim Report 69-3, Technical Report, ECOM 2-68, I-6, Cornell University, Ithaca, New York. (Available from National Technical Information Service, Springfield, Virginia.

Tanner, C. B., and Sinclair, T. R. (1983). Efficient water use in crop production. *In* "Limitations to Efficient Water Use in Crop Production" (H. M. Taylor, W. R. Jordan, and T. R. Sinclair, eds.), pp. 1–27. American Society of Agronomy, Madison, Wisconsin.

Thom, A. S. (1971). Momentum absorption by vegetation. *Q. J. R. Meteorol. Soc.* **97,** 414–428.

Thornley, J. H. M. (1976). "Mathematical Models in Plant Physiology." Academic Press, New York.

Vennard, J. K., and Street, R. L. (1976). "Elementary Fluid Mechanics," 5th Ed., Chap. 8. Wiley, New York.

Verma, S. B. (1989). Aerodynamic resistance to transfers of heat, mass and momentum. *In* "Estimation of Areal Evapotranspiration" (T. A. Black, D. L. Splittlehouse, M. D. Novak, and D. T. Price, eds.), pp. 13–20. IAHS Publ. No. 177, International Society of Hydrological Science, Wallingford, U.K.

Wallace, H. A. (1920). Mathematical inquiry into the effect of weather on crop yield in the eight corn belt states. *Month. Weather Rev.* **48,** 439–446.

Warrick, A. W., Mullen, G. J., and Nielson, D. R. (1977). Scaling field measured soil hydraulic properties using a similar media concept. *Water Resource Res.* **13,** 355–362.

Weiss, A., and Norman, J. M. (1985). Partitioning solar radiation into direct and diffuse, visible and near-infrared components. *Agric. For. Meteorol.* **34,** 205–213.

5

Scaling Water Vapor and Carbon Dioxide Exchange from Leaves to a Canopy: Rules and Tools

Dennis D. Baldocchi

I stand upon a hill so green
And spy upon a leaf.
It's occupied with chlorophyll,
Beyond the mind's belief.

Microchasms everywhere!!
Stomatal holes of Hell.
Cascades of carbon swallowed,
For vapor they expel.

It's this molecular ballet,
That makes chestnuts grow so tall.
Breathing air so silently,
'Til expiring in Fall.

I. Introduction

Through the ages a poetic picture of nature has been painted by artists and authors. Scientists do not have the latitude to describe nature poetically. Instead, scientists must describe nature as it is and understand how and why it operates. How scientists describe biological systems depends on their educational background. Scientists trained in the physical and mathematical sciences are inclined to describe biological processes and systems in terms of elegant analytical solutions to differential equations.

To describe carbon and water exchange of leaves mathematically—the subject of this essay—and to extend this information to the canopy scale, we must consider the roles of photosynthesis, stomatal mechanics, leaf and root respiration, transpiration, soil evaporation, decomposition, turbulence, diffusion, and radiative transfer. The diverse nature of this topic precludes defining and unifying a set of elegant mathematical equations for scaling carbon dioxide and water vapor fluxes from leaves to canopies, because environmental and evolutionary pressures force plants to accomodate carbon gain, water loss, nutrient uptake, growth, and reproduction through an assortment of routes and strategies. These pressures also cause plant processes to exhibit complex variations in time and space. Yet the reader should not despair and conclude that any attempt to integrate leaf-scale physiological processes to the canopy scale will be fruitless. Scaling frameworks, based on linkages between micrometeorological, phytoactinometric, physiological, and biochemical theories, have existed for over 25 years and are being improved continually.

In this essay, I review past work on the scaling of carbon dioxide and water vapor exchanges. Next, I discuss rules and theories required to integrate information on leaf carbon dioxide and water vapor exchange to the canopy scale. Finally, I demonstrate our ability to transfer flux information from leaf to canopy scale by considering two cases: a homogeneous crop canopy and a complex broadleaf forest canopy.

II. Literature Overview

The earliest attempts to scale carbon and water vapor exchange from leaves to canopies focused on horizontally homogeneous crop canopies. Scaling was conducted by linking light-dependent transpiration, photosynthesis, and stomatal conductance models for leaves with canopy radiative transfer models (de Wit, 1965; Duncan *et al.*, 1967; Miller, 1967,1970; Horn, 1971). These initial efforts were made possible by contemporary advances in modeling radiative transfer in closed plant canopies (Monsi and Saeki, 1953; de Wit, 1965; Monteith, 1965; Anderson, 1966; Ross, 1981). Soon the research community recognized that ideal plant canopies were more often an exception rather than the rule. Subsequently, more complex geometrical and statistical canopy radiative transfer models were developed. Particular models treated the orientation and spacing of row crops (Jackson and Palmer, 1972; Allen, 1974; Fukai and Loomis, 1976; Mann *et al.*, 1980), the nonrandom spatial distribution of leaves (Acock *et al.*, 1970 Nilson, 1971; Oker-Blom and Kellomäki, 1983), arrays of plants and leaves (Roberts and Miller, 1977; Norman and Welles, 1983; Myneni and Impens, 1985), the shapes of crowns (Oker-Blom and Kellomäki, 1983; Grace *et al.*, 1987; Wang and Jarvis, 1990), the distribution of needles on stems (Norman and Jarvis, 1975; Oker-Blom *et al.*, 1983), and penumbra (Miller and Norman, 1971; Denholm, 1981a; Myneni and Impens, 1985; Oker-Blom, 1985). From these advances in canopy radiative transfer modeling came the next generation of light-dependent photosynthesis, transpiration, and stomatal conductance scaling models; these models were adapted for the specific needs of widely spaced row crops (Fukai and Loomis, 1976; Gijzen and Goudriaan, 1989), orchards (Cohen and Fuchs, 1987; Cohen *et al.*, 1987), grasslands (Norman and Polley, 1989), desert cactus (Garcia *et al.*, 1985), broadleaf forests (Baldocchi and Hutchison, 1986; Caldwell *et al.*, 1986; Baldocchi, 1989), and conifer forests (Grace *et al.*, 19; Wang and Jarvis, 1990).

Meanwhile, a parallel effort was being conducted by the micrometeorological community to develop scaling models that focused on microenvironmental variables that control photosynthesis, transpiration, and stomatal conductance. The earliest micrometeorological scaling models were based on K-theory and the resistance-analog scheme (Cowan, 1968; Waggoner *et al.*, 1969; Miller, 1971; Allen *et al.*, 1974; Shawcroft *et al.*, 1974; Goudriaan, 1977; Norman, 1979,1982; Jarvis *et al.*, 1985; Caldwell *et al.*, 1986). In the past decade, experimental and theoretical advances have shown that K-theory models are subject to some important fundamental weaknesses (Denmead and Bradley, 1985; Wilson, 1989). Recent scaling efforts instead have focused on using higher-order closure (Meyers and Paw U, 1987; Naot and Mahrer, 1989) and Lagrangian (Raupach, 1988; van den Hurk and Baldocchi, 1990; Baldocchi, 1992) models to link leaf carbon dioxide and water vapor exchange rates with the canopy microenvironment.

In the past decade, carbon exchange models based on biochemical and physiological principles have been developed (Farquhar et al., 1980; Farquhar and von Caemmerer, 1982). This information is being used now to model stomatal mechanics and transpiration of leaves (Collatz *et al.*, 1991) and gas exchange in crops and broadleaf and coniferous forests (Baldocchi, 1989,1992; Grant *et al.*, 1989; Norman and Polley, 1989; Price and Black, 1990; Wang and Jarvis, 1990).

III. Basic Scaling Rules

Carbon dioxide and water vapor exchange of biological systems occurs across a spectrum of time and space scales (Osmond, 1989). These scales range from micrometers and microseconds (for cellular processes) to kilometers and centuries (for ecosystems). Two key challenges arise when transferring information from one scale to the next. One challenge involves determining what processes control the system at the scale of interest and the other entails evaluating key regulating processes at the scale being probed. The first challenge is relevant because processes that are important at one scale may not be at another, or vice versa. For example, the energy balance of a canopy is affected by soil and vegetation heat storage and soil evaporation, whereas the energy balance of a leaf is not affected by these processes. The second challenge is noteworthy because interactions between plant canopies and their microenvironment often cause governing variables to vary in time and space. Hence, the status of a governing variable above a plant canopy can be distinct from its state at the leaf surface. Consequently, it is best to assess leaf gas exchange rates by evaluating key governing variables at the surface of leaves (Grantz and Meinzer, 1990; Collatz *et al.*, 1991).

From a practical standpoint, one should only be concerned with information from adjacent time and space scales. Consequently, a system should be described using information from across a hierarchy of three scales: (1) reductionist, (2) operational, and (3) macro. The mechanistics and the dynamics at the operational scale are described at the lower or reductionist scale. Operational-scale information can be obtained by integrating reductionist-scale information in both time and space. The state variables that drive the operational-scale and, consequently, the reductionist-scale processes, are determined at the higher, or macro scale. Since we are interested in processes operating at the canopy scale, the leaf is assigned as the fundamental scale unit. We typically do not concern ourselves with super-fine scale information because its influence typically does not transfer linearly across larger time and space scales. (This point is discussed in more detail in Section V,B.) This scaling recommendation is conservative, however, as it has been used successfully as a modeling framework for many years (de Wit, 1970; Osmond, 1989).

IV. Leaf to Canopy Scaling: Linking Transpiration and Photosynthesis with Their Microenvironment

Water vapor transfer depends on the flows of energy that convert liquid to vapor and on molecular and turbulent diffusion that transfers water vapor molecules between the leaf and the free atmosphere (Campbell, 1981). Net leaf carbon exchange comprises a balance between carbon gain, through photosynthetic carbon reduction, and carbon losses, through photorespiratory carbon oxidation and mitochondrial (dark) respiration (Farquhar and Sharkey, 1982; Farquhar and von Caemmerer, 1982; Farquhar, 1989). The rate of carbon assimilation depends on the abundance of substrates [CO_2 and ribulose bisphosphate (RuBP)] and the degree to which photosynthetic carbon reduction outcompetes photorespiratory carbon oxidation for the enzyme RuBP carboxylase–oxygenase. The supply of CO_2 to the leaf chloroplast is regulated by diffusion through the leaf boundary layer and stomatal aperture (see Collatz *et al.*, 1991). At low intercellular CO_2 levels, RuBP carboxylase–oxygenase is saturated with respect to the RuBP, so photosynthetic activity increases with additional CO_2. When intercellular CO_2 is ample, carbon assimilation becomes limited by the capacity of the leaf electron transport system (a light dependent process) to produce ATP and NADPH; these compounds are needed to regenerate RuBP (see Farquhar and Sharkey, 1982).

To evaluate water vapor and carbon exchange rates of leaves in plant canopy, linkages between the strengths of respective sources and sinks (i.e., transpiration and photosynthesis) and scalar concentrations in the canopy air space must be considered. These linkages arise because the rate at which material is released, or taken up, affects the local scalar concentration in the canopy, and the rate of leaf emission, or uptake, depends on the local scalar concentration. The conservation budget for a passive scalar provides the foundation for computing scalar fluxes and their local ambient concentrations. The budget equation expresses how a scalar (c) varies with time in a controlled volume. The concentration in a volume will change if more material flows into a given volume than out or if material is being produced or consumed within the confines of the volume.

If a canopy is horizontally homogeneous and environmental conditions are steady, the scalar conservation equation can be expressed as an equality between the change, with height, of the vertical turbulent flux and the diffusive source/sink strength. $S(c,z)$:

$$\frac{\partial F(c,z)}{\partial z} = S(c,z). \tag{1}$$

The diffusive source/sink strength of a scalar in a unit volume of leaves is proportional to the concentration gradient normal to individual leaves, the surface area (A) of individual leaves, and the number (M) of leaves in the volume (Finnigan, 1985). The diffusive source strength can be expressed in the form of a resistance-analog relationship, using equations originated by Gaastra (1959):

$$S(c,z) = -\rho_a a(z) \frac{[c(z) - c_i]}{r_{bc}(z) + r_{sc}(z)}, \tag{2}$$

where $a(z)$ is the leaf area density, $[c(z) - c_i]$ is the concentration difference between air outside the laminar boundary layer of leaves and the air within the stomatal cavity, r_{bc} is the boundary layer resistance to molecular diffusion, r_{sc} is the stomatal resistance, and ρ_a is air density.

Components of Eq. (2) are regulated by abiotic variables, such as solar and terrestrial radiation, temperature, humidity, wind speed, and soil moisture. The state of the regulating variables is determined by turbulent mixing, the physiological status of the vegetation, and the fate of incoming and outgoing streams of radiative energy above and within a plant canopy (see Norman, 1979; Campbell, 1981). The micrometeorological conditions within and above a plant canopy can be described with models that are derived from the equations describing the conservation of energy, mass, momentum, and turbulent kinetic energy (Finnigan, 1985; Meyers and Paw U, 1987; Raupach, 1988; Wilson, 1989) and the short and long wave radiation balance (Norman, 1979; Ross, 1981; Myneni *et al.*, 1989). Leaf temperature and the vapor concentration inside the leaf must be assessed by evaluating the leaf energy balance (Campbell, 1981). The leaf boundary layer resistance commonly is computed using fluid dynamic theories for flow over flat plates (Grace, 1980). The intercellular CO_2 concentration and stomatal resistance can be evaluated using physiologically and biochemically based models (Jarvis, 1976; Farquhar *et al.*, 1980; Farquhar, 1989; Collatz *et al.*, 1991).

A. Evaluating the Conservation Budget Equation

A multilayer framework typically is used to assess the conservation equation [Eq. (1)] because spatial variation in vegetation and the abiotic variables that control S is greatest in the vertical dimension. Estimates of the net turbulent flux of material between a plant canopy and the overlying atmosphere are determined by summing contributions of individual layers. This summing (or integration) also requires specification of the flux at the lowest boundary. When considering CO_2 and water vapor exchange, soil/root respiration and soil evaporation must be assessed. The magnitudes of soil evaporation and soil/root respiration often account for 10 to 30% of the net exchange in closed canopies (Denmead, 1984; Baldocchi *et al.*, 1987a; Black and Kelliher, 1989) and can exceed these values in open canopies. If the canopy is horizontally heterogeneous, a three-dimensional gridding should be employed.

Two basic reference frames exist for evaluating the conservation budget: Eulerian and Lagrangian. The Eulerian framework describes the scalar concentration at a fixed point and time, as occurs when measuring the concentration of a given scalar from a tower. The Lagrangian approach analyzes the conservation equation by following parcels of fluid as they move with the wind, much like the trajectory of a neutrally buoyant balloon. The principles behind these frameworks are explored in the following sections.

1. Eulerian Models The conservation budget equation for a scalar cannot be solved readily because it does not form a closed set of equations and unknowns. The equation defining the time rate of change in c contains a higher order moment that is also a function of c. This higher-order moment is the vertical turbulent flux (F), which is defined as the covariance between vertical velocity (w) and scaler concentration fluctuations ($\overline{w'c'}$) (primes denote fluctuations from the mean and the overbar represents time averaging). The simplest, and earliest, Eulerian models on turbulent exchange in plant canopies adopted a first order closure scheme called K-theory (see Legg and Monteith, 1975; Thom, 1975). The appeal of this model is in its simple reduction of the number of unknown variables. K-theory models assume that turbulent transfer and molecular diffusion are analogs; thus, the vertical velocity-scalar covariance is represented as the product of the scalar concentration gradient and a turbulent diffusivity (K):

$$F_c(z = \overline{w'c'} = K\frac{\partial c}{\partial z}.$$
(3)

The concept of along-gradient transfer is valid only when the turbulence length scales are fine-scaled in comparison to the curvature of the scalar concentration gradient (Corrsin, 1974). An accumulating body of evidence now shows that many of the assumptions supporting K-theory are false in plant canopies. Turbulent transport is dominated by large scale and intermittent eddies, which can cause counter gradient transfer (see Finnigan, 1985; Denmead and Bradley, 1985; Baldocchi and Meyers, 1988a,b; Raupach, 1988; Wilson, 1989). Since K-theory is invalid in a plant canopy, one cannot derive canopy-level exchange rates from concentration gradients measured inside the canopy, as was discovered unsuspectingly by Johnson *et al.* (1976).

Higher-order closure models have been proposed as a means of circumventing the inherent limitation of first-order closure models (Wilson and Shaw, 1977; Meyers and Paw U, 1986,1987; Wilson, 1988). The appeal of this method is its mechanistic base and an ability to simulate counter gradient transport (Wilson and Shaw, 1977). Higher-order closure models introduce formal budget equations for higher-order moments such as $\overline{w'c'}$. Equations that describe mean wind speed and turbulence are also introduced to evaluate dependent terms in the second moment equation and in the source-sink function [i.e., r_b and $c(z)$]. For example, wind speed and turbulence in a plant canopy are described by the budget equations for mean horizontal wind velocity ($\overline{u}$), tangential momentum stress ($\overline{w'u'}$), and the turbulent kinetic energy components ($\overline{u'u'}$, $\overline{v'v'}$, $\overline{w'w'}$).

The budget equations for the second-order moments, unfortunately, include additional unknowns of the third order (such as $\overline{w'w'u'}$, $\overline{w'w'c'}$). Deriving budget equations for these unknown terms introduces more unknowns, consisting of the next-order moment. Hence, an equal set of equations and unknowns can be obtained only through parameterizing the highest-order moment with an effective eddy exchange coefficient (Wilson and Shaw, 1977; Meyers and Paw U, 1986, 1987). The logic of attaining closure at orders of two or three assumes that errors introduced at higher orders will have a miminal effect on the estimate of the flux and concentration field. Deardorff (1978) perceptively criticizes the use of effective exchange coefficients to close budget equations of higher-order moments because of an ultimate reliance on down-gradient diffusion. Deardorff argues that effective exchange coefficients are inadequate for near-field flows which occur in the vicinity of sources and sinks, because any turbulent diffusivity, K, in the vicinity of a source or sink is related linearly to the time period that fluid parcels have traveled. Only after a long travel distance is the time-independent "far-field" limit of K reached. The dispersion of a scalar released by sources at different distances upwind from an observer (as in a plant canopy) cannot be described by a single effective diffusivity (Wilson, 1989), as is attempted in higher-order closure schemes. Other criticisms of higher-order closure models revolve around the use of certain laboratory-based model parameters and parameterization schemes in the natural environment (Wyngaard, 1988). Yet despite the criticisms listed, Eulerian higher-order closure models have simulated temperature and wind speed profiles and fluxes of momentum, heat, and moisture within and above crop canopies successfully (Meyers and Paw U, 1986, 1987; Naot and Mahrer, 1989).

2. Lagrangian Models In the Lagrangian frame, a concentration field and its turbulent flux are defined by the statistics of an ensemble of dispersing marked fluid parcels and the strength and spatial distribution of sources and sinks (see Lamb, 1980; Sawford, 1985; Raupach, 1988; Wilson, 1989, for detailed discussions.). This approach is valid as long as the molecular diffusion with a fluid parcel is negligible in comparison with its turbulent diffusion through the atmosphere. The concentration of a scalar can be defined at a particular location and after a given travel time using the principle of superposition. For the case of vertical diffusion, the concentration observed at point z and time t is

$$c(z,t) = \int_0^t \int_0^z P(z, t|z_0, t_0) \, S(z_0, t_0) dz_0 dt_0 \, . \tag{4}$$

S is the same diffusive source/sink function defined in Eq. (1). $P(z,t|z_0,t_0)$ is a joint probability density function, which describes the probability that a fluid parcel released from a point in space z_0 at time t_0 will be observed at another location and time (z,t). The strength of the Lagrangian approach is an ability to stimulate the different time dependencies of near- and far-field diffusion, as noted earlier.

The probability density function for the diffusion of fluid parcels depends only on the properties of the turbulent wind field, which must be prescribed. Inside a plant canopy, turbulence is inhomogeneous and non-Gaussian (Wilson *et al.*, 1982; Raupach, 1988; Baldocchi and Meyers, 1988a,b); length, time, and velocity scales of turbulence vary with height and the probability density functions of the three vector velocity components are skewed and kurtotic. The joint probability density function, $P(z,t|z_0,t_0)$, is difficult to define analytically in such conditions, yet $P(z,t|z_0 t_0)$ can be specified using a Markovian "random-walk" approach. This method computes the trajectory of a large number of fluid parcels. Methods for determining fluid parcel movement are derived from the Langevin equation and are discussed by Raupach (1988) and Wilson (1989); the Langevin equation defines the acceleration of a fluid parcel as a function of the memory of its initial value and a random forcing.

Lagrangian models based on the Langevin equation will fail where the vertical velocity variance increases with height, as occurs inside plant canopies. The vertical gradient in the vertical velocity variance imposes a downward drift on a Markovian random flight model. This net downward drift occurs because downward directed fluid parcels from above enter a lower region with a decreased vertical velocity scale and a reduced probability of leaving that region (Sawford, 1985; Raupach, 1988; Wilson, 1989). Accumulation of matter near the surface would otherwise occur in nature if it were not for the intermittent gusts that disproportionately transfer matter and maintain continuity. Heuristic arguments have been used to develop approaches to remove the unrealistic accumulation of matter that would otherwise occur when using Markov sequence models in a field of inhomogeneous turbulence. One approach introduces an additional force term into the Langevin equation. This method yields a mean upward drift velocity in the solution of the differential equation (Wilson *et al.*, 1981; Legg and Raupach, 1982). Another method bypasses the addition of a bias velocity by reflecting marked particles according to a probability calculated from the gradient in the vertical velocity variance (Leclerc *et al.*, 1988). In the event of horizontal variability, the Lagrangian frame must account for horizontal parcel movement and the horizontal distribution of sources and sinks.

In practice, the source-sink function, S, is dependent on local concentration, c, and vice versa. Raupach (1988) devised a scheme that computes the interdependence between S and c. The canopy can be expressed by summing the contributions of material diffusing to or from layers in the canopy (denoted by the subscript j):

$$c_i\text{-}c_r = \sum_{j=1}^{N} S_j(c_j)D_{i,j}\Delta Z_j .$$ (5)

The dispersion matrix $(D_{i,j})$ is solely a function of the turbulence statistics and can be computed by uniformly releasing fluid parcels from each canopy level. Equation (5) can be solved by iteration or by solving simultaneous equations.

3. How Well Do Eulerian and Lagrangian Models Simulate the Canopy Microenvironment?   As discussed earlier, both Eulerian and Lagrangian models have inherent closure limitations. Eulerian models depend on questionable assumptions to parameterize higher-order moments, whereas Lagrangian models require the turbulence regime to be specified a priori. The obvious question that must be posed in this chapter on scaling is, "In spite of inherent theoretical flaws, do Eulerian and Lagrangian models simulate the microenvironment well enough to be used as a framework for scaling water vapor exchange and CO_2 exchange?"

Figure 5.1A compares normalized measurements of water vapor concentration against calculations derived from the higher-order closure model of Meyers and Paw U (1987) and a Lagrangian random-walk model. The Lagrangian model was based on the particle trajectory algorithm from Legg and Raupach (1982) and prescribed the sources of water vapor through the canopy as a function of leaf area density and the flux density of net radiation (see van den Hurk and Baldocchi, 1990). Above the canopy, both models simulate the normalized humidity profile reasonably well. Model computations derived from the Lagrangian model mimic the distinct shape of the vertical water vapor profile (which is caused by the strong diffusive source in the upper canopy) reasonably well, but place the nose in the concentration profile too high. Computations based on the Eulerian model, on the other hand, simulate water vapor profiles well in the upper half of the canopy, but overestimate measured values in the lower half of the canopy.

Figure 5.1B compares simulations of water vapor flux densities against the value measured above the canopy. Despite different abilities of the two modeling frames to compute the concentration field, both models compute comparable flux densities of canopy water vapor exchange. Further, these values are in reasonable agreement with the measured values, underestimating them by only 10 to 12%. Consequently, it may be argued that errors attributed to computing water vapor profiles in the lower half of the canopy with the Eulerian model have little consequence on leaf-to-canopy scaling; the contribution of sources in the lower canopy

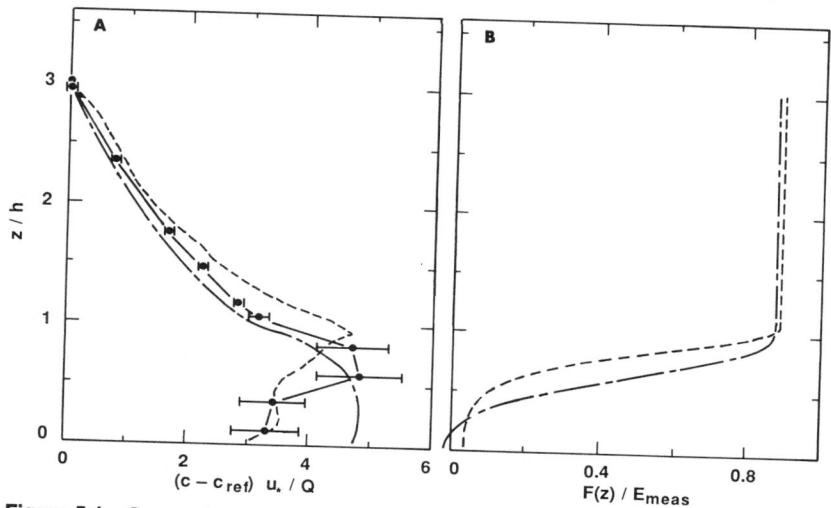

Figure 5.1 Comparison of water vapor concentration (A) and flux density (B) computations based on the Lagrangian random walk model of van den Hurk and Baldocchi (1990) (---) and the Eulerian model of Meyers and Paw U (1987) (—·—) against measurements made in a mature soybean canopy (—●—). The Lagrangian model used the particle trajectory algorithm of Legg and Raupach (1982). The ordinate axis denotes elevation (z) normalized by canopy height (h). On the abscissa, concentrations are normalized by friction velocity (u^*) and the relevant canopy water vapor source strength (Q), whereas calculated flux densities [$F(z)$] are normalized by measured values. Test data were collected August 4, 1979, between 1100 and 1400 hr.

to the integrated canopy water vapor flux density is relatively small because foilage is sparse and the turbulence and available energy that drive transpiration are low (Fig. 5.1B).

Comparing Figs. 5.1A and 5.1B reveals that the Lagrangian model aptly calculates the countergradient transfer of water vapor inside a soybean canopy, whereas the Eulerian model does not. This failure of the Eulerian model to compute water vapor concentrations in the lower canopy accurately substantiates the theoretical concerns of Deardorff (1978) mentioned earlier.

B. Radiative Transfer in Plant Canopies

Information on the radiation balance of leaves is needed to assess many processes that contribute to CO_2 and water vapor exchange, for example, photosynthesis, respiration, stomatal conductance, transpiration, and leaf temperature. The flux density of radiation in a plant canopy, at a given wavelength, is the sum of the beam and diffuse sky radiation that penetrates through the foilage gaps and the complementary radiation that is generated as radiation is intercepted and scattered by leaves and soil surface (see Ross, 1981; Myneni *et al.*, 1989). The net transfer of photons in a plant canopy is described by the differential photon transport equation. A change in the intensity of a stream of monochromatic light through an optical depth of vegetation results from a balance between sinks (photon flux attentuation) and sources (phase scattering) of radiation along that path. The photon transport equation and its application in plant canopies is discussed thoroughly in Ross (1981) and Myneni *et al.* (1989).

Statistical models are used classically to estimate radiative transfer in ideal closed canopies. The simplest models assume that (1) the canopy is a plane-parallel turbid medium; (2) the sun is a point source that emits parallel beam radiation; (3) foliage is distributed randomly in space; (4) the azimuthal distribution of foliage is symmetrical; and (5) the leaf inclination angle distribution can be defined, but is invariant with height (Lemeur and Blad, 1974; Ross, 1981; Myneni *et al.*, 1989). When these assumptions are valid, the probability of beam penetration (P_b) can be described by the Poisson probability function:

$$P_b = \exp(-\frac{GL}{\sin \beta}), \tag{6}$$

where L is the cumulative leaf area per unit ground area (starting at the top of the canopy), β is the solar elevation angle, and G is the mean direction cosine between the solar zenith angle and the leaf normals. G is interpreted as the fraction of leaf area projected in a plane normal to the source of radiation. G varies as a function of the canopy leaf inclination angle distribution and the solar elevation angle. If the inclination angles of leaves distributed uniformly over the surface of a sphere, G takes a value of 0.5 for all solar elevation angles (Norman, 1979; Ross, 1981). The probability that diffuse radiation penetrates into a plant canopy can be described by integrating Eq. (6) over the surface of a hemisphere, as long as the diffuse sky exhibits uniform brightness (Norman, 1979; Ross, 1981).

Scattering of radiation depths depends on the bidirectional optical properties of leaves (their reflectivity and transmissivity), their orientation relative to an incident light beam, and the wavelength of the radiation (Myneni *et al.*, 1989). Detailed treatment of scattering is not critical when considering photosynthesis and stomatal conductance because leaf absorbtance in the critical wavebands is high. On the other hand, careful attention to scattering is necessary when computing leaf energy balances because scattering in the near-infrared (NIR) waveband is appreciable (Norman, 1979).

The assumptions for computing radiative transfer in ideal homogeneous canopies are often not valid in broadleaf and conifer forest canopies and in open vegetation, so modifications to the differential equation describing photon transfer through a turbid medium must be made. In broadleaf forest stands, leaf inclination angle distributions vary with depth (Hutchison *et al.*, 1986; Hollinger, 1989) and the spatial distribution of foilage can be clumped (Baldocchi, 1989; Kruijt, 1989). For closed broadleaf stands with clumped foilage, a negative binomial probability function provides a better estimate of the probability of beam penetration than the Poisson distribution (Acock *et al.*, 1970; Baldocchi, 1989).

If the sun was a point source, then leaves in a plant canopy would be exposed either to full sunlight (if the solar disk is not obscured by leaves above) or umbral shade (if the solar disk is fully obscured by upper leaves). Since the sun has a finite radius of 0.0046 radians, it does not emit parallel radiation. Consequently, penumbral shade can occur if a plant element, as seen from a point below, partially obscures the solar disk; this happens when the angular radius of a leaf is less than that of the sun (see Miller and Norman, 1971; Denholm, 1981a,b; Oker-Blom and Kellomäki, 1983; Myneni *et al.*, 1989). The flux density of radiation on a leaf in penumbral shade is between that of a leaf in a sunfleck and one in full shade. The probability of penumbra is a function of leaf size and the distance between a plant element that is partially obscuring the solar disk and a reference point below. The probability of penumbral shade will be greatest in tall canopies and in vegetation with small needles and leaves, such as conifer (Oker-Blom and Kellomäki, 1983) and broadleaf (Baldocchi, 1989) forest stands.

In conifer stands, needles on individual shoots, shoots on branches, branches on whorls, and whorls on stems often are grouped in distinct geometrical patterns (Leverenz and Hinckley, 1990; Whitehead *et al.*, 1991; Wang *et al.*, 1990), causing the assumption of a plane-parallel turbid medium to be invalid (Norman and Jarvis, 1975; Oker-Blom and Kellomäki, 1983; Oker-Blom, 1986). The geometric arrangement of needles on a shoot causes the projected area of a shoot to be less than the projected area that would otherwise occur if individual needles on the shoot were removed and distributed randomly in space (Carter and Smith, 1985; Leverenz and Hinckley, 1990) because needles arranged on a shoot mutually shade each other (Oker-Blom, 1986). The radiation environment within a shoot depends on the arrangement and size of needles, which cause significant umbral and penumbral shading (Norman and Jarvis, 1975; Oker-Blom, 1986). The simplest way to calculate the probability of beam penetration in a conifer stand is to substitute the leaf area index (*L*) in Eq. (6) with the projected shoot area index (see

Oker-Blom, 1986). The reader should be advised that this concept is valid only if the *shoots* are distributed randomly in space.

In plant stands with distinct crown geometry, we must define the crown shape and the three-dimensional distribution of foilage to compute probabilities of light penetration correctly. If the spatial distribution of branches, whorls, and tree crowns has a preferred geometry, gaps will arise through which pencils of radiation can travel without interception (Roberts and Miller, 1977; Norman and Welles, 1983; Grace *et al.*, 1987; Wang and Jarvis, 1990). The probability of beam penetration through a distinct plant crown is a function of the path length that a pencil of radiation passes through the canopy, r, so

$$P = \exp[-GA(z)r], \tag{7}$$

where $A(z)$ is leaf area density in broadleaf canopies (Norman and Welles, 1983) and can be assumed to equal the projected shoot area density in conifer canopies (Oker-Blom, 1986). The probability of beam penetration through a plant crown increases with the degree of shoot grouping (Norman and Jarvis, 1975; Oker-Blom, 1986), so caution should be used when applying Eq. (7) under such conditions.

Radiative transfer in isolated plant stands requires specific detail on crown geometry and foilage distribution within the foilage envelope. Models by Roberts and Miller (1977), Norman and Welles (1983), Grace *et al.* (1987), and Wang and Jarvis (1990) are adapted for these circumstances and are the most comprehensive ones available. The procedural approach, advocated by Myneni and Impens (1985), also is well suited for computing radiative transfer in heterogeneous canopies. The data requirement, however, is overwhelming: the three-dimensional distribution of individual leaves is needed to execute the model. This limitation may be overcome if it is coupled to a statistical Monte Carlo approach, which prescribes the foilage dispersion in space.

C. Surface Energy Balance

The temperature and humidity at the leaf surface are determined from the balance between net incoming short- and longwave radiation and its partitioning into sensible (H_f) and latent (LE_f) heat exchange. Leaf temperature can be estimated using an iterative scheme based on Newton's method (Bristow, 1987) or an analytical solution (Paw U and Gao, 1988), or by linearizing the leaf energy balance (Campbell, 1981). When considering the energy partitioning of the whole canopy, consideration must also be made for advection and soil and canopy heat storage.

V. What Information Is Needed to Scale CO_2 and Water Vapor Exchange from a Leaf to a Canopy?

The type and amount of information needed to apply the equations discussed earlier to scale water and carbon exchange from leaves to canopies depends on the form and functional features of a plant canopy. One may surmise initially that an overwhelming amount of information is needed for scaling because plant canopies are tremendously complex. Instead, workable frameworks for scaling gas fluxes from leaves to canopies are possible, in a seemingly imposing landscape, because convergent evolution has conspired favorably to minimize the physiognomic detail of plant communities (Crawley, 1984). It is my opinion that many aspects of scaling can be accomplished by addressing the scaling needs of four broad categories of homogeneous plant canopies and their heterogeneous counterparts. These categories are short vegetation (crops and grasslands), shrublands (sagebrush, savannah, and chaparral), broadleaf forests (temperate and tropical), and conifer forests.

Support for adopting a limited number of scaling categories comes, in part, from the attributes that are required to simulate radiative transfer through vegetation (see Section V,A). Other support for adopting a limited number of broad scaling categories is based on the omega theory of Jarvis and McNaughton (1986). This theory describes the role that the bulk aerodynamic and surface properties have in controlling water vapor exchange of otherwise dissimilar canopies. For example, the evaporation from crops and grasslands is similar because both canopy types are aerodynamically smooth and exert a weak surface resistance to vapor transfer. Conifer forests fall into another class because they are aerodynamically rough and exert a significant surface resistance to water vapor transfer. Broadleaf forests and shrublands are moderately rough and exert a moderate resistance to water vapor transfer, forcing these vegetation types to be categorized into yet another class. Empirical evidence supporting this call for broad scaling categories comes from the observed similarity in the evaporative fluxes between otherwise disparate temperate deciduous and tropical forests (Baldocchi, 1989; Shuttleworth, 1989; Fitzjarrald *et al.*, 1990).

A. Defining the Canopy

Information on the physical attributes of a plant canopy is needed to assess the diffusive source/sink and the transmission of radiation directly. We need to know how much leaf area a canopy has, how the leaves are distributed in space, their elevational and angular orientation, and the dominant species that are present (see Ross, 1981). The information needed to describe the characteristics of the four noted categories of plant canopies is not the same. A hierarchical list of information that is needed to describe a plant canopy follows.

Short closed crop canopies are the simplest to describe. Their description only requires information pertaining to the leaf area index, the vertical profile of leaves, and the leaf inclination angle distribution. The description of the spatial and leaf inclination angle distribution is often simplified because leaves tend to be placed randomly and their inclination angles often are distributed spherically (Norman, 1979). Information on the azimuthal orientation of leaves is needed only if the plants are heliotropic or if their leaf distribution is asymmetrical (Lemeur and Blad, 1974). Closed broadleaf forest canopies require the information just cited as well as data on leaf size, the spatial dispersion of foilage, and the amount of woody biomass (Baldocchi, 1989). A description of conifer stands require all the information mentioned so far, plus a description of the crown envelope and the geometry of shoots (Oker-Blom, 1986; Wang and Jarvis, 1990). The most complicated canopies to describe are heterogeneous counterparts. They require information on the three-dimensional distribution of plant crowns and the foliage inside (Roberts and Miller, 1977; Norman and Welles, 1983; Whitehead *et al.*, 1991). The three-dimensional distribution of plant foliage can be quantified using nested ellipsoidal shells that contain randomly dispersed foliage of a given foliage density (Norman and Welles, 1983; Grace *et al.*, 1987; Wang and Jarvis, 1990). This approach provides a reasonable representation of most plant and tree crowns, except when whorl and branch geometry allows distinct foliage gaps between internodes (see Whitehead *et al.*, 1992) and the foliage is clumped (Oker-Blom, 1986).

Leaf area distribution measurements can be made easily in agricultural crops by direct destructive sampling or indirect remote sensing means (Lang, 1986; Norman and Campbell, 1989). On the other hand, leaf area information often is not widely available for natural ecosystems because of logistical difficulties. However, a growing database is becoming available for many broadleaf and coniferous forest canopies (Massman, 1982; Hutchison *et al.*, 1986; Hollinger, 1989; Kruijt, 1989; Wang *et al.*, 1990; Whitehead *et al.*, 1991). Remote sensing methods also are becoming available for use in forest stands, but they often must be calibrated in stands of known structure (Chason *et al.*, 1991).

One criticism of the multilayer modeling approach just described is that vertical leaf area distribution data are not always available. A satisfactory solution to the problem has been proposed by Meyers and Paw U (1986). They observed that the vertical leaf area distribution, for a wide range of crop and forest canopies, falls into several categories. Continuous and normalized leaf area distributions for these canopy categories can be generated with a β distribution (Massman, 1982). Subsequently, an estimate of the actual leaf area distribution can be obtained by simply selecting the appropriate normalized leaf area distribution and providing an estimate of canopy height and the leaf area index (using destructive or remotely sensed means; Lang, 1986; Norman and Campbell, 1989).

B. Evaluating Leaf Scaling Parameters and Processes

Leaf-to-canopy scaling requires estimating the expected value of a dependent function $[E[f(x)]]$ in terms of independent variables that vary in time and space. This scaling is complicated by the nonlinearity of many key dependent processes and by the non-Gaussian temporal and spatial distributions of many driving independent variables, such as light, wind, temperature, humidity, and CO_2, within a canopy. The expected value of a function $E[f(x)]$ can be assessed, at a given level in a canopy, by integrating the product of the dependent function $[f(x)]$ and its probability density function $[p(x)]$ over the domain of the independent variable:

$$E[f(x)] = \int_{x_{min}}^{x_{max}} f(x)p(x)dx. \tag{8}$$

The evaluation of nonlinear light-dependent functions merits careful attention because the probability distribution for radiation can be bimodal. In practice, a simple discretized version of Eq. (8) can be used for this purpose (see Norman, 1980; Smolander, 1984; Jarvis *et al.*, 1985; Koen, 1987; Baldocchi, 1989):

$$E[f(I,L)] = f(I_s)P_s(L) + f(I_u)P_u(L) + f(I_p)P_p(L). \tag{9}$$

Equation (9) weights the functional dependence on solar radiation (I) at a given cumulative leaf area (L) according to the fractions of leaf area that are associated with the sunlit (P_s), umbral (P_u), and penumbral (P_p) classes; P_s, P_u, and P_p sum to 1 and can be estimated with a canopy radiative transfer model.

Leaf photosynthesis, transpiration, and stomatal conductance of leaves also are known to respond differently to a given stimulus due to differences in age, physiology, species, and acclimation to the local environment (Field, 1983, 1991; Jurik *et al.*, 1985; Gutschick and Weigel, 1988; Marek *et al.*, 1989). For example, leaf photosynthetic capacity diminishes with depth into the canopy because less nitrogen needs to be invested to fix carbon in this low light environment (Field and Mooney, 1986; Gutschick and Weigel, 1988; Evans, 1989; Schimel *et al.*, 1991). The strength of the multilayer approach is its flexibility to allow parameter values to vary with height as such information becomes available.

Knowledge about how gas exchange model parameters are determined is crucial in evaluating scaling functions. For example, biochemically based photosynthesis models are derived according to the quanta *absorbed* by the chloroplast, yet model parameters often are determined by gas exchange experiments in terms of light *incident* on a leaf (Harley *et al.*, 1985). Photosynthesis–light response curves for conifers sometimes are derived for shoots and sometimes for needles. Because shoot geometry causes needles to shade each other, the amount of irradiance incident on a shoot is not equal to that on individual needles. Consequently, one should not use parameter values derived for shoots and needles interchangeably.

Physiological studies show that CO_2, light, and photosynthetic capacity gradients exist through the cross section of a leaf (Terashima and Saeki 1983; Terashima and Inouye, 1985; Parkhurst *et al.*, 1988; Vogelmann *et al.*, 1989). Other studies reveal that stomatal aperture varies widely across the surface of a leaf (Farquhar and Sharkey, 1982). Do we need to include super-fine scale information when scaling CO_2 and water vapor exchange from a leaf to a canopy? Farquhar (1989) suggests that the distribution of light across a leaf may not complicate the computation of leaf photosynthesis because a leaf optimizes the distribution of nitrogen (a surrogate for photosynthetic substrate) across a leaf. The variability in CO_2 throughout a leaf also may be too small to be of concern in the scaling of leaf carbon exchange. For example, the spatial variability in C_1 across a leaf causes the internal CO_2, generally estimated in the substomatal cavity, to overestimate the mean CO_2 concentration throughout the mesophyll by only 2 to 12 μbars for amphistomatous leaves (Parkhurst *et al.*, 1988).

If super-fine scale information is deemed critical to scaling, it is possible to assess such information by using a nested hierarchy of models. A nested model can be used to lump finer-scaled biophysical information and provide information that is compatible with the needs of the higher scale. One example is to develop a model that computes photosynthesis according to light and chlorophyll gradients across a leaf (Terashima and Saeki, 1985) and accounts for CO_2 diffusion through the three-dimensional intercellular spaces (Parkhurst, 1986). A problem with this approach deals with correctly evaluating forcing variables at such fine scales. The outlined procedure requires information on the spatial distribution of cells and differential scattering properties of light in narrow wavebands due to pigment absorption. To obtain this information for leaves throughout a plant canopy is impractical and increases the potential for error propagation. Yet, despite these problems, nesting cellular-based models into canopy models can be viewed as a means of guiding research to obtain more mechanistic information on canopy-scale exchange rates.

VI. Can Information on Leaf CO_2 and Water Vapor Exchange Rates Be Extended to the Canopy Scale?

General frameworks for scaling water vapor and CO_2 exchange have been outlined and discussed. In this section I demonstrate our ability to integrate this information from the leaf to the canopy scale. To do so, I examine the scaling of leaf CO_2 exchange for two different cases. One case involves testing scaling themes for a uniform and aerodynamically smooth crop canopy with high photosynthesis capacity (soybeans). The other case tests scaling schemes for a clumped and aerodynamically rough canopy, with a lower photosynthetic capacity (temperate deciduous forest).

A. Homogeneous Closed Canopy: Soybeans

To scale photosynthesis from the leaf to the canopy of an aerodynamically smooth soybean canopy, I propose the following hypothesis.

The homogeneous features of a soybean canopy allow us to simplify the treatment of radiative transfer processes; one may assume that the leaf inclination angle distribution is spherical and the probability of beam penetration is Poisson. On the other hand, the combined effects of a high photosynthetic capacity, a similarity between leaf boundary layer and stomatal resistances, and an aerodynamically smooth canopy can cause a substantial drawdown of CO_2. Such an occurrence would necessitate the incorporation of a detailed turbulent-diffusion model to compute feedback between the local scalar regime and the source/sink strength rigorously.

To test this hypothesis I link together (1) a Lagrangian random-walk turbulent-diffusion model; (2) a Poisson radiative transfer model; and (3) a biochemical/physiological photosynthesis model to estimate canopy CO_2 exchange. The attributes of this coupled canopy CO_2/water vapor exchange model are listed in Table 5.1. A detailed description of the model is provided by Baldocchi (1992).

The coupled micrometeorological/physiological canopy gas exchange model is tested against measurements of canopy CO_2 exchange in Fig. 5.2. Calculations of net CO_2 exchange (canopy photosynthesis minus soil/root respiration) yield values that are well correlated ($r^2 = 0.68$) with field measurements over a wide range of environmental conditions. Model calculations account for 68% of the variance between calculated and measured fluxes and a paired Student's t test reveals that there is no significant difference between calculated and measured fluxes at the 5% probability level ($t = 1.12$, $t_{0.05} = 1.68$). Based on these calculations, I conclude that it is possible to scale leaf-level fluxes in uniform canopies reasonably well, based on the posited assumptions. However, it must be acknowledged that calculated fluxes systematically overestimate measured values at the low end and underestimate them at the high end.

Differences between computed and measured values are attributed

Table 5.1   Data and Sources of Models and Parameters Used to Compute CO_2 Flux Densities over Soybeans

Model or parameter	Data and source
Leaf area index	4.1
Species/cultivar	*Glycine max*, Clark cv.
Canopy height	1.0 m
Photosynthesis model	Biochemical, Farquhar *et al.* (1980), Harley *et al.* (1985)
Stomatal conductance model	Phenomenological, Jarvis (1976)
Radiative transfer model	Random spatial distribution, spherical leaf angle distribution, Norman (1979)
Turbulent transfer model	Lagrangian random walk model, Legg and Raupach (1982), Raupach (1988)
Surface energy balance model	Bristow (1987)
Photosynthesis parameters	Harley *et al.* (1985)
Stomatal conductance parameters	Baldocchi *et al.* (1987b)
Field measurements: Fluxes and driving environmental variables	Bowen ratio energy balance, Baldocchi (1992)
Soil/root respiration parameters	da Costa *et al.* (1986)

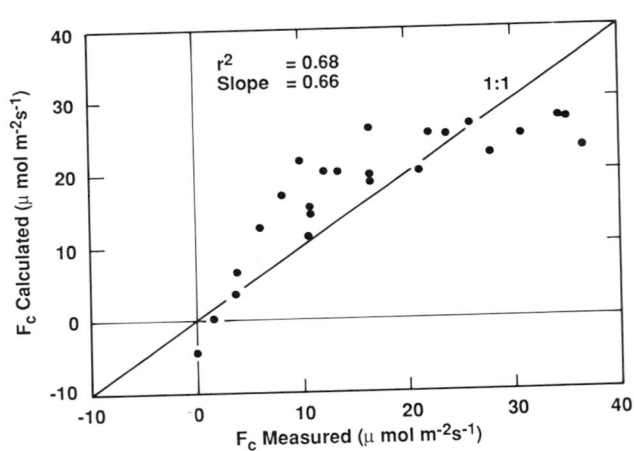

Figure 5.2   Comparison of net CO_2 exchange rates computations (canopy photosynthesis minus soil/root respiration) with micrometeorological flux measurements over a soybean canopy. The data are from a well-watered soybean canopy growing in Nebraska during August and September, 1979. Model specifications are listed in Table 5.1. A wide range of environmental conditions was encountered. PAR ranged between 100 and 2100 μmol · m^{-2} · sec^{-1}, air temperature ranged between 15 and 32°C, humidity ranged between 15 and 28 mbar, and friction velocity ranged between 0.10 and 0.65 m·sec^{-1}.

to a variety of sources. Some sources of error are associated with the specification of model parameters and the measurement of turbulent fluxes and leaf area index. Better model performance could be expected if the photosynthetic model parameters were derived from the foliage of the canopy under study. Measurement and sampling errors attributed to leaf area index and canopy CO_2 exchange are each on the order of 20%. Consequently, the accuracy of these variables exerts a limit on the accuracy that can be expected in such a model comparison. Bias errors associated with computing canopy radiative transfer, stomatal conductance, and surface energy balance are small since the submodels have been validated successfully by tests against field measurements (Norman, 1979; Meyers and Paw U, 1987; Baldocchi, 1992).

The tested canopy model can be used to ask some interesting scaling questions. Physiological ecologists often ask, "What limits carbon gain of plant canopies? Is it photosynthetic capacity or the capacity of atmospheric turbulence to deliver enough carbon dioxide to the leaves?" (Field, 1991). We can address this question by examining the vertical distribution of W_c and W_j associated with the sunlit and shaded leaf fractions in a soybean canopy. Remember, the photosynthetic carboxylation velocity, V_c, equals the minimum between the RuBP-saturated rate of carboxylation (W_c) and the carboxylation rate allowed the electron transport (W_j). W_c is a function of intercellular CO_2 and oxygen concentrations, and W_j is primarily a function of photon flux density, but also has a minor dependence on CO_2. Figure 5.3 shows that W_c of sunlit leaves

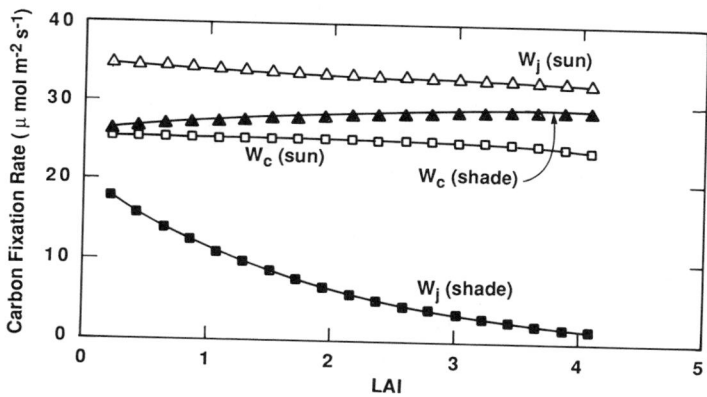

Figure 5.3 Computations of the RuBP-saturated carboxylation rate (W_c) (squares) and the electron transport-limited rate (W_j) (triangles) on sunlit (open) and shaded (filled) leaves in a soybean canopy. Model subroutines are listed in Table 5.1. These computations assume that photosynthesis model parameters are invariant with height. Input variables were PAR of 2000 μmol · m^{-2} · sec^{01}, air temperature of 25°C, wind speed of 3 m · sec^{-1}, and [CO_2] of 350 ppm.

is less than W_j. This result suggests that the supply of CO_2 to the leaves through the turbulent mixing and diffusion through the leaf boundary layer and stomata is the limiting factor for this leaf class. Shaded leaves, on the other hand, are always light-limited because W_j is less than W_c. These results suggest that scaling of CO_2 uptake by an aerodynamically smooth crop with a high photosynthetic capacity requires an accurate radiative transfer model to calculate W_j on the shaded leaves and an accurate model of turbulent transfer to calculate W_c well on the sunlit leaves.

Since photosynthesis of sunlit soybean leaves is limited by the supply of CO_2 to drive W_c, to what extent does a reduction in turbulent mixing affect photosynthetic CO_2 uptake of an aerodynamically smooth canopy? Hypothetically, we would expect a reduction in turbulence to cause a drawdown in the scalar concentration field, with a subsequent reduction in canopy photosynthesis. Figure 5.4A shows that reducing turbulent mixing (as occurs when friction velocity u^* decreases from 0.55 to 0.1 m·sec^{-1}) limits P_c by only 7% at high light levels ($Q=$ 2000 μE · m^{-2}·sec^{-1}). When photosynthetic capacity is low due to light levels ($Q=500$ μE·m^{-2}·sec^{-1}), variations in turbulent mixing have an even weaker influence on canopy photosynthesis.

The reduction in canopy photosynthesis is primarily a consequence of increasing boundary layer resistances because the drawdown in CO_2 is relatively small in the upper third of the canopy (Fig. 5.4B). The potential drawdown in CO_2, due to diminished turbulence, was offset by a significant buildup of CO_2 in the lower two-thirds of the canopy. CO_2 builds up in the lower canopy because the ability to transfer CO_2 respired by roots and the soil out of the canopy is diminished as turbulence decreases.

Since little drawdown in CO_2 occurred in the canopy crown, can we use this information to derive a simplified parameterization scheme for modeling canopy photosynthesis? For example, can we derive estimates of soybean canopy photosynthesis by assuming a constant CO_2 profile, thereby circumventing the need to apply a detailed micrometeorological model? Assuming a constant CO_2 profile reduces the mean estimate of canopy photosynthesis by less than 1% compared with values computed on the basis of a turbulence and diffusion model that allows concentration–source/sink feedbacks (Fig. 5.5). It must be recognized that the conclusion drawn from this analysis is valid for the windy conditions of Nebraska, from which the data set was derived. On the other hand, whether or not it is important to consider the explicit effects of turbulence and diffusion on feedbacks between sources and the scalar concentration field when simulating photosynthesis of high capacity and aerodynamically smooth vegetation in regions of limited turbulent mixing, e.g., C_4 crops in the southeastern United States during the summer must be

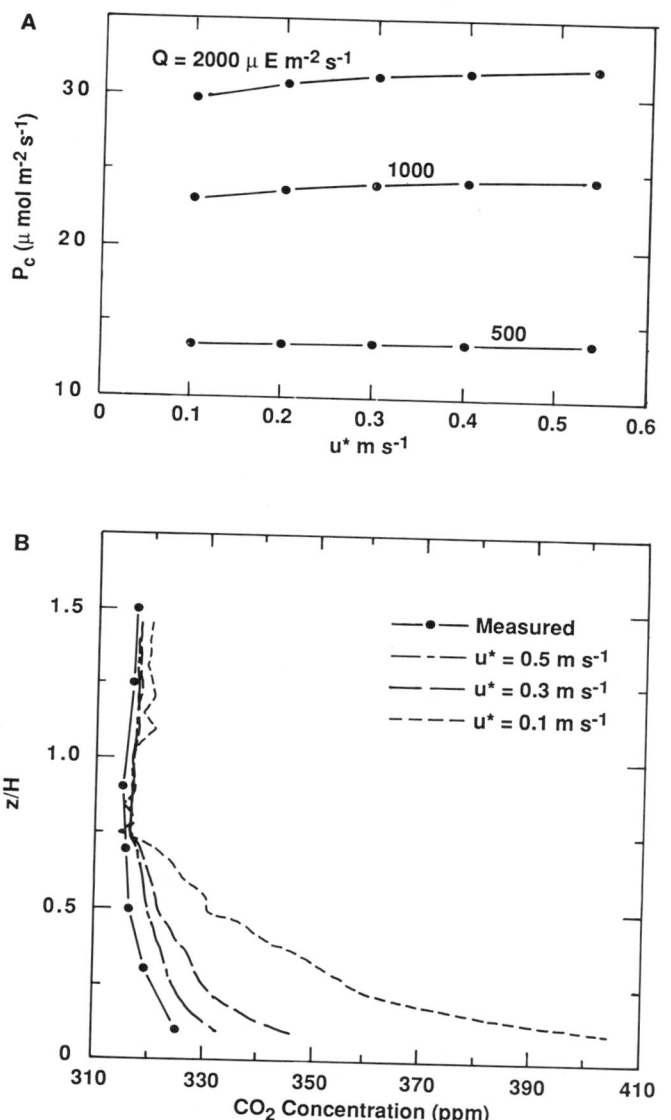

Figure 5.4 (A) Response of canopy photosynthesis of soybeans to variations in friction velocity ($u*$) at various photosynthetic photon flux densities (Q): 500, 1000, and 2000 $\mu E \cdot m^{-2} \cdot sec^{-1}$. (B) Measured (●) and computed vertical profiles of CO_2 concentration in a soybean canopy. The computed concentrations were derived from a Lagrangian random walk model (Table 5.1). $u*$: 0.5 m · sec⁻¹ (—·—); 0.3 m · sec⁻¹(——); 0.1 m · sec⁻¹ (---).

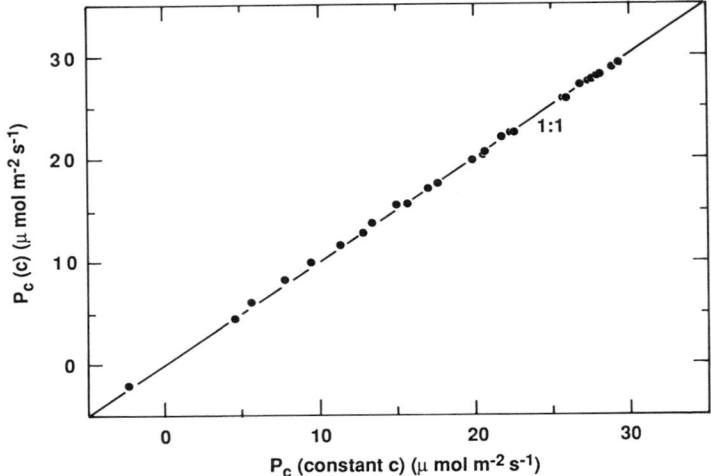

Figure 5.5 Comparison of canopy photosynthesis in soybeans, computed using a model that accounts for sink-scalar feedbacks [$(P_c(c))$] and one assuming a constant CO_2 profile [$(P_c(\text{constant } c))$]. The data set used in Figure 5.2 is tested here.

tested. This note of caution also holds for cases of evaporation and sensible heat exchange of aerodynamically smooth canopies. Here, the driving potential above the canopy can differ greatly from that in the vicinity of leaves, as was shown in the case of water vapor exchange in soybeans (Fig. 5.1), and has been reported for sugarcane (Grantz and Meinzer, 1990), alfalfa, and water hyacinth (Monteith, 1990).

B. Aerodynamically Rough and Complex Plant Stand: Broadleaf Forest

Scaling carbon and water vapor exchange in an aerodynamically rough and complex forest causes us to alter the scaling hypothesis that was proposed earlier. The distinct features of a deciduous forest canopy force us to (1) account for leaf clumping, penumbra, and vertical variations in leaf inclination angles when treating radiative transfer processes; (2) account for vertical variations in photosynthetic capacity (Jurik *et al.,* 1985); and (3) consider bole and branch respiration (Ryan, 1991). The features of the coupled micrometeorological/physiological canopy CO_2 exchange model are listed in Table 5.2.

Model estimates of canopy photosynthesis are relatively well correlated ($r=0.77$, $r^2=0.59$) with field measurements of canopy photosynthesis over a range of light, temperature, and wind conditions (Fig. 5.6). A paired Student's *t* test indicates no significant difference between measured and

Table 5.2 Data and Sources of Models and Parameters Used to Compute CO_2 Flux Densities over a Deciduous Forest Canopy

Model or parameter	Data and source
Species	*Quercus alba, Acer rubrum, Carya glabra, Lirodendrum tulipfera*
Canopy height	22.5 m
Photosynthesis model	Biochemical, Farquhar *et al.* (1980), Harley *et al.* (1985)
Stomatal conductance model	Phenomenological, Jarvis (1976)
Radiative transfer model	Negative binomial probability distribution, variable leaf angle distribution, penumbra, Baldocchi (1989)
Turbulent transfer model	Lagrangian random walk model, Legg and Raupach (1982), Raupach (1988); turbulence parameters, Baldocchi and Meyers (1988b)
Photosynthesis parameters	*Quercus rubra*, vertical variation in leaf capacity, Jurik *et al.* 1985)
Stomatal conductance parameters	*Quercus alba*, Baldocchi *et al.* (1987b)
Bole respiration	Model, Ryan (1991); stand parameters, Johnson and van Hook (1989)
Field flux data and driving environmental variables	Fluxes: eddy correlation, Baldocchi *et al.* (1987a)

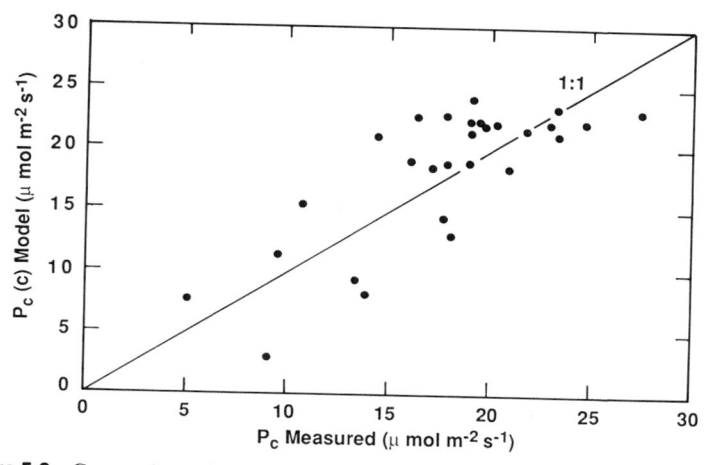

Figure 5.6 Comparison of canopy photosynthesis computations against micrometeorological flux measurements. The experimental data are from a study over a deciduous forest (Baldocchi *et al.*, 1987a). Model specifications are listed in Table 5.2.

calculated flux densities on the 5% probability level ($t = 0.764$, 27df). On the other hand, had bole respiration been ignored, model calculations would have overestimated measured fluxes significantly.

Errors in Fig. 5.6 can be attributed to using model parameters from the literature and errors in the field flux measurements. Errors ascribed to the calculation of photosynthetically active radiation (PAR), photosynthesis, and stomatal conductance can be discounted. This conclusion is drawn because model tests reveal that computations of these variables improved markedly by considering penumbra, leaf clumping, and vertical distributions in leaf inclination angle instead of using a spherical Poisson radiative-transfer model (see Fig. 5.7).

As with the soybean case, the tested model can be used to ask whether canopy photosynthesis of a broadleaf forest is limited by the supply of CO_2 (W_c) or by electron transport (W_j). In contrast to the soybean canopy, W_j is less than W_c for the sunlit, penumbral, and shaded leaf classes at all levels in the canopy (Fig. 5.8). These results suggest that all leaves in a broadleaf forest canopy are light limited rather than CO_2 limited. Discontinuities in the vertical profiles of W_c and W_j also are observed in

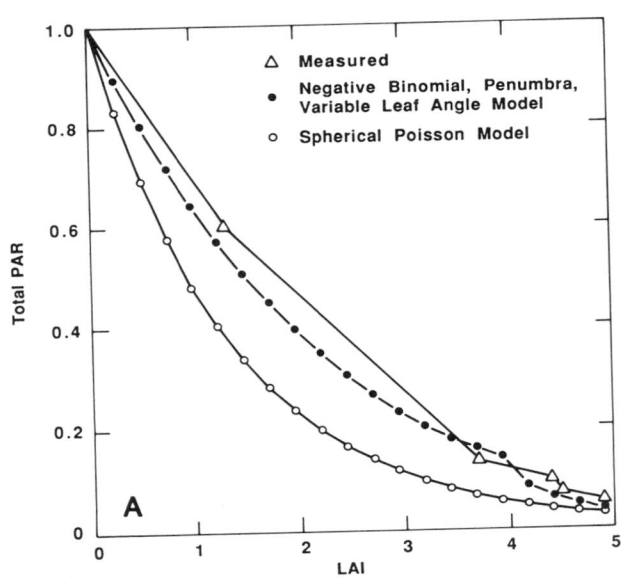

Figure 5.7   (A) Comparison of photosynthetically active radiation flux density computations against values measured in a deciduous forest. Details on the experimental data and field site are reported in Baldocchi and Hutchison (1986). The radiative transfer model accounts for leaf clumping, penumbra, and vertical variations in leaf inclination angles. Model specifications are listed in Table 5.2.

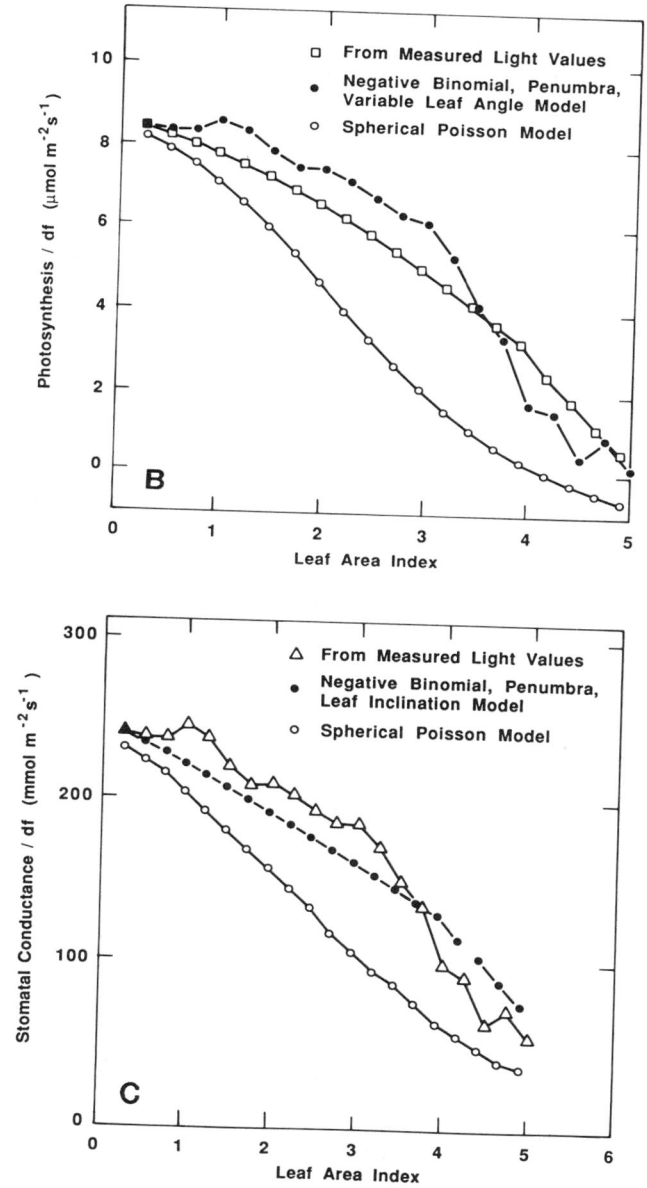

(B) Comparison of photosynthesis computations, based on the radiation model, against photosynthesis values derived from PAR measured in a broadleaf deciduous forest (after Baldocchi, 1989). (C) Comparison of stomatal conductance computations, based on the radiation model, against stomatal conductance values derived from PAR measured in a broadleaf deciduous forest (after Baldocchi, 1989).

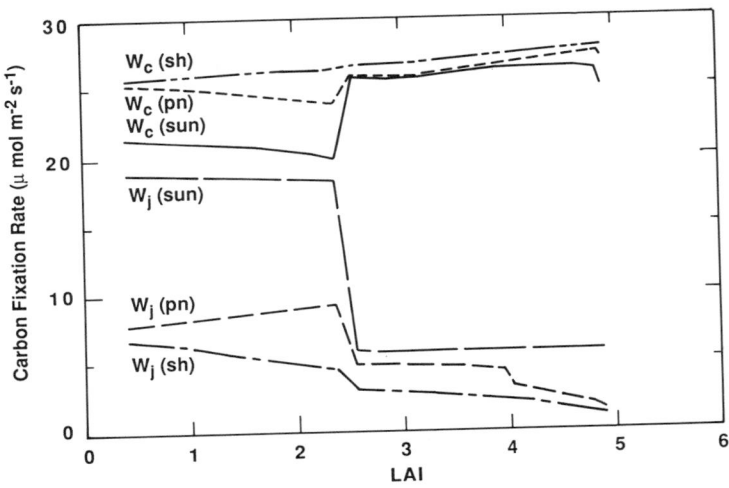

Figure 5.8 Computations of the RuBP-saturated carboxylation rate (W_c) and the electron transport-limited rate (W_j) on sunlit (sun), shaded (sh), and penumbra (pn) leaves in a broadleaf forest canopy. These carboxylation rates were computed with the model of Farquhar *et al.* (1980). Input variables were PAR of 2000 μmol $\cdot$ m^{-2} $\cdot$ sec^{-1}, air temperature of 25°C, wind speed of 3 m $\cdot$ sec^{-1}, and [CO_2] of 350 ppm.

Fig. 5.8. Jurik *et al.* (1985) report that the photosynthetic capacity of a shade leaf is less than that of leaves exposed to full sun. I attempted to account for this artifact by varying leaf photosynthetic parameters at a leaf area index where most leaves are shaded. A smoother transition could be expected had a more extensive data set been available. For example, one could scale maximum photosynthesis vertically according to leaf nitrogen measurements (see Field and Mooney, 1986).

What is the consequence of forest photosynthesis being limited exclusively by electron transport? It can be argued that the aerodynamic roughness of the canopy allows a simplified parameterization of feedbacks between the local scalar regime and the source/sink strength. In other words, one can assume a uniform scalar concentration profile without introducing significant error. This proposition is tested in Fig. 5.9. Data reveal that reductions in turbulent mixing have a small impact on canopy photosynthesis of a broadleaf forest. For example, a 5% reduction in P_c occurs as friction velocity ($u*$ decreases from 0.5 to 0.1 m $\cdot$ sec^{-1} (Fig. 5.9A). Computations of vertical CO_2 profiles support the contention that concentrations are relatively uniform within the canopy (Fig. 5.9B). Only at very low turbulent mixing levels ($u*=0.1$ m $\cdot$ sec^{-1}) does a moderate buildup of CO_2 respired from the soil occur. Since the photosynthetic

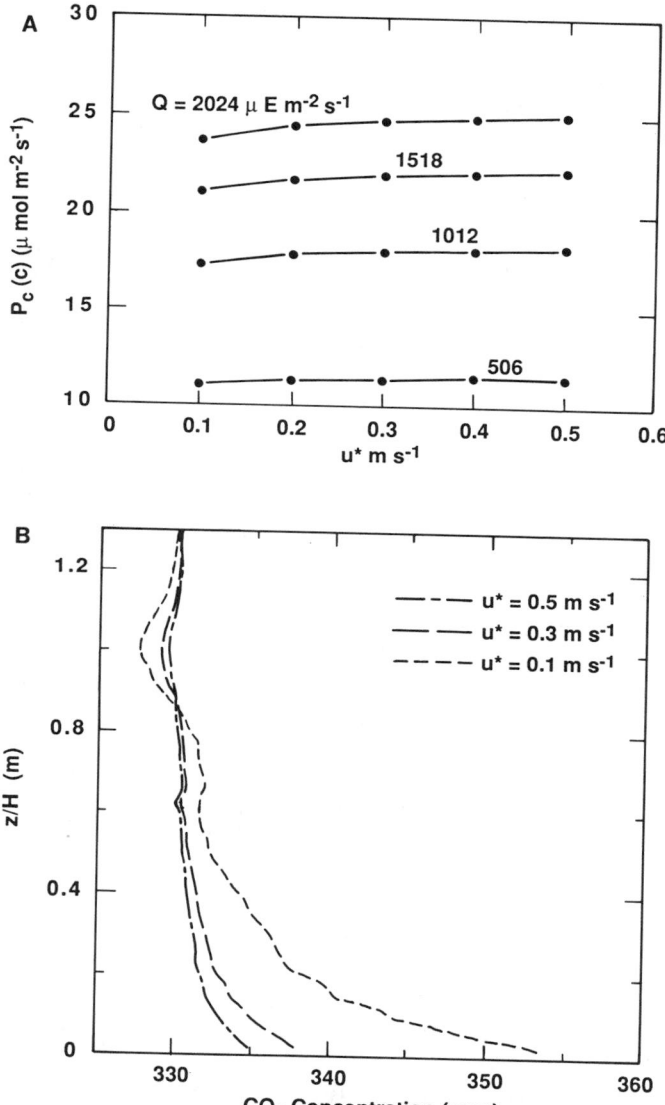

Figure 5.9 (A) Response of canopy photosynthesis in a deciduous forest to variations in friction velocity ($u*$) at various light levels ($Q = 2024, 1518, 1012$, or $506\ \mu E \cdot m^2 \cdot sec^{-1}$). (B) Measured and computed vertical profiles of CO_2 concentration in a broadleaf forest canopy. The computed concentrations were derived from a Lagrangian random-walk model (Table 5.2). $u*$, $0.5\ m \cdot sec^{-1}$(——); $0.3\ m \cdot sec^{-1}$ (——); $0.1\ m \cdot sec^{-1}$ (---).

capacity of the understory leaves is low, these higher CO_2 levels will not contribute to the net canopy exchange significantly.

The model results shown here lend support to the hypothesis that it is more important to model radiative transfer, hence, electron transport rates, well than to model turbulent transfer in great detail when calculating dioxide exchange in a broadleaf forest canopy. Additional support for allowing simplified micrometeorological models to be used within aerodynamically rough forests comes from studies by Tan *et al.* (1978), Jarvis *et al.* (1985), and Wang *et al.* (1991). These research groups were able to make reasonable estimates of canopy transpiration and photosynthesis of conifer forests by assuming constant scalar profiles within aerodynamically rough conifer canopies.

VIII. Concluding Comments

Although successful scaling of leaf-level gas exchange rates to the canopy scale was demonstrated, the reader must remember that the case studies examined here were from ideal situations, the canopies were one dimensional and conditions were steady. Obviously, many scaling problems remain because the natural environment often fluctuates and plant canopies are rarely horizontally uniform.

The roles of dynamic variations in turbulence and physiology are active fields of research. Unfortunately, many of these efforts are only at a descriptive stage. For example, we know that turbulent transfer occurs primarily through a succession of rapid sweep and ejection events that are followed by a prolonged quiescent period (Gao *et al.*, 1989); the whole sequence occurs within a time period of 100 to 300 sec. However, no mechanistic model exists to describe dynamic turbulence and so it may be linked to a dynamic description of evaporation and photosynthesis. Physiological studies show that photosynthesis and the rate of stomatal opening are also dynamic (Pearcy, 1990). Theoretical work has been conducted to model photosynthesis and stomatal dynamics on the leaf level (Kirschbaum *et al.*, 1988), but we do not know how significant these dynamic leaf-level responses are in the field and if significant, no one has proposed how to scale up such information to the canopy level under fluctuating field conditions. One possible approach would involve (1) prescribing the three-dimensional distribution of leaves according to probability statistics with a Monte Carlo approach; (2) evaluating radiation balance of each leaf using a procedural model (e.g., Myneni and Impens, 1985); and (3) using the calculated radiation field to calculate and integrate the dynamic leaf model of Kirschbaum *et al.* (1988). Although this approach is computationally intensive, it is not beyond the capability of available computers.

Scaling carbon dioxide and water vapor exchange in conifers and heterogeneous canopies in uniform and complex terrain remains a tough challenge. Excellent work is being done regarding radiative transfer of isolated and closed conifer canopies, but no single effort is being made to unite all the features cited in this chapter. For example, the MAESTRO model of Wang and Jarvis (1990) accounts for crown geometry and age differences, but assumes that foliage is distributed randomly within ellipsoidal envelopes and ignores penumbra. Oker-Blom and colleagues have elegant radiative transfer models that consider penumbra, shoot clumping, and shoot geometry, but these models ignore scattering.

Regarding complex terrain or advection across surface discontinuities, no comprehensive model addresses controls by both aerodynamics and physiology. Raupach *et al.* (1992) attempt to evaluate moisture fluxes over simples hills with a detailed micrometeorological model, but to achieve an analytical solution they neglect to consider variations in the radiation budget and stomatal conductance on different faces of the hill. It is interesting to note that Raupach *et al.* (1992) found that streamline deformation over simple hills had a minor impact on latent and sensible heat flux densities. On the other hand, Running and colleagues (1987) account for the effects of slope and azimuth on radiation balance to compute physiological variables, but neglect aerodynamic effects associated with advection, streamline deformation, cold air drainage, and atmospheric inversions.

Leaf-to-canopy scaling efforts require modeling of soil gas exchange rates. At present, state-of-art CO_2 exchange models rely on empirical links to soil temperature, moisture, and litter characteristics. Better predictive models and a larger experimental database on which to build these models are needed. Regarding soil evaporation, many models of varying sophistication exist and give comparable results for daytime evaporation (see Mahfouf and Noilhan, 1991).

Finally, the detailed theory presented in this chapter may not be amenable to routine applications, as in assessing global or ecosystem productivity and hydrology. More work is needed to assess if and how detailed scaling models can be simplified for routine use.

Acknowledgments

I dedicate this paper to the late Dr. Phil Miller. He was a pioneer and leading contributor in the field of leaf-to-canopy scaling long before it was fashionable, and his work inspired me to study this field.

This study is supported partially by the U.S. Department of Energy and the National Oceanic and Atmospheric Administration. The author acknowledges and appreciates the contributions made through conversations, collaborations, and correspondence with Tilden Meyers, Wang Han Jie, Bart van den Hurk, and Mike Ryan. Editorial comments by Chris Field and Jim Ehleringer are appreciated also.

References

Acock, B., Thornley, J. H. M., and Wilson, J. W. (1970). Spatial variation of light in the canopy. *In* "Proceedings of the IBP/PP Technical Meeting," pp. 91–102. Pudoc, Wageningen, The Netherlands.

Allen, L. H., Jr. (1974). Model of light penetration into a wide-row crop. *Agron. J.* **66,** 41–47.

Allen, L. H., Jr., Stewart, D. W., and Lemon, E. R. (1974). Photosynthesis in plant canopies effect of light response curves and radiation source geometry. *Photosynthetica* **3,** 184–207.

Anderson, M. C. (1966). Stand structure and light penetration. II. A theoretical analysis. *J. Appl. Ecol.* **3**(1), 41–54.

Baldocchi, D. D. (1989). Turbulent transfer in a deciduous forest. *Tree Physiol.* **5,** 357–377.

Baldocchi, D. D. (1992). A Lagrangian random-walk model for simulating water vapor, CO_2, and sensible heat flux densities and scalar profiles over and within a soybean canopy. *Boundary Layer Meteorol.* (in press).

Baldocchi, D. D., and Meyers, T. P. (1988a). A spectral and lag-correlation analysis of turbulence in a deciduous forest. *Boundary Layer Meteorol.* **45,** 31–58.

Baldocchi, D. D., and Meyers, T. P. (1988b). Turbulence structure in a deciduous forest. *Boundary Layer Meteorol.* **43,** 345–364.

Baldocchi, D. D., and Hutchison, B. A. (1986). On estimating canopy photosynthesis, and stomatal conductance in a deciduous forest with clumped foliage. *Tree Physiol.* **2,** 155–165.

Baldocchi, D. D., Verma, S. B., and Anderson, D. E. (1987a). Canopy photosynthesis and water use efficiency in a deciduous forest. *J. Appl. Ecol.* **24,** 251–260.

Baldocchi, D. D., Hicks, B. B., and Camara, P. (1987b). A canopy stomatal resistance model for gaseous deposition to vegetated surfaces. *Atmos. Environ.* **21,** 91–101.

Batchelor, G. K. (1949). Diffusion in a field of homogeneous turbulence. I. Eulerian analysis. *Aust. J. Sci. Res.* **2,** 437–450.

Black, T. A., and Kelliher, F. M. (1989). Process controlling understorey evapotranspiration. *Proc. R. Soc. London B* **324,** 207–231.

Bristow, K. L. (1987). On solving the surface energy balance equation for surface temperature. *Agric. Forest Meteorol.* **39,** 49–54.

Caldwell, M. M., Meister, H. P., Tenhunen, J. D., and Lange, O. L. (1986). Canopy structure, light microclimate and leaf gas exchange of *Quercus coccifera* L. in a Portuguese macchia: Measurements in different canopy layers and simulations with a canopy model. *Trees* **1,** 25–41.

Campbell, G. S. (1981). Fundamentals of radiation and temperature relations. (O. L. Lange *et al.,* ed.), *In* "Encyclopedia of Plant Physiology" Vol. 12A, pp. 11–40. Springer-Verlag, Berlin.

Carter, G. A., and Smith, W. K. (1985). Influence of shoot structure on light interception and photosynthesis in conifers. *Plant Physiol.* **79,** 1038–1043.

Chason, J., Huston, M., and Baldocchi, D. (1991). A comparison of direct and indirect methods for estimating forest canopy leaf area. *Agric. For. Meteorol.* **57,** 107–128.

Cohen, S., and Fuchs, M. (1987). The distribution of leaf area, radiation, photosynthesis and transpiration in a Shamouti orange hedgerow orchard. I. Leaf area and radiation. *Agric. For. Meteorol.* **40,** 123–144.

Cohen, S., Fuchs, M., Moreshet, S., and Cohen, Y. (1987). The distribution of leaf area, radiation, photosynthesis and transpiration in a Shamouti orange hedgerow orchard. II. Photosynthesis, transpiration and the effect of row shape and direction. *Agric. For. Meteorol.* **40,** 145–162.

Collatz, G. J., Ball, J. T., Grivet, C., and Berry, J. A. (1991). Regulation of stomatal conductance and transpiration: A physiological model of canopy processes. *Agric. For. Meteorol.* **54,** 107–136.

Corrsin, S. (1974). Limitations of gradient transport models in random walks and in turbulence. *Adv. Geophys.* **18A,** 25–60.

Cowan, I. R. (1968). Mass, heat and momentum exchange between stands of plants and their atmospheric environment. *Q. J. R. Meteorol. Soc.* **94,** 523–544.

Crawley, M. J. (1984). "The Structure of Plant Communities," pp. 1–50.

Da Costa, J. M. N., Rosenberg, N. J., and Verma, S. B. (1986). Respiratory release of CO_2 in alfalfa and soybean under field conditions. *Agric. For. Meteorol.* **37,** 143–157.

Deardorff, J. W. (1978). Closure of second- and third-moment rate equations for diffusion in homogeneous turbulence. *Phys. Fluids* **21,** 525–530.

Denholm, J. V. (1981a). The influence of penumbra on canopy photosynthesis. I. Theoretical considerations. *Agric. Meteorol.* **25,** 145–161.

Denholm, J. V. (1981b). Influence of penumbra on canopy photosynthesis. II. Canopy of horizontal circular leaves. *Agric. Meteorol.* **25,** 167–194.

Denmead, O. T. (1984). Plant physiological methods for studying evapotranspiration: Problems of telling the forest from the trees. *Agric. Water Mgmt.* **8,** 167–189.

Denmead, O. T., and Bradley, E. F. (1985). Flux-gradient relationships in a forest canopy. *In* Forest-Atmosphere Interactions" (B. A. Hutchison and B. B. Hicks, ed.), pp. 421–442. Reidel, Dordrecht, The Netherlands.

de Wit, C. T. (1965). "Photosynthesis of Leaf Canopies," pp 1–57. Institute for Biological and Chemical Research on Field Crops and Herbage, Wageningen.

de Wit, C. T. (1970). Dynamic concepts in biology. *In* "Prediction and Measurement of Photosynthetic Productivity," pp. 17–23. Agricultural Publ. Documentation, Wageningen.

Duncan, W. G., Loomis, R. S., Williams, W. A., and Hanau, R. (1967). A model for simulating photosynthesis in plant communities. *Hilgardia* **38,** 181–206.

Evans, J. R. (1989). Photosynthesis and nitrogen relationships in leaves of C_3 plants. *Oecologia* **78,** 9–19.

Farquhar, G. D. (1989). Models of integrated photosynthesis of cells and leaves. *Plant Environ. Biol.* **323,** 357–367.

Farquhar, G. D., and Sharkey, T. D. (1982). Stomatal conductance and photosynthesis. *Annu. Rev. Plant Physiol.* **33,** 317–45.

Farquhar, G. D., von Caemmerer, S., and Berry, J. A. (1980). A biochemical model of photosynthetic CO_2 assimilation in leaves of C_3 species. *Planta* **149,** 78–90.

Farquhar, G. D., and von Caemmerer, S. (1982). Modeling photosynthetic response to environmental conditions. *In* "Encyclopedia of Plant Physiology" (O. L. Lange *et al.,* ed.), Vol. 12B, pp. 549–587. Springer-Verlag, Berlin.

Field, C. (1983). Allocating leaf nitrogen for the maximization of carbon gain: Leaf age as a control on the allocation program. *Oecologia* **56,** 341–347.

Field, C. B. (1991). Ecological scaling of carbon gain to stress and resource availability. *In* "Integrated Responses of Plants to Stress" (H. A. Mooney, W. E. Winner, and E. J. Pell, eds.), pp. 1–32. Academic Press, San Diego.

Field, C., and Mooney, H. A. (1986). The photosynthesis–nitrogen relationship in wild plants. *In* "On the Economy of Plant Form and Function" (T. J. Giunish, ed.), pp. 25–55. Cambridge Univ. Press, Cambridge.

Finnigan, J. J. (1985). Turbulent transport in flexible plant canopies. *In* "The Forest-Atmosphere Interaction" (B. A. Hutchison and B. B. Hicks, eds.), pp. 443–480. Reidel, Dordrecht, The Netherlands.

Fitzjarrald, D. R., Moore, K. E., Cabral, O. M. R., Scolar, J., Manzi, A. O., and de Abreu Sá, L. D. (1990). Daytime turbulent exchange between the Amazon Forest and the atmosphere. *J. Geophys. Res.* **95,** 16,825–16,838.

Fukai, S., and Loomis, R. S. (1976). Leaf display and light environments in row-planted cotton communities. *Agric. Meteorol.* **17,** 353–379.

Gaastra, P. (1959). Photosynthesis of crop plants as influenced by light, carbon dioxide, temperature, and stomatal diffusion resistance. *Meded. Landbouwhogesch. Wageningen* **59,** 1–68.

Gao, W., Shaw, R. H., and Paw U, K. T. (1989). Observation of organized structure in turbulent flow within and above a forest canopy. *Boundary Layer Meteorol.* **47,** 349–377.

Garcia de Cortazar, V., Acevedo, E., and Nobel, P. S. (1985). Modeling of PAR interception and productivity by *Opuntia ficus-indica. Agric. Forest Meteorol.* **34,** 145–162.

Gijzen, H., and Goudriaan, J. (1989). A flexible and explanatory model of light distribution and photosynthesis in row crops. *Agric. Forest Meteorol.* **48,** 1–20.

Goudriaan, J. (1977). "Crop Micrometeorology: A Simulation Study." Centre for Agricultural Publishing and Documentation, Wageningen, The Netherlands.

Grace, J. (1980). Some effects of wind on plants. *In* "Plants and Their Atmospheric Environment" (J. Grace *et al.*, eds.), pp. 31–56. Blackwell Scientific Publications, Oxford.

Grace, J. C., Jarvis, P. G., and Norman, J. M. (1987). Modelling the interception of solar radiant energy in intensively managed stands. *N. Z. J. For. Sci.* **17,** 193–209.

Grant, R. F., Peters, D. B., Larson, E. M., and Huck, M. G. (1989). Stimulation of canopy photosynthesis in maize and soybean. *Agric. For. Meteorol.* **48,** 75–92.

Grantz, D. A., and Meinzer, F. C. (1990). Stomatal response to humidity in a surgarcane field: Simultaneous porometric and micrometeorological measurements. *Plant Cell Environ.* **13,** 27–37.

Gutschick, V. P., and Wiegel, F. W. (1988). Optimizing the canopy photosynthetic rate by patterns of investment in specific leaf mass. *Am. Nat.* **132,** 67–86.

Harley, P. C., Weber, J. A., and Gates, D. M. (1985) Interactive effects of light, leaf temperature, CO_2 and O_2 on photosynthesis in soybean. *Planta* **165,** 249–263.

Hollinger, D. (1989). Canopy organization and foliage photosynthetic capacity in a broad-leaved evergreen montane forest. *Func. Ecol.* **3,** 53–62.

Horn, H. S. (1971). "The Adaptive Geometry of Trees." Princeton University Press, Princeton, New Jersey.

Hutchison, B. A., Matt, D. R., McMillen, R. T., Gross, L. J., Tajchman, S. J., and Norman, J. M. (1986). The architecture of a deciduous forest canopy in eastern Tennessee. *J. Ecol.* **74,** 635–646.

Jackson, J. E., and Palmer, J. W. (1972). Interception of light by model hedgerow orchards in relation to latitude, time of year and hedgerow configuration and orientation. *J. Appl. Ecol.* **9,** 341–357.

Jarvis, P. G. (1976). The interpretation of the variations in leaf water potential and stomatal conductance found in canopies in the field. *Phil. Trans. R. Soc. London B* **273,** 593–610.

Jarvis, P. G., and McNaughton, K. G. (1986). Stomatal control of transpiration: Scaling up from leaf to region. *Adv. Ecol. Res.* **15,** 1–48.

Jarvis, P. G., Miranda, H. S., and Muetzelfeldt, R. I. (1985). Modelling canopy exchanges of water vapor and carbon dioxide in coniferous forest plantations. *In* "The Forest-Atmosphere Interaction" (B. A. Hutchison and B. B. Hicks, eds.), pp. 521–542. Reidel, Dordrecht, The Netherlands.

Johnson, C. E., Biscoe, P. V., Clark, J. A., and Littleton, E. J. (1976). Turbulent transfer in a barley canopy. *Agric. Meteorol.* **16**, 17–35.

Johnson, D. W., and van Hook, R. I. (eds) (1989). "Analysis of Biogeochemical Cycling Processes in Walker Branch Watershed." Springer-Verlag. New York.

Jurik, T. W., Briggs, G. M., and Gates, D. M. (1985). Carbon dynamics of northern hardwood forests: Gas exchange characteristics. *Carbon Dioxide Res. Div.* **19**, 1–63.

Kirschbaum, M. U. F., Gross, L. J., and Pearcy, R. W. (1988). Observed and modelled stomatal responses to dynamic light environments in the shade plant *Alocasia macrorrhiza*. *Plant Cell Environ.* **11**, 111–121.

Koen, C. (1987). A note on estimating the mean level of photosynthesis from radiation measurements. *Agric. For. Meteorol.* **40**, 207–211.

Kruijt, B. (1989). Estimating canopy structure of an oak forest at several scales. *Forestry* **62**(3), 269–284.

Lamb, R. G. (1980). Mathematical principles of turbulent diffusion modeling. "Atmospheric Boundary Layer Physics" (A. Longhetto, ed.), pp. 173–210. Elsevier Science, Amsterdam.

Lang, A. R. G. (1986). Leaf area and average leaf angle from transmission of direct sunlight. *Aust. J. Bot.* **34**, 349–355.

Leclerc, M. Y., Thurtell, G. W., and Kidd, G. E. (1988). Measurements and Langevih simulations of mean tracer concentration fields downwind from a circular line source inside an alfalfa canopy. *Boundary Layer Meteorol.* **43**, 287–308.

Legg, B. J., and Monteith, J. L. (1975). Heat and mass transfer in plant canopies. *In* "Heat and Mass Transfer in the Biosphere" (D. A. de Vries and H. H. Afgan, eds.), pp. 167–186. Wiley, New York.

Legg, B. J., and Raupach, M. R. (1982). Markov-chain simulation of particle dispersion in inhomogeneous flows: The mean drift velocity inducted by a gradient in Eulerian velocity variance. *Boundary Layer Meteorol.* **24**, 3–13.

Lemeur, R., and Blad, B. L. (1974). A critical review of light models for estimating the shortwave radiation regime of plant canopies. *Agric. Meteorol.* **14**, 255–286.

Leverenz, J. W., and Hinckley, T. M. (1990). Shoot structure, leaf area index and productivity of evergreen conifer stands. *Tree Physiol.* **6**, 135–149.

Mahfouf, J. F., and Noilhan, J. (1991). Comparative study of various formulations of evaporation from bare soil using *in situ* data. *J. Appl. Meteorol.* **30**, 1354–1365.

Mann, J. E., Curry, G. L., DeMichele, D. W., and Baker, D. N. (1980). Light penetration in a row-crop with random plant spacing. *Agron. J.* **72**, 131–142.

Marek, M., Masarovicova, E., Kratochvilova, I., Elias, P., and Janous, D. (1989). Stand microclimate and physiological activity of tree leaves in an oak-hornbeam forest. II. Leaf photosynthetic activity. *Trees* **4**, 234–240.

Massman, W. J. (1982). Foliage distribution in old-growth coniferous tree canopies. *Can. J. For. Res.* **12**, 10–17.

McNaughton, K. G., and Spriggs, T. W. (1989). An evaluation of the Priestly-Taylor equation and the complementary relationship using results from a mixed layer model of the convective boundary layer. *In* "Estimation of Areal Evapotranspiration" (Black *et al*, eds.), pp. 89–104. IAHS Press, Wallingford, UK.

Meyers, T. P., and Paw U, K. T. (1986). Testing of a higher-order closure model for modeling airflow within and above plant canopies. *Boundary Layer Meteorol.* **37**, 297–311.

Meyers, T. P., and Paw U, K. T. (1987). Modelling the plant canopy micrometeorology with higher-order closure principles. *Agric. For. Meteorol.* **41**, 143–163.

Miller, E. E., and Norman, J. M. (1971). A sunfleck theory for plant canopies. II. Penumbra effect: Intensity distribution along sunfleck segments. *Agron. J.* **63**, 739–743.

Miller, P. C. (1967). Leaf temperatures, leaf orientation and energy exchange in quaking aspen (*Populus tremuloides*) and Gambell's oak (*Quercus gambellii*) in central Colorado. *Oecologia Plantarum* **2**, 241–270.

Miller, P. C. (1970). A model of temperatures, transpiration rates and photosynthesis of sunlit and shaded leaves in vegetation canopies. *Unesco* **4**, 427–434.

Miller, P. C. (1971). Bioclimate, leaf temperature, and primary production in red mangrove canopies in south Florida. *Ecology* **53**, 22–45.

Monsi, M., and Saeki, T. (1953). Über den Lichtfaktor in den Pflanzengesellschaften und seine Bedeutung für die Stoffproduktion. *Jpn. J. Bot.* **14**, 22–52.

Monteith, J. L. (1965). Light distribution and photosynthesis in field crops. *Ann. Bot.* **29**, 17–37.

Monteith, J. L. (1990). Porometry and baseline analysis: The case for compatibility. *Agric. For. Meteorol.* **49**, 155–167.

Myneni, R. B., and Impens, I. (1985). A procedural approach for studying the radiation regime of infinite and truncated foliage spaces. I. Theoretical considerations. *Agric. For. Meteorol.* **34**, 3–16.

Myneni, R. B., Ross, J., and Asrar, G. (1989). A review on the theory of photon transport in leaf canopies. *Agric. For. Meteorol.* **45**, 1–153.

Naot, O., and Mahrer, Y. (1989). Modeling microclimtae environments: A verification study. *Boundary Layer Meteorol.* **46**, 333–354.

Nilson, T. (1971). A theoretical analysis of the frequency of gaps in plant stands. *Agric. Meteorol.* **8**, 25–38.

Norman, J. M. (1979). Modeling the complete crop canopy. *In* "Modification of the Aerial Environment of Plants" (B. J. Barfield and J. F. Gerber, eds.), pp. 249–277. American Society of Agricultural Engineers, St. Joseph, Michigan.

Norman, J. M. (1982). Simulation of microclimates *In* "Biometeorology in Integrated Pest Management" (J. L. Hatfield and I. Thompson, eds.). Academic Press, New York.

Norman, J. M., and Campbell, G. S. (1989). Canopy structure. *In* Physiological Plant ecology: Field Methods and Instrumentation" (Pearcy *et al.*, eds.), pp. 301–325. Chapman Hall, New York.

Norman, J. M., and Jarvis, P. G. (1975). Photosynthesis in sitka spruce (Picea sitchensis (bong.) carr.). V. Radiation penetration theory and a test case. *J. Appl. Ecol.* **12**, 839–878.

Norman, J. M., and Polley, W. (1989). Canopy photosynthesis. *In* "Photosynthesis" (W. R. Briggs, ed.). pp. 227–241. A. R. Liss, New York.

Norman, J. M. and Welles, J. M. (1983). Radiative transfer in an array of canopies. *Agron. J.* **75**, 481–488.

Oker-Blom, P. (1985). The influence of penumbra on the distribution of direct solar radiation in a canopy of Scots pine. *Photosynthetica* **19**, 312–317.

Oker-Blom, P. (1986). Photosynthetic radiation regime and canopy structure in modeled forest stands. *Acta Forestalia Fennica* **197**, 1–44.

Oker-Blom, P., and Kellomäki, S. (1983). Effect of grouping foliage on the within-stand and within-crown light regime: Comparison of random and grouping canopy models. *Agric. Meteorol.* **28**, 143–155.

Oker-Blom, P., Kellomäki, S., and Smolander, H. (1983). Photosynthesis of a Scots pine shoot: The effect of shoot inclination on the photosynthetic response of a shoot subjected to direct radiation. *Agric. Meteorol.* **29**, 191–206.

Osmond, C. B. (1989). Photosynthesis from the molecule to the biosphere: A challenge for integration. *In* "Photosynthesis" (W. R. Briggs, ed.), pp. 5–17. A. R. Liss, New York.

Parkhurst, D. F. (1986). Internal leaf structure: A three-dimensional perspective. "On the Economy of Plant Form and Function" (T. J. Givnish, ed.), pp. 215–249. Cambridge Press, Cambridge.

Parkhurst, D. F., Wong, S., Farquhar, G. D., and Cowan, I. R. (1988). Gradients of intercellular CO_2 levels across the leaf mesophyll. *Plant Physiol.* **86**, 1032–1037.

Paw U, K. T., and Gao, W. (1988). Applications of solutions to non-linear energy budget equations. *Agric. For. Meteorol.* **43**, 121–145.

Pearcy, R. W. (1990). Sunflecks and photosynthesis in plant canopies. *Annu. Rev. Plant Physiol. Plant Mol. Biol.* **41**, 421–453.

Price, D. T., and Black, T. A. (1990). Effects of short-term variation in weather on diurnal canopy CO_2 flux and evaporation of a juvenile Douglas-fir stand. *Agric. For. Meteorol.* **50**, 139–158.

Raupach, M. R. (1988). Canopy transport processes. *In* "Flow and Transport in the Natural Environment: Advances and Applications" (W. L. Steffen and O. T. Denmead, eds.), pp. 1–33. Springer-Verlag, Berlin.

Raupach, M. R., Weng, W. S., Carruthers, D. J., and Hunt, J. C. R. (1992). Temperature and humidity fields and fluxes in the air flow over low hills. *Q. J. R. Meteor. Soc.* **118**, 191–226.

Roberts, S. W., and Miller, P. C. (1977). Interception of solar radiation as affected by canopy organization in two mediterranean shrubs. *Ecol. Plant.* **12**(3),273–290.

Ross, J. (1981). "The Radiation Regime and Architecture of Plant Stands." Junk Publisher, The Hague.

Running, S. W., Nemani, R. R., and Hungerford, R. D. (1987). Extrapolation of synoptic meteorological data in mountainous terrain and its use for simulating forest evapotranspiration and photosynthesis. *Can. J. For. Res.* **17**, 472–483.

Ryan, M. G. (1991). Effects of climate change on plant respiration. *Ecol. Appl.* **1**, 157–167.

Sanford, A., and Jarvis, P. G. (1986). Stomatal responses to humidity in selected conifers. *Tree Physiol.* **2**, 89–103.

Sawford, B. L. (1985). Lagrangian statistical simulation of concentration mean and fluctuation fields. *J. Climate Appl. Meteorol.* **24**, 1152–1166.

Schimel, D. S., Kittel, T. G. F., Knapp, A. K., Seastedt, T. R., Parton, W. J., and Brown, V. B. (1991). Physiological interactions along resource gradients in a tallgrass prairie. *Ecology* **72**, 672–684.

Shawcroft, R. W., Lemon, E. R., Allen, L. H., Jr., Stewart, D. W., and Jensen, S. E. (1974). The soil-plant-atmosphere model and some of its predictions. *Agric. Meteorol.* **14**, 287–307.

Shuttleworth, W. J. (1989). Micrometeorology of temperate and tropical forest. *Phil. Trans. R. Soc. London B* **324**, 299–334.

Smolander, H. (1984). Measurement of fluctuating irradiance in field studies of photosynthesis. *Acta Forestalia Fennica* **187**, 1–56.

Tan, C. S., Black, T. A., and Nynamah, J. U. (1978). A simple diffusion model of transpiration applied to a thinned Douglas-fir stand. *Ecology* **59**, 1221–1229.

Terashima, I., and Saeki, T. (1983). Light environment within leaf. I. Optical properties of peridermal sections of camellia leaves with special reference to differences in the optical properties of palisade and spongy tissues. *Plant Cell Physiol.* **24**,(8), 1493–1501.

Terashima, I., and Saeki, T. (1985). A new model for leaf photosynthesis incorporating the gradients of light environment and of photosynthetic properties of chloroplasts within a leaf. *Ann. Bot.* **56**, 489–499.

Terashima, I., and Inoue, Y. (1985). Vertical gradient in photosynthetic properties of spinach chloroplasts dependent on intra-leaf light environment. *Plant Cell Physiol.* **26**, 781–785.

Thom, A. S. (1975). Momentum, mass and heat exchange of plant communities. "Vegetation and the Atmosphere" (J. L. Monteith, ed.), pp. 57–109. Academic Press, New York.

Van den Hurk, B., and Baldocchi, D. D. (1990). "A Random Walk Model for Simulating Water Vapor Exchange in a Soybean Canopy." NOAA Technical Report, U.S. Department of Commerce, Silver Spring, Maryland.

Vogelmann, T. C., Bornman, J. F., and Josserand, S. (1989). Photosynthetic light gradients and spectral regime within leaves of *Medicago sativa*. *Phil. Trans. R. Soc. London B* **323**, 411–421.

Waggoner, P. E., Furnival, G. M., and Reifsnyder, W. E. (1969). Simulation of the microclimate in a forest. *For. Sci.* **15**, 37–45.

Wang, Y. P., and Jarvis, P. G. (1990). Description and validation of an array model - MAESTRO. *Agric. For. Meteorol.* **51**, 257–280.

Wang, Y. P., Jarvis, P. G., and Benson, M. L. (1990). Two-dimensional needle-area density distribution within crowns of Pinus radiata. *For. Ecol. Mgmt.* **32**, 217–237.

Wang, Y. P., McMurtrie, R. E., and Landsberg, J. J. (1992). Modelling canopy photosynthetic productivity. *In* "Spatial and Temporal Determinants of Photosynthesis" (N. R. Baker, ed.) (in press).

Whitehead, D., Grace, J. C., and Godfrey, M. J. S. (1991). Architectural distribution of foliage in individual *Pinus radiata* D. Don crowns and the effect of clumping on radiation interception. *Tree Physiol.* **7**, 135–155.

Wilson, J. D. (1988). A second-order closure model for flow through vegetation. *Boundary Layer Meteorol.* **42**, 371–392.

Wilson, J. D. (1989). Turbulent transport within the plant canopy. *In* "Estimation of Areal Evaporation" (T. A. Black *et al.*, eds.). IAHS Press, Wallingford, UK.

Wilson, J. D., Thurtell, G. W., and Kidd, G. E. (1981). Numerical simulation of particle trajectories in inhomogeneous turbulence. I. Systems with constant turbulent velocity scale. *Boundary Layer Meteorol.* **21**, 295–313.

Wilson, J. D., Ward, D. P., Thurtell, G. W., and Kidd, G. E. (1982). Statistics of atmospheric turbulence within and above a corn canopy. *Boundary Layer Meteorol.* **24**, 495–519.

Wilson, N. R., and Shaw, R. H. (1977). A higher order closure model for canopy flow. *J. Appl. Meteorol.* **16**, 1197–1205.

Wyngaard, J. C. (1988). Convective processes in the lower atmosphere. *In* "Flow and Transport in the Natural Environment: Advances and Applications" (W. L. Steffen and O. T. Denmead, eds.), pp. 240–260. Springer-Verlag, Berlin.

6

Prospects for
Bottom-Up Models

Paul G. Jarvis

It has become widely accepted that process-based models provide a useful way forward, and indeed a necessary way forward because there is no reasonable alternative, when endeavoring to predict the consequences, sometime in the future, of environmental changes. As a result, a wide spectrum of more or less mechanistic models, in which processes are represented explicitly, has been developed on a range of spatial scales (e.g., leaf, plant, stand, region) for particular purposes. Ideally, these models contain a consistent, interlocking, interdependent set of processes that represent a coherent description of the way the system functions. In particular, these models must contain explicit representation of the processes on which the variables of particular interest have been shown to act.

Because the processes usually are defined on a smaller spatial scale and a shorter time scale than the scale of the output (e.g., processes at leaf scale, predictions at canopy scale), these models often have been called "bottom-up" models, although "disaggregated" model might be a better term.

The earliest bottom-up models were developed for agricultural and forest crops to provide an objective basis for the analysis of the relative performance of a range of yield-determining attributes (e.g., de Wit, 1965; Duncan *et al.*, 1967; Waggoner *et al.*, 1969; Lemon *et al.*, 1971). Since those early days, a wide range of bottom-up models has been developed for different crops for a wider range of purposes (see de Wit *et al.*, 1978; Loomis *et al.*, 1979; Norman, 1979; Hesketh and Jones, 1980) and, more recently, for forest plantations (e.g., Jarvis *et al.*, 1985; Mohren, 1987; McMurtrie *et al.*, 1988; Thornley, 1991; Wang *et al.*,

115

1992), agroforests (e.g., Grace *et al.*, 1987; Wang and Jarvis, 1990), energy forests (Perttu and Kowalik, 1989), and other seminatural and natural vegetation areas of minimal complexity, for example, grasslands and prairie (e.g., Johnson and Thornley, 1985).

Bottom-up models encompassing the complexity of natural ecosystems are, however, in their infancy. Some processes may be represented adequately (e.g., de Wit and Goudriaan, 1978), but a balanced representation of the whole system remains to be achieved (see Landsberg *et al.*, 1991).

Despite widespread acceptance of this trend, many rigorous scientists who work on the individual processes are suspicious of bottom-up modeling because they are aware of the deficiencies in understanding that exist with respect to some of the processes. They see only too well the short cuts and empiricism required because basic knowledge about certain processes is lacking.

There are other problems, too, relating to the usefulness of such models in practice. Increasingly, we are being asked to predict the likely responses of vegetation to environmental change on spatial and temporal scales appropriate to hydrological and climate models. Over what range of scales is it reasonable to take output from a set of processes on one scale and use it as input for another set of processes on a larger scale? The range of scales from leaf to region may cover seven or eight orders of magnitude. Can one with impunity model over more than adjacent scales? In discussing this question, de Wit (1970) concluded that two-stage models linking three adjoining scales were the limit beyond which it was impractical to go. He went on to say that

> seven-stage simulation models by means of which ecosystems may be explained on the basis of the molecular sciences are impossibly large and detailed and it is naive to pursue their construction. Likewise, it is naive to pursue construction of simulation models which are supposed to simulate complex ecosystems like forests as a whole.

Today, however, serious attempts are being made to model forests and other ecosystems over scales ranging from the leaf to the region and from minutes to centuries. Are we being hopelessly naive, as perceived by de Wit in 1970, or has the situation changed since then?

To what extent is bottom-up modeling open ended, possibly leading to predictions that may be seriously in error, because of inadequate representation of the processes and because of transfer over too large a range of spatial scales or too long a range of temporal scales? Does the modeling process suppress or exaggerate errors in the initial assumptions? Can bottom-up modeling yield realistic solutions to practical problems or is it primarily an academic exercise?

To address these kinds of questions, a discussion was held on bottom-up modeling. This chapter is an attempt to synthesize the views expressed.

I. What Are Bottom-Up Models?

In general, bottom-up models are deterministic, state-of-the-art, mechanistic, process-based models that provide a statement of how a system may function on its spatial and temporal scale, on the basis of knowledge acquired on smaller spatial and shorter temporal scales.

As such, a bottom-up model is a *tool* that may be used to provide answers to a range of questions, solve a range of problems, and, particularly, make predictions about the likely response of the system to future changes in a range of environmental variables.

As far as possible, bottom-up models are based on knowledge that has been established by experiment and is generally accepted, but because they may contain hypotheses for which empirical knowledge does not exist or is inadequate, they may be teleonomic, that is, they may arrive at a solution on the basis that the plant has been optimized through evolution to operate in a certain way, as conceived by the modeler.

The most familiar bottom-up model takes knowledge at leaf or subleaf scale, combines this with environmental information, particularly weather and stand sructural information, and derives descriptions of how a stand functions in a range of circumstances. Less familiar, but also very relevant, are bottom-up models that take stand scale information and derive descriptions of how a region or landscape may function in relation to climate and weather, and yield predictions of likely responses to changes in management.

Bottom-up models have usually been conceived by one person or a small group without regard to any particular objective such as land management, climate change, or general circulation models (GCMs). They represent an individual view of how the components of the sysem function and, by bringing the components together, how the system functions.

The components vary in degree of empirical and mechanistic content, depending on the current state of knowledge. When opinions differ, alternative formulations of the same process may be available. As new knowledge is acquired (e.g., the root-produced messenger hypothesis for effects of water stress on stomata; Johnson *et al.*, 1991), new routines may be added and old ones removed or put in storage.

Although a bottom-up model has usually been constructed by one person or a small group and represents their ideas about how the system functions, there is a large degree of similarity in the way processes are represented in different bottom-up models. There is general agreement on many of the processes that need to be represented and on the current state of knowledge about these processes. In other cases, however, there is substantial lack of agreement and alternative ways forward are being practiced (e.g., allocation of carbon within the plant, phenology of leaf growth). Nonetheless, a substantial amount of unproductive reinvention and duplication of effort goes on that could be reduced or avoided with better communication and exchange of ideas.

The basic properties of different kinds of vegetation are essentially similar; consequently the modules (i.e., functions and subroutines) needed in a bottom-up model to describe the processes of stand functioning are general. The differences between a grassland and a woodland, for example, are superficial; essentially the same bottom-up model, with only minor changes, can be used for both (e.g., Thornley, 1991). In principle, there is no major reason why we should not have a generic bottom-up model, at least for many types of C_3 vegetation. Although different users may wish to represent certain processes in particular, individual ways, this becomes completely feasible given an appropriate model structure, as exemplified in Fig. 6.1.

Because most bottom-up models have been created by an individual, they are generally unbalanced with respect to the modules contained and the amount of detail in the modules. To a considerable extent, a bottom-up model may express the prejudices of the modeler, with an emphasis on the processes that the modeler considers to be important and, usually, an excessive amount of detail about these processes. As such, the bottom-up model is an indulgence, but an understandable one.

A bottom-up model can contain only those interactions and feedbacks that are known explicitly or about which there are tenable hypotheses. Because many of the interactions and feedbacks may not be known, the bottom-up model is necessarily incomplete. Nonetheless, a bottom-up model provides a framework within which new knowledge about the processes and their interactions can be added as it accrues. For many, a bottom-up model is an archive of all that is known about the system on which it works, an archive that can serve as an indication of gaps and deficits in knowledge about the system, and one that can be updated readily.

For a modeler to have faith in what he or she does and for the community at large to accept the predictions made with the model, all bottom-up models should be tested and validated. Critical assessment of the assumptions in a bottom-up model by others working in the field is an important part of the validation process, but is frequently difficult because the assumptions are not defined explicitly, the documentation is poor, and the model itself is insufficiently transparent. Empirical calibration of the model to force agreement between predictions and measurements by adjustment of parameter values cannot be regarded as a proper procedure, but may be necessary when either functions or parameters are defined poorly.

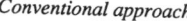

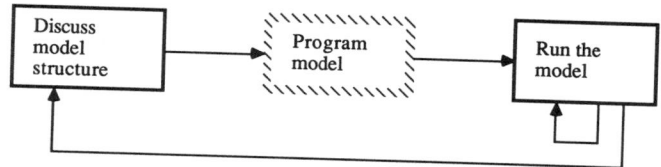

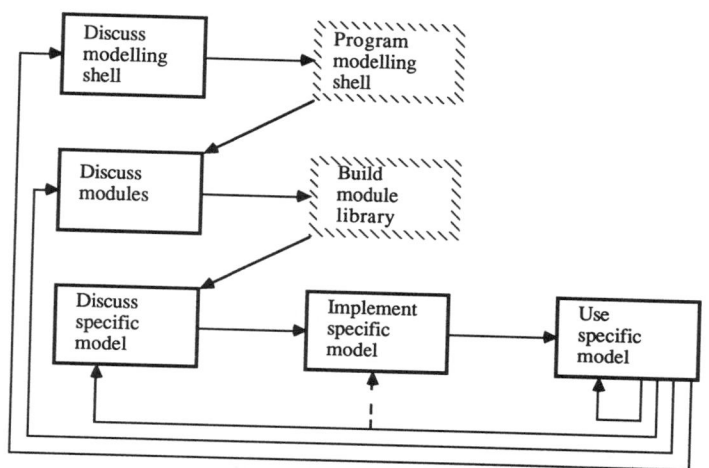

Figure 6.1 Comparison of conventional (top) and proposed (bottom) approaches to implementing computer simulation models of ecological problems. In the conventional approach, any changes to the model require a programming phase, which is an impediment to rapid development and refinement of a model. In the proposed approach, the major (bottom) loop involves scientists in all the phases of model design, implementation, and use, without the need to undertake reprogramming of the model. Addition of new modules and extensions to the model-building shell do involve programming, but much less frequently (R. I. Muetzelfeldt, personal communication).

Optimization is not an appropriate procedure to use in running a bottom-up model, but it certainly has its uses in deriving parameters from data. A common procedure is to remove from the model a set of functions or a subroutine, invert it, apply it to a set of measurements of a property (e.g., surface conductance) obtained over a range of conditions of one of more variables (e.g., quantum flux density and saturation deficit), and extract the parameters relating the property to the variables by optimization using "maximum likelihood" or "nonlinear least squares" procedures. The parameter values then can be used in the model.

Not all the complexity embedded in a typical bottom-up model is needed to answer particular questions and may increase the running time unnecessarily. A streamlined cut-down version of the model, in which assumptions and approximations replace some of the full rigor of the original, is a desirable option for many applications. Intellectual analysis of the system, intuition, and sensitivity analysis can be used to determine which detail can be removed or approximated without introducing significant bias or error. The performance of the cut-down model can be evaluated against the full model, as well as against independent data. Such a cut-down model can be run over long periods of simulated time at reasonable cost in computer time. However, such a model adds nothing new to our ideas about what is important or needs to be included in simplified models on larger and longer spatial and temporal scales.

II. Problems

Heterogeneity of the vegetation raises problems with respect to the description of structure and the parameterization of physiological processes, and is a major problem in the application of bottom-up models to natural and seminatural vegetation. Heterogeneity of vegetation also places a greater emphasis on soil properties in many situations. The bottom-up models in current use for modeling ecological systems have, in general, been based on the successful use of bottom-up models over many years in modeling agricultural crops but the degree of uniformity found in crops is in marked contrast to the heterogeneity in natural and seminatural vegetation.

As more and more routines or submodels are added to a bottom-up model, and more detail is included, the computation and running time tends to increase. Consequently, comprehensive bottom-up models are frequently only run for short periods of simulated time, for example, days rather than years. If the simulation time is to be one or more years, the running time becomes unacceptably long, if the whole model must be run on every occasion. The development of appropriate software structures that will allow only those modules needed to answer a particular question to be pulled down from a library and be run is required urgently.

Schemes for this purpose have been proposed and prototypes developed, but it seems that no major bottom-up model can be operated in this way. Consequently, to obtain solutions to problems with any reasonably comprehensive bottom-up model, a programmer is required to be constantly on hand to modify the model for a particular purpose. Figure 6.1

shows one approach that avoids having to run the whole model to answer a simple question and, thus, both economizes on running time and increases the practical usefulness of the model to the user.

III. Top-Down Models: An Alternative Approach

Because the output from bottom-up models is open ended and may be very sensitive to error in certain inputs such as leaf area, an alternative more empirical approach has achieved considerable practical usefulness in recent years. In essence, the output from the model is constrained totally through an experimentally determined relationship with a crucial driving variable. The empirical relationship that is derived puts a lid on the system, constraining predictions to lie within the realm of observation, and preventing extreme predictions that may occur from the open-ended bottom-up approach.

Two well-known examples of such a lumped parameter approach are

- the Monteith model of stand growth as a function of absorbed radiation at the stand scale (Monteith, 1977)
- the Priestley–Taylor model on evapotranspiration as a function of net radiation at the regional scale (McNaughton and Spriggs, 1986)

In both examples, much detail evident in bottom-up models at the scale below is no longer included in the formulation of the lumped parameter model. However, such empirical relationships may be decomposed to "explain" their success or to provide modifying coefficients for exceptional circumstances; some of that detail ultimately may be included as routines that are essentially similar to those used in coming up from below (Fig. 6.2) (e.g., Wang *et al.*, 1991). Because explanation and generality are achieved by fingering down into the processes from above, such models frequently are known as "top-down" models rather than lumped parameter models.

An advantage of a top-down model is that it can be used immediately to make predictions, whereas it is likely to take a number of years to formulate the functions and to determine the values of the parameters required in a bottom-up method. A major disadvantage of a top-down model is that predictions cannot be made safely outside the range of the variables encountered in the derivation of the lumped parameter function. Thus, without supporting, process-based routines like those in a bottom-up model, a top-down model cannot be used for predicting the consequences of global change.

By using the two approaches of bottom-up modeling and top-down modeling at the same time, one can derive *and* explain an appropriate

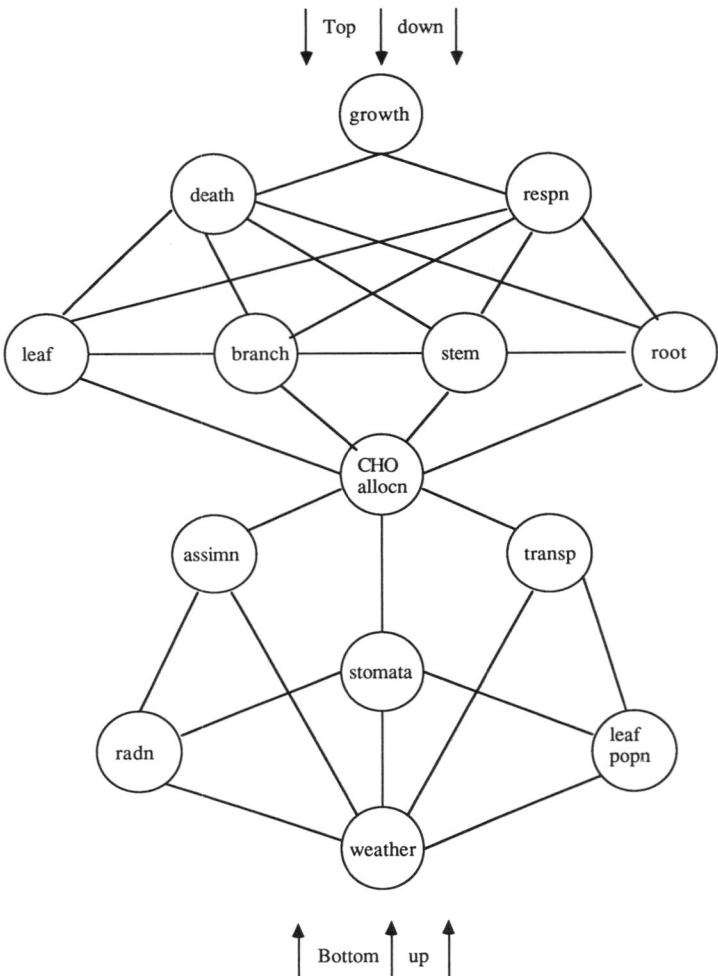

Figure 6.2 A diagram of a plant or stand growth model to illustrate the convergence of bottom-up and top-down approaches to the prediction of growth. Assembly of the disaggregated processes in a bottom-up model may lead to predictions of growth. In a top-down model, growth may be related to absorbed radiation (radn) by an empirical function, the lumped parameter(s) in that function being explained or derived by identifiable processes at successive levels of detail.

top-down model (e.g., Wang *et al.*, 1991). It should be emphasized that, if both modeling approaches go on together, the best of both worlds is obtained.

IV. Bottom-Up Models and Scaling

It is neither necessary nor desirable to carry all the detail in a bottom-up model conceived for a particular spatial and temporal scale up through a succession of larger and larger spatial and temporal scales. Further, it is inevitable that bottom-up models will be made for each scale by modelers working on that scale and will, as a result, contain less detail than models conceived on smaller scales.

For example, a model that scales-up from leaf to canopy with a time scale of minutes will inevitably be much richer in the detail of leaf-scale processes than a model that scales up from canopies to region with a time scale of days. In fact, what is conceived as a bottom-up approach on the larger scale (e.g., the growth/absorbed PAR function) may well be regarded as a top-down approach on the smaller scale (Fig. 6.3).

It was suggested that top-down and bottom-up models between leaf and stand scales should converge on the individual plant, because it is the individual that is the repository of genetic information and is susceptible to change by breeding or evolution. Although this statement is certainly correct, the evolutionary time scale is far too long in relation to the predicted rate of global change to be of practical relevance in this context (Huntley, 1991). As far as crops are concerned, stands of new genotypes are a regular consequence of plant breeding. It has been shown many times that the success of a genotype depends very strongly on the structure of the crop in which it is grown. Further, the genotypes in a crop are all closely similar, if not the same. However, this is far from the case in natural and seminatural vegetation, where the genotypes are much more varied and the time scale for evolutionary change is much longer than the time scale for climatic change. Thus, the degree of acclimation and the speed of migration are processes more likely to determine the composition of vegetation in the future (Bradshaw and McNeilly, 1991).

Negative feedbacks stabilize the response of a system to change in a variable and make the system less sensitive to a change in that variable. On larger spatial scales, more negative feedbacks come into play and the gains in the feedback paths are generally larger, meaning that the model can be simpler. Identification of the feedback paths on smaller scales through the use of a bottom-up model is, therefore, a major step in scaling up and can lead to logically derived conclusions about what can be left out of the model without significant detriment (e.g., McNaughton

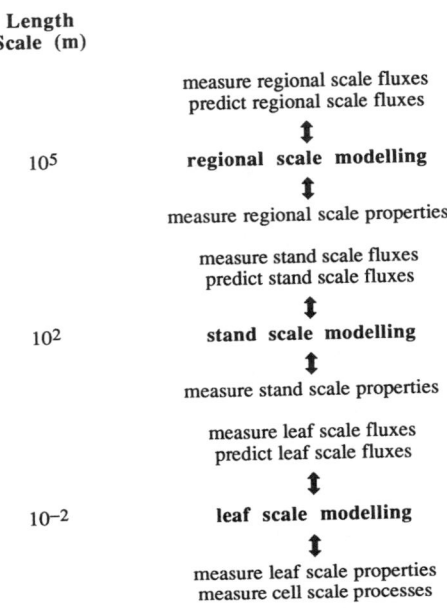

**Length
Scale (m)**

measure regional scale fluxes
predict regional scale fluxes
$\updownarrow$
10^5 **regional scale modelling**
$\updownarrow$
measure regional scale properties

measure stand scale fluxes
predict stand scale fluxes
$\updownarrow$
10^2 **stand scale modelling**
$\updownarrow$
measure stand scale properties

measure leaf scale fluxes
predict leaf scale fluxes
$\updownarrow$
10^{-2} **leaf scale modelling**
$\updownarrow$
measure leaf scale properties
measure cell scale processes

Figure 6.3   A strategy for upscaling from leaf to region by bottom-up modeling. Models
are developed at each major identifiable scale and tested by comparison of model predictions with measurements made at the scale above. A bottom-up model developed at one
scale may resemble a top-down model for the scale beneath. In principle, a bottom-up
model at the smallest scale may be cascaded up to the largest scale; in practice this may be
a very risky thing to do.

and Jarvis, 1991). Simplification should not be left to the unassailable
instincts of experts but should result from analysis of the system, leading
to justifiable conclusions. A bottom-up model, simplified as a result of
the stabilizing influence of negative feedbacks on the larger scales, may
differ little from the appropriate top-down function.

V. Conclusions

We are under considerable pressure to come up with answers to questions
put by the public and by politicians concerning our future well-being.
For this purpose, we need to be able to predict the likely impacts of
global change on ecosystems and biomes, as well as on plants and crops.
Conversely, we also need to be able to predict the effects of terrestrial
vegetation, ecosystems, and biomes on the global atmosphere, and the
meteorological and vegetational consequences of the feedbacks between

vegetation and atmosphere. Naive or not, we must make use of the knowledge we have about the functioning of plants and terrestrial ecosystems in models to make these predictions. Nonetheless, it behooves us to bear in mind de Wit's (1970) final remark: "Success in this field is only possible when we have the common sense to recognize that we know only bits and pieces of nature around us and restrict ourselves to quantitative and dynamic analyses of the simplest ecological systems that can be made or found in biology." We know more of the bits and pieces now than we did in 1970, but there are still very substantial gaps. We should be cautious in our use of bottom-up models and in the conclusions we reach.

References

Bradshaw, A. D., and McNeilly, T. (1991). Evolutionary response to global climatic change. *Ann. Bot.* **67**, 5–14.

de Wit, C. T. (1965). "Photosynthesis of Leaf Canopies," pp. 1–57. Argic. Res. Rep. 663, Wageningen Centre for Agricultural Publications and Documentation.

de Wit, C. T. (1970). *In* "Prediction and Measurement of Photosynthetic Productivity" (I. Setlik, ed.), pp. 17–23. Pudoc, Wageningen.

de Wit, C. T., *et al.* (1978). "Simulation of Assimilation, Respiration and Transpiration of Leaf Surfaces." Pudoc, Wageningen.

de Wit, C. T., and Goudriaan, J. (1978). "Simulation of Ecological Processes." Pudoc, Wageningen.

Duncan, W. G., Loomis, R. S., Williams, W. A., and Hanau, R. (1967). A model for simulating photosynthesis in plant communities. *Hilgardia* **38**, 181–203.

Grace, J. C., Jarvis, P. G., and Norman, J. M. (1987). Modelling the interception of solar radiant energy in intensively managed stands. *N. Z. J. For. Sci.* **17**, 193–209.

Hesketh, J. D., and Jones, J. W. (1980). "Predicting Photosynthesis for Ecosystem Models," Vol. II. CRC Press, Boca Raton. Florida.

Huntley, B. (1991). How plants respond to climate change: Migration rates, individualism and the consequences for plant communities. *Ann. Bot.* **67**, 15–22.

Jarvis, P. G., Miranda, H. S., and Muetzelfeldt, R. I. (1985). Modelling canopy exchanges of water vapour and carbon dioxide in coniferous forest plantations. *In* "The Forest-Atmosphere Interaction" (B. A. Hutchison and B. B. Hicks, eds.), pp. 521–542. Reidel, Dordrecht.

Johnson, I. R., and Thornley, J. H. M. (1985). Dynamic model of the response of a vegetative grass crop to light, temperature and nitrogen. *Plant Cell Environ.* **8**, 485–499.

Johnson, I. R., Melkonian, J. J., Thornley, J. H. M., and Riha, S. J. (1991). A model of water flow through plants incorporating shoot/root 'message' control of stomatal conductance. *Plant Cell Environ.* **14**, 531–544.

Landsberg, J. J., Kaufmann, M. R., Binkley, D., Isebrands, J., and Jarvis, P. G. (1991). Evaluating progress towards closed forest models based on fluxes of carbon, water and nutrients. *Tree Physiol.* **9**, 1–15.

Lemon, E., Stewart, D. W., and Shawcroft, R. W. (1971). The sun's work in a cornfield. *Science* **174**, 371–378.

Loomis, R. W., Rabbinge, R., and Ng, E. (1979). Explanatory models in crop physiology. *Annu. Rev. Plant Physiol.* **30**, 339–367.

McMurtrie, R. E., Rook, D. A., and Kelliher, F. M. (1988). Modeling the yield of *Pinus radiata* on a site limited by water and nitrogen. *For. Ecol. Mgmt.* **30**, 381–413.

McNaughton, K. G., and Jarvis, P. G. (1991). Effects of spatial scale on stomatal control of transpiration. *Agric. For. Meteorol.* **54**, 279–301.

McNaughton, K. G., and Spriggs, T. W. (1986). A mixed-layer model for regional evaporation. *Boundary Layer Meteorol.* **34**, 243–262.

Mohren, G. M. J. (1987). "Simulation of Forest Growth, Applied to Douglas-Fir Stands in The Netherlands." Pudoc, Wageningen.

Monteith, J. L. (1977). Climate and the efficiency of crop production in Britain. *Phil. Trans. Roy. Soc. London Ser. B* **281**, 277–294.

Norman, J. M. (1979). Modeling the complete crop canopy. *In* "Modification of the Aerial Environment of Crops" (B. J. Barfield and J. F. Gerber, eds.), pp. 249–277. Am. Soc. Agric. Engin. Monograph No. 2, ASAE, St Joseph, Michigan.

Perttu, K. L., and Kowalik, P. J. (1989). "Modeling of Energy Forestry: Growth, Water Relations and Economics." Pudoc, Wageningen.

Thornley, J. H. M. (1991). A transport-resistance model of forest growth and partitioning. *Ann. Bot.* **68**, 211–226.

Waggoner, P. E., Furnival, G. M., and Reifsnyder, W. E. (1969). Simulation of the microclimate in a forest. *For. Sci.* **15**, 37–45.

Wang, Y.-P., and Jarvis, P. G. (1990). Description and validation of an array model: MAESTRO. *Agric. For. Meteorol.* **51**, 257–280.

Wang, Y.-P., Jarvis, P. G., and Taylor, C. M. A. (1991). PAR absorption and its relation to above-ground dry matter production of Sitka spruce. *J. Appl. Ecol.* **28**, 547–560.

Wang, Y.-P., McMurtrie, R. E., and Landsberg, J. J. (1992). Modeling canopy photosynthetic productivity. *In* "Crop Photosynthesis: Spatial and Temporal Determinants" (N. R. Baker and H. Thomas, eds.), Elsevier Science, Amsterdam.

7

Scaling Ecophysiology from the Plant to the Ecosystem: A Conceptual Framework

James F. Reynolds, David W. Hilbert, and Paul R. Kemp

I. Introduction

There is a growing consensus among atmospheric scientists that global temperature and precipitation will increase due to increasing levels of atmospheric CO_2, although the exact magnitudes are unclear (Chen and Drake, 1986; Schlesinger, 1986; Kellogg, 1991). Although the primary cause of increased atmospheric CO_2 appears to be the combustion of fossil fuels (Rotty and Marland, 1986), the increase represents a disequilibrium among all the sources and sinks in the bio–geosphere carbon cycle (Woodwell *et al.*, 1983; Chen and Drake, 1986; Mooney *et al.*, 1987). In addition to its potential for inducing climatic changes, increased CO_2 also has a direct effect on plants through the physiological processes of photosynthesis, respiration, and stomatal closure (Strain, 1985). Thus, global climate change, in concert with these direct effects of CO_2 on plants, could have a significant impact on both natural and agricultural ecosystems (Committee on Global Change, 1988). The ability of our society to prepare for, and respond to, such changes depends largely on the ability of climate and ecosystem researchers to provide predictions of regional-level ecosystem responses with sufficient confidence and adequate lead time (Dahlman, 1985).

The effects that increased CO_2 will have on ecosystems—and even broader questions such as how much CO_2 will increase and how the various bio–geospheric pools of carbon will change—are related directly to plants because of the fundamental role plants play in the global carbon cycle. Plants are a significant sink for CO_2 from the atmosphere; at least for individual plants, their capacity as sinks (i.e., for CO_2 uptake) is dependent on the atmospheric concentration of CO_2. Plants form the base of terrestrial and aquatic food chains; all carbon available to the other organisms is linked directly or indirectly to productivity and carbon flow through plants. Thus, predicting the changes in vegetation on Earth in response to elevated CO_2 and climate change is crucial to understanding the future behavior of the atmosphere–biosphere–geosphere cycles of carbon (Woodwell *et al.*, 1983; National Research Council, 1986; Mooney *et al.*, 1987).

In this chapter, we address various issues related to using plant ecophysiological information to understand and predict how ecosystems may respond to elevated CO_2 and climate change. First, we discuss the role of modeling, emphasizing its value as a extrapolation tool. Second, problems of scaling are considered. A hierarchical scheme for model aggregation is presented and the role of various types of models within this conceptual framework are considered. We give examples of (1) aggregating leaf-level physiological information to predict ecosystem dynamics and (2) the danger of transposing scales, that is, using fine-scale data to predict larger scale phenomena directly.

II. Role of Modeling

It is currently impossible to determine the long-term outcome of elevated CO_2 and climate change on ecosystems experimentally. Consequently, increasing atmospheric CO_2 is essentially an Earth-scale "experiment" (Ramanathan, 1988). To predict the outcome of this experiment is a daunting challenge, particularly across the various levels of organization in the bio–geosphere. For example, consider the multiple factors involved in predicting the response of single plants to elevated CO_2. Research has shown that not all plants are affected in the same way and that the way in which individual plants respond to CO_2 is affected greatly by the levels of other environmental resources (e.g., nutrients and water) and regulators (e.g., temperature and salinity) (see Strain and Cure, 1985; Mooney *et al.*, 1991). Since these resources and regulators are spatially and temporally variable over the Earth, large differences can be expected among regions or biomes in their response to elevated CO_2. These factors, coupled with the potential for climate change that will produce further changes in environmental factors, make it extremely difficult to design experiments that test more than a small fraction of the many different combinations of environmental conditions that may be involved in causing ecosystem change with elevated CO_2 (Reynolds and Acock, 1985). Even at very small scales (e.g., 1 m^2), there are many factors that make it difficult to determine the effects of elevated CO_2 on whole ecosystems experimentally, including long-term changes that may lead to a new homeostasis (Hilbert *et al.*, 1987), potential herbivory feedbacks (Fajer *et al.*, 1991), and multiple environmental stresses (Chapin *et al.*, 1987).

Modeling has an essential role in developing an understanding of potential plant, community, and ecosystem responses to simultaneous changes in atmospheric CO_2 and climate (Dahlman, 1985; Reynolds and Acock, 1985). The judicious use of modeling will allow a limited body of empirical knowledge to be extended greatly through simulation of complex combinations of environmental–biotic interactions. The major challenge is how to build mechanistic models that are based on short-term or small-scale studies but are useful for predicting phenomena on larger temporal and spatial scales (Reynolds and Leadley, 1992).

III. Scaling Issues and Hierarchy Theory

A. Scaling and Aggregation Problems

Scaling issues greatly complicate how we investigate the response of ecosystems to climate change (Rosswall *et al.*, 1988). The spatial and temporal scales on which the plant carbon pool (e.g., tropical forests, coniferous forests, and grasslands) reaches equilibrium with other carbon pools of the bio–geosphere are vastly greater than the scale on which carbon dioxide directly affects plant growth through its effects on photosynthesis and stomatal aperture.

The effects of enriched atmospheric CO_2 on ecosystems can be viewed in a hierarchical manner. Carbon dioxide directly enters the biosphere through plant leaves, affecting photosynthesis, respiration, stomatal conductance, and transpiration. Because these cellular and physiological processes are reasonably well understood, the direct effects of elevated CO_2 concentrations can be predicted with a fairly high degree of confidence. However, the indirect effects of CO_2 are complex, as illustrated in Fig. 7.1. Changes in leaf physiological processes lead to changes in allocation patterns, growth, and other whole-plant properties (Strain, 1985; Hilbert *et al.*, 1991). In turn, these effects feed back on lower hierarchical levels (e.g., altering leaf photosynthetic capacity) and are translated to higher hierarchical levels, for example, plant–plant interactions (Bazzaz and Garbutt, 1988) and interactions among organisms of different trophic levels (Committee on Global Change, 1988). The plethora of interactions likely to occur will produce changes in the distribution of resources and, ultimately, ecosystem structure and function (Mooney *et al.*, 1991), thus affecting the global carbon balance (Post *et al.*, 1990).

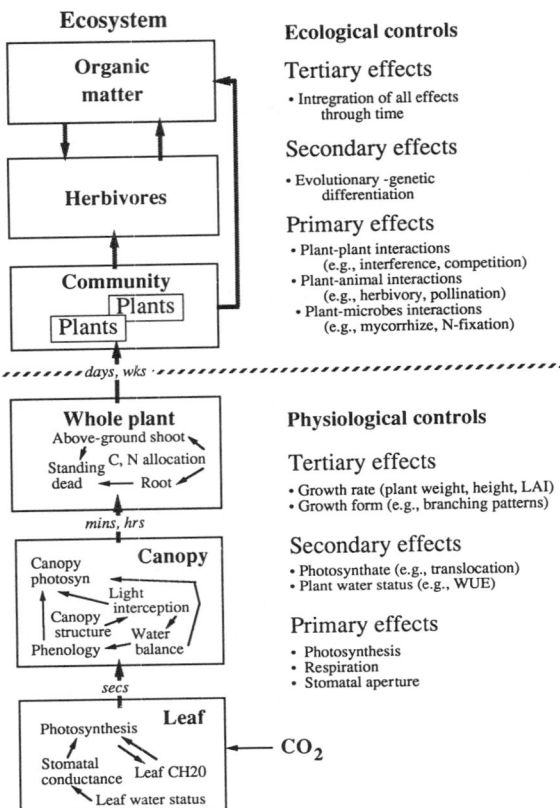

Figure 7.1 Physiological and ecological controls on translating effects of elevated CO_2 from leaf-level processes up to ecosystems.

How will elevated CO_2 and climate change affect systems and processes at specific hierarchical levels? Importantly, how will lower-level effects (e.g., physiology) be translated to higher levels in the hierarchy (Fig. 7.1)? O'Neill (1988) calls this the "aggregation problem." Addressing this problem will involve a careful match of scales of observation (pattern), theory (explanation), and modeling (description) to develop models that are capable of prediction on the scale of interest (e.g., leaf) while avoiding errors that occur when a model attempts to scale up processes and mechanisms over too great a range (e.g., leaf to ecosystem) (Patten, 1983; O'Neill, 1988; Woodmansee, 1988). Thus, model aggregation is the parallel development of (1) the theory of how CO_2 and climate effects actually are scaled up from levels at which they have direct effects, for example, on plant physiology and growth, to higher levels in the ecosystem at which they act indirectly; and (2) the technical aspects of linkage, simplification, and information flow between models developed on different scales. These two issues are discussed in the following sections.

B. Role of Mechanistic Models

We suggest that mechanistic descriptions of systems are necessary to address the aggregation problem (Reynolds and Acock, 1985; Reynolds and Leadley, 1992). We use the term mechanistic for models that decompose a system into its component parts and describe the behavior of the whole through the interaction of those parts. Such models are considered to be based on underlying mechanisms that give rise to the observed behavior of interest. (Such models also frequently are referred to as bottom-up, explanatory, or process models.) Of course, "mechanism" is itself a scale-dependent concept: One person's "mechanism" is another person's empiricism!

The special characteristic of mechanistic models in our usage is that behavior of the system at any chosen level in a hierarchy (e.g., organ, whole plant, community, ecosystem) is modeled by representing the interactions of the component systems at lower hierarchical levels, for example, modeling whole plant dynamics based on organ descriptions or modeling community dynamics based on whole plants. Generally, mechanistic models are composed of both empirical and phenomenological equations that describe the component parts of the system (Thornley, 1976). Empirical models are based on statistical formulations (e.g., multiple regression) that best correlate a set of observed variables to some response variable of interest; these models are simple, readily built, and easy to apply, but lack generality, so their use must be restricted to the range of data on which they are based (see Fig. 7.2A). Phenomenological equations also describe the behavior of a system (the phenomena of interest) without decomposition into lower-level subsystems (see Fig. 7.2B). Phenomenological models or equations do not "plunge beneath the surface," but describe the process or system property directly at the level of interest (Spanner, 1964; Thornley, 1976).[1] In contrast to empirical models, the form of a phenomenological model is derived from a general understanding of the process or system of interest (e.g., logistic growth). Hence, the equation used for the model is meaningful and describes some logical or expected relationship between the variables.

An example of the use of phenomenological equations to build a mechanistic model is given by Reynolds and Thornley (1982). To develop a whole-plant growth model, they decompose plants into four compo-

---

[1] An example is the ideal gas law, $PV = nRT$, which accurately describes the behavior of gases over a wide range of conditions *without* any reference to (or even knowledge of) the molecules of which the gases are composed.

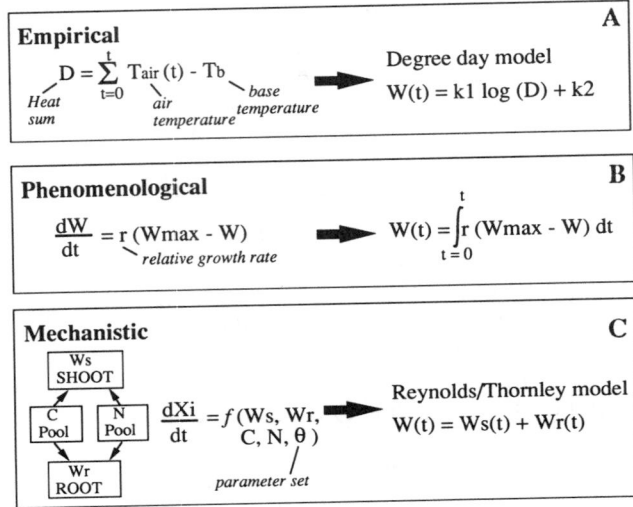

Figure 7.2 Three types of models to describe the change in whole plant weight over time, $W(t)$. (A) Empirical (degree day), in which plant is treated as a "black box" (D, heat sum; T_{air}, air temperature; T_b, basic temperature). (B) Phenomenological, in which functional form (logistic equation) is based on knowledge of plant growth (r, relative growth rate). (C) Mechanistic, in which plant is decomposed into interacting components (root, shoot, carbon pool, and nitrogen pools) (θ, parameter set).

nents (shoot biomass, root biomass, carbon substrate pool, and nitrogen substrate pool) and model growth as the interaction of these components (see Fig. 7.2C). The equations they use to represent the behavior of each component in this mechanistic model are phenomenological. For example, the differential equation for the carbon substrate pool is formulated from a knowledge of the possible inputs to this component and the law of mass balance.

C. Hierarchical Framework for Model Aggregation

Hierarchy theory (Allen and Starr, 1982; O'Neill *et al.*, 1986) suggests that it is unnecessary to look further down than one level in the search for "mechanistic" explanations of the behavior of a system. Ecological models that explicitly include processes and structures at several hierarchical levels (i.e., composed of numerous coupled process models, having many interactions) tend to be very complex and are frequently hypersensitive to slight changes in parameters (Allen and Starr, 1982). Although such models have been successful in simulating specific short-term processes, they are not well suited for long-term simulations (Allen and Starr, 1982). Models that include "mechanism" across a large range of hierarchical levels necessarily become very complex, are difficult to maintain and verify, tend to be unstable, and usually are not amenable to modification for new situations (Reynolds and Leadley, 1992). Ultimately, such models "lose" their mechanism as noise (Rexstad and Innis, 1985; Landsberg, 1986; O'Neill, 1988).

In our effort to build models that will enable us to predict ecosystem responses to global change, we argue that only mechanistic models will permit us to extrapolate beyond our data with any degree of confidence, yet we must avoid building complex bottom-up ecosystem models that attempt to include processes operating at many different levels of hierarchy directly. The conceptual approach we present is a hierarchical scheme in which the level of focus (L) is modeled by describing the interactions among components that exist at any hierarchical level below L (typically one level below, i.e., $L-1$) but are themselves represented in the model as phenomenological relationships (Fig. 7.3A). Thus, a fundamental focus is to understand the ways in which processes and system properties are aggregated and translated among and within subsystems. The mechanistic modeling of each hierarchical level is emphasized via the mechanism that occurs at the lower level, coupled with constraints from the level above (O'Neill, 1988; Salthe, 1985). We emphasize the hierarchical nature of our mechanistic modeling approach whereby modeling on each scale is based on information from lower scales (Fig. 7.3). The model is parameterized with data collected at the subordinate level scale, validated against data collected at the scale of focus, and constrained by data or information collected at the superior scale. In this way, we continually test our ability to extrapolate from one scale to the next.

The importance and utility of mechanistic modeling has been discussed in the context of predictive system-level response models. Mechanistic modeling seems particularly applicable to global change questions. Perhaps most importantly, the examination of the effects of global change at several hierarchical levels will give us a greater understanding of how processes that operate at one level in a system are translated to or influence other levels in the system. A second justification for this approach is that most of the data regarding effects of CO_2 and climate change (mainly temperature, moisture, and nutrients) are for individual plant responses or responses of very small and simple "systems." It is necessary to scale up these observations to make predictions for ecosystems, landscapes, regions, and, ultimately, the biomes of the Earth.

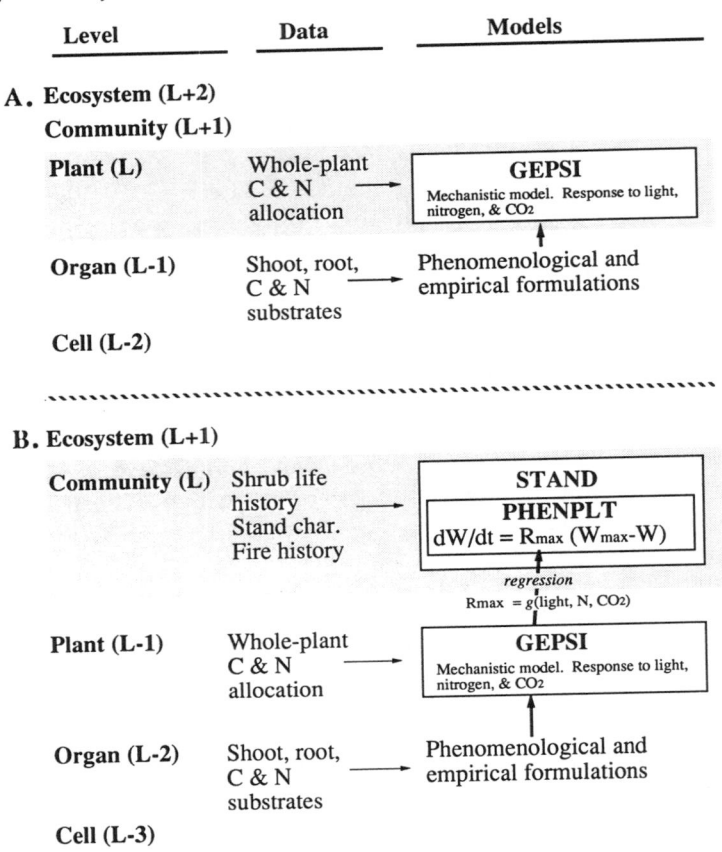

Figure 7.3 Scheme to develop mechanistic models at Level *L*. (A) Whole-plant model GEPSI (*L*) is based on the dynamics of organs (*L* − 1) and whole-plant allocation of carbon and nitrogen substrates (*L*). (B) Community model STAND (*L*) is based on relationships among components of community, for example, shrub life histories (*L*), GEPSI (*L* − 1), and organ dynamics (*L* − 2).

IV. Examples of Model Aggregation

A. Effects of Elevated CO₂ and Soil Nitrogen on Stand Dynamics

A model of long-term stand dynamics in chaparral communities was developed by Hilbert and Larigauderie (1990). This model (STAND, Fig. 7.3B), which operates on an annual time step, includes a simple phenomenological submodel of plant growth (PHENPLT, Fig. 7.3B). PHENPLT describes the change in weight (*W*) of the average plant in the community as a function of a maximum growth rate (R_{max}) and a maximum attainable weight (W_{max}). In this example, we examine the effects of variable soil nitrogen (N), solar radiation, and elevated CO_2 on stand dynamics by modifying PHENPLT. This modification consists of the use of lower-level information (i.e., plant ecophysiology) to predict how higher levels (i.e., plant community) will respond. This is a classic aggregation "problem:" because of the complex interactions likely to exist in predicting plant and stand response to soil N, solar radiation, and elevated CO_2, we wish to take advantage of existing ecophysiological information on these species.

To accomplish this, we used GEPSI (Fig. 7.3A; J. F. Reynolds, unpublished observations), a detailed mechanistic plant growth model that runs on a daily basis, to predict R_{max} for chaparral plants as a function of N, solar radiation, and CO_2. GEPSI was run under a wide range of these environmental variables to calculate a response surface for R_{max}. A simple hyperbolic model (function g, Fig. 7.3B) was then fit to this response surface and substituted for the fixed parameter R_{max} in PHENPLT. We now can run STAND and investigate stand-level consequences of altered N, solar radiation, and CO_2 availabilities.

The results are summarized in Fig. 7.4. The mortality rate of shrubs is accelerated under nitrogen stress and affects the ultimate stand density and biomass. Interesting interactions between CO_2 and nitrogen availability emerge that would not have been predicted from GEPSI models alone. For example, total biomass and leaf area index are increased significantly by CO_2 only when the stand is strongly nitrogen limited. This constrasts with the behavior of individual plants predicted from GEPSI, which suggests that growth rates are most stimulated by CO_2 when nitrogen is abundant. Although this example is preliminary and requires more analysis, it appears that the more open stands that develop under nutrient limitation are more responsive to CO_2 because they are less light limited. Stands with greater nutrient availability quickly attain a closed canopy where production is limited by competition for light.

In this example, fine-scale ecophysiological information has been "aggregated" up the hierarchy by substituting the empirical regression equation for R_{max}, developed from output data generated by GEPSI (a model with high extrapolation potential), into the phenomenological submodel PHENPLT of the STAND model. Stand-level effects of altered resource availability could not have been investigated in the original version of STAND because plant growth responses to resources were not represented. On the other hand, population and community effects of resource availability cannot be addressed by GEPSI since community level processes like mortality and competition among plants are not included. Each mechanistic model (GEPSI and STAND) is best at answering questions at its own level of focus. However, the method described here allows

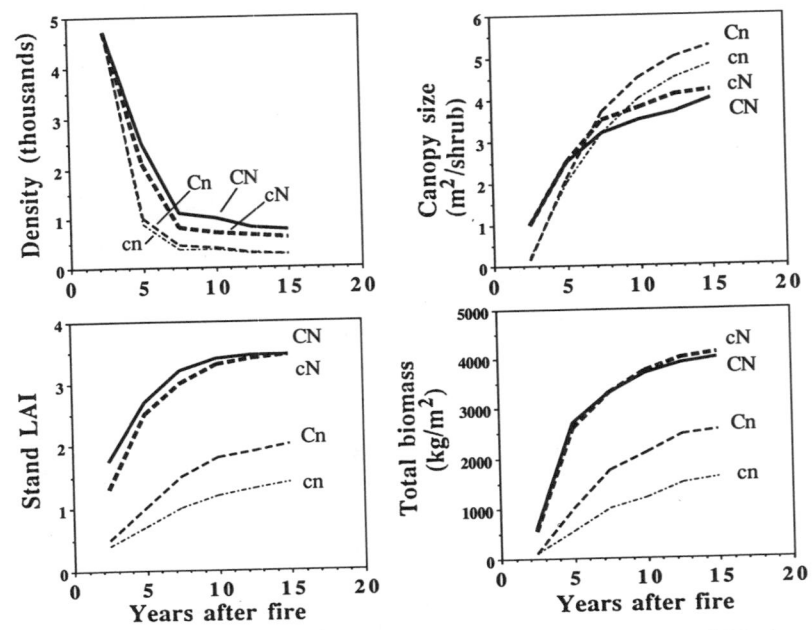

Figure 7.4 Results from running STAND model at two levels of internal CO_2 (c = 240 and C = 500 μbar) and low (n) and high (N) soil nitrogen availability in four combinations: CN (~), Cn (- - -), CN (- - -), and cn (-··-).

us to include information from a mechanistic model at level *L*-1 (GEPSI) into a mechanistic model at level *L* (STAND) without directly incorporating the complete model structure.

B. Danger of Direct Scaling: Transposition of Scale

Significant errors may be introduced if information obtained at a lower level in the hierarchy is used directly to make predictions at a higher level, without incorporating the interactions between the components of an increasingly complex system. O'Neill (1988) calls this a "transposition of scale." We present a simple example to illustrate this effect.

In Fig. 7.5, we show the percentage increase in critical states or processes at three hierarchical levels (leaf, whole-plant, and stand) in response to increasing atmospheric CO_2, where the responses are predictions based on the models GEPSI, PHENPLT, and STAND shown in Fig. 7.3. For the sake of this example, we assume constant climatic conditions. At the leaf level, under saturating light and high nutrient status, the

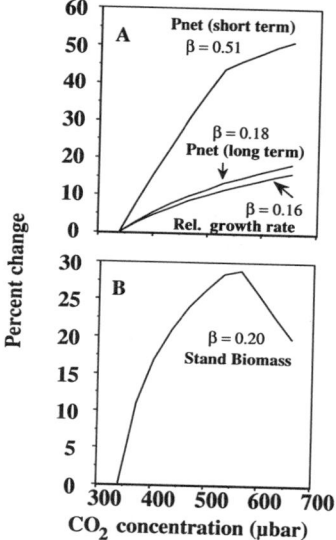

Figure 7.5 Predicted change in response to increasing atmospheric CO_2. Beta is the proportional change in each variable following a doubling of CO_2 concentration. (A) P_{net} over short term ($\beta = 0.51$; *top*); P_{net} over long term ($\beta = 0.18$; *middle*); relative growth rate ($\beta = 0.16$, *bottom*). (B) Stand biomass ($\beta = 0.20$).

immediate short-term effect of increasing atmospheric CO_2 is substantial, resulting in a 51% increase in net photosynthesis when CO_2 rises from 340 to 680 μbar (Fig. 7.5A). During long-term exposure to elevated CO_2, leaf photosynthetic capacity generally declines and the modeled increase in net photosynthesis at 680 μbar is only 18% (Fig. 7.5A).

The next step is to consider the results of the leaf simulations in a whole-plant context using GEPSI (which includes feedback to the leaves resulting from changes in carbon and nitrogen allocation to root and shoot). The whole-plant relative growth rate was calculated (R_{max}, Fig. 7.3) and the effect of doubling CO_2 is an increase in R_{max} of 16% (Fig. 7.5A). Next, these results were included at the stand level following the scheme outlined in Fig. 7.3. At the stand level, there is a 20% increase in biomass predicted by STAND for a doubling of CO_2 (Fig. 7.5B). (Note that this effect is contingent on several other factors, including soil nitrogen availability, which we do not consider here.)

The percentage stimulation of these leaf, whole-plant, and stand processes at 680 μbar is equivalent to the biotic growth factor (β-factor) of Bacastow and Keeling (1973), values of which are given in Fig. 7.5.

Although values of the β-factor obtained at the level of the individual leaf or small crop stand have been used to infer large-scale effects of increasing CO_2 (e.g., Gates, 1985), we caution against such an approach. Information obtained from low levels of the hierarchy is certainly of value in understanding potential ecosystem responses, but if such information is to be incorporated into models at the next higher level, the potential interactions and feedbacks (positive and negative) that arise at the new level must be recognized and dealt with explicitly. To ignore them will result in potentially large errors in prediction. Jarvis and Mc-Naughton (1986) present an analogous argument in discussing the problems of scaling measurements of evapotranspiration made at the leaf or plant level to make predictions on a regional or global scale.

V. Summary

In summary, we have argued that complex global change issues, such as elevated atmospheric CO_2, are especially difficult to address since they involve translating information across a variety of spatial and temporal scales. Although the direct effects of a particular change may be focused primarily on a single scale, the potential ramifications of the change are likely to be seen on many higher, and possibly lower, scales. Understanding these interactions and making predictions on larger scales will require modeling. However, in our view, a single model cannot be expected to span all scales, so an effort must be made to develop mechanistic models that are effective on particular scales and can be used also to provide information to models on larger scales.

We also must progress beyond the naive view that "mechanisms" exist solely at lower levels of organization. The process and elements that constitute the mechanism of a model are dictated by the level (scale) of focus of the model. Mechanistic interactions of system components can be represented best at one, or at most, two levels lower than the focal level. Developing a hierarchy of models in this way, each of which is suited to a particular scale, and drawing on insights and output from lower levels is, in our opinion, the most effective way of accounting for the translation of effects from the physiological level of organization to the ecosystem level and the ramifications of those effects.

Acknowledgment

This research was supported by U.S. Department of Energy Grant DE-FG03-86ER60490.

References

Allen, T. F. H., and Starr, T. B. (1982). "Hierarchy: Perspectives for Ecological Complexity." University of Chicago Press, Chicago.

Bacastow, R., and Keeling, C. D. (1973). *In* "Progress in Photosynthesis Research" (I. Briggins, ed.), pp. 221–224. Martinus Nijhoff, Dordrecht, Netherlands.

Bazazz, F. A., and Garbutt, K. (1988). The response of annuals in competitive neighborhoods: Effects of elevated CO_2. *Ecology* **69**, 937–946.

Chapin, F. S., III, Bloom, A. J., Field, C. B., and Waring, R. H. (1987). Plant responses to multiple environmental factors. *BioScience* **37**, 49–57.

Chen, C. T. A., and Drake, E. T. (1986). Carbon dioxide increase in the atmosphere and oceans and possible effects on climate. *Annu. Rev. Earth Planet. Sci.* **14**, 201–235.

Committee on Global Change (1988). "Toward an Understanding of Global Change—Initial Priorities for U.S. Contributions to the International Geosphere-Biosphere Program." National Academy Press, Washington, D.C.

Dahlman, R. C. (1985). Modeling needs for predicting responses to CO_2 enrichment; plants, communities and ecosystems. *Ecol. Model.* **29**, 77–106.

Fajer, E., Bowers, M. D., and Bazazz, F. A. (1991). The effects of enriched CO_2 atmospheres on the buckeye butterfly, *Junonia coenia*. *Ecology* **72**, 751–754.

Gates, D. M. (1985). Global biospheric response to increasing atmospheric carbon dioxide concentration. *In* "Direct Effects of Increasing Carbon Dioxide on Vegetation" (B. R. Strain and J. D. Cure, eds.), pp. 171–184. U.S. DOE/ER-0238, National Technical Information Service, Springfield, Virginia.

Hilbert, D. W., and Larigauderie, A. (1990). A modelling approach to the concept of stand senescence in chaparral. *Acta Oecologica* **11**, 181–190.

Hilbert, D. W., Larigauderie, A., and Reynolds, J. F. (1991). Influence of carbon dioxide and daily photon-flux density on optimal leaf nitrogen concentration and root:shoot ratio. *Ann. Bot.* **68**, 365–376.

Hilbert, D. W., Prudhomme, T., and Oechel, W. C. (1987). Response of tussock tundra to elevated carbon dioxide regimes: Analysis of ecosystem CO_2 flux through nonlinear modeling. *Oecologia* **72**, 466–472.

Jarvis, P. G., and McNaughton, K. G. (1986). Stomatal control of transpiration: Scaling up from leaf to region. *Adv. Ecol. Res.* **15**, 1–49.

Kellogg, W. M. (1991). Response to skeptics of global warming. *Bull. Amer. Meteor. Soc.* **72**, 499–511.

Landsberg, J. J. (1986). "Physiological Ecology of Forest Production." Academic Press, New York.

Mooney, H. A., Vitousek, P. M., and Matson, P. A. (1987). Exchange of materials between terrestrial ecosystems and the atmosphere. *Science* **238**, 926–932.

Mooney, H. A., Drake, B. G., Luxmoore, R. J., Oechel, W. C., and Pitelka, L. R. (1991). Predicting ecosystem responses to elevated CO_2 concentrations. *BioScience* **41**, 96–104.

National Research Council (1986). "Global Change in the Geosphere-Biosphere. Initial Priorities for an IGBP." National Academy Press, Washington, D.C.

O'Neill, R. V. (1988). Hierarchy theory and global change. *In* "Scales and Global Change" (T. Rosswall, R. G. Woodmansee, and P. G. Risser, eds.), pp. 29–45. John Wiley & Sons, New York, New York.

O'Neill, R. V., DeAngelis, D. L., Wade, J. B., and Allen, T. F. H. (1986). "A Hierarchical Concept of Ecosystems." Princeton University Press, Princeton, New Jersey.

Patten, B. C., ed. (1983). "System Analysis and Simulation in Ecology," Vol. 1. Academic Press, New York.

Post, W. M., Peng, T.-H., Emanuel, W. R., King, A. W., Dale, V. H., and DeAngelis, D. L. (1990). The global carbon cycle. *Am. Sci.* **78**, 310–326.

Ramanathan, V. (1988). The greenhouse theory of climate change: a test by an inadvertent global experiment. *Science* **240**, 293–299.

Rexstad, E., and Innis, G. S. (1985). Model simplification—Three applications. *Ecol. Model.* **27**, 1–13.

Reynolds, J. F., and Acock, B. (1985). Predicting the response of plants to increasing carbon dioxide: A critique of plant growth models. *Ecol. Model.* **29**, 107–129.

Reynolds, J. F., and Thornley, J. H. M. (1982). A shoot:root partitioning model. *Ann. Botany* **49**, 585–97.

Reynolds, J. F., and Leadley, P. W. (1992). Modeling the response of arctic plants to changing climate. *In* "Arctic Ecosystems in a Changing Climate" (F. S. Chapin, R. Jefferies, J. F. Reynolds, G. Shaver, and J. Svoboda, eds.), pp. 413–438. Academic Press, San Diego.

Rosswall, T., Woodmansee, R. G., and Risser, P. G. (eds.) (1988). "Scales and Global Change." Wiley, New York.

Rotty, R. M., and Marland, G. (1986). Fossil fuel combustion: Recent amounts, patterns, and trends of CO_2. *In* "The Changing Carbon Cycle—A Global Analysis" (J. R. Trabalka and D. E. Reichle, eds.), pp. 474–490. Springer-Verlag, New York.

Salthe, S. N. (1985). "Evolving Hierarchical Systems." Columbia University Press, New York.

Schlesinger, M. E. (1986). Equilibrium and transient climatic warming induced by increased atmospheric CO_2. *Climate Dyn.* **1**, 35–51.

Spanner, D. C. (1964). "Introduction to Thermodynamics." Academic Press, New York.

Strain, B. R. (1985). Physiological and ecological controls on carbon sequestering in ecosystems. *Biogeochemistry* **1**, 219–232.

Strain, B. R., and Cure, J. D. (eds.) (1985). "Direct Effects of Increasing Carbon Dioxide on Vegetation." U.S. DOE/ER-0238, National Technical Information Service, Springfield, Virginia.

Thornley, J. H. M. (1976). "Mathematical Models in Plant Physiology: A Quantitative Approach to Problems in Plant and Crop Physiology." Academic Press, New York.

Woodmansee, R. G. (1988). Ecosystem processes and global change. *In* "Scales and Global Change" (T. Rosswall, Woodmansee, and P. G. Risser, eds.), pp. 11–27. Wiley, New York.

Woodwell, G. M., Hobbie, J. E., Houghton, R. A., Melillo, J. M., Moore, B., Peterson, B. J., and Shaver, G. R. (1983). Global deforestation: Contribution to atmospheric carbon dioxide. *Science* **222**, 1081–1086.

8

Generalization of a Forest Ecosystem Process Model for Other Biomes, BIOME-BGC, and an Application for Global-Scale Models

Steven W. Running and E. Raymond Hunt, Jr.

I. Introduction

Modeling terrestrial ecosystems on the global scale demands the development of simple, generalized representations of important plant processes that can be used in different biomes with minimal change. However, the sophistication and complexity of this task requires models of various domains of space and time resolution which, in turn, requires emphasis on different ecological and biophysical processes. We have developed a family of coniferous forest process models over the last several years and have used them on a variety of spatial and temporal scales to address important ecological questions. What began as a single-tree daily water-balance model run for 1 year is now an integrated carbon, nitrogen, and water cycle model with dual timestep resolution run for 100 years. With FOREST-BGC (biogeochemical cycles) embedded in our Regional Ecosystem Simulation System (RESSys) with a microclimate simulator (MT-CLIM) and satellite definition of the vegetation, we now map ecosystem processes such as photosynthesis, respiration, evapotranspiration, decomposition, and nitrogen mineralization over landscapes of hundreds of square kilometers (Running et al., 1989). We now use these models to calibrate simple models for global implementation using satellite data.

However, the transition from the essentially one-dimensional calculations of FOREST-BGC to the three-dimensional landscape-scale simulation of RESSys has required a great deal of effort in nonbiological topics. Defining the "site" requires knowledge of meteorology, climatology, geomorphology, soil physics, hydrology, and remote sensing, which requires broader perspectives and interactions with colleagues in other disciplines.

This chapter has three objectives. The first explores the scaling lessons learned in the development of our forest ecosystem model from the 1970s to the present. The second uses our forest models to simulate two other ecosystems, grasslands and deciduous forests, by changing only model parameters. The goal is to determine if generic ecosystem processes can simulate a range of ecosystems. We call this new model BIOME-BGC and, for our purposes here, we define a biome as a combination of life-form type and climate. For example, the differences in process rates between cottongrass–sedge tundra and temperate grasslands result from climate, not life-form. Third, we show how BIOME-BGC can be used to scale biospheric processes globally from the remote sensing of absorbed photosynthetically active radiation (APAR) by satellites.

II. Lessons Learned in the Evolution of Forest-BGC and RESSys

A. Precursors of FOREST-BGC and RESSys

The spatial and temporal scaling implicit in RESSys has had a long historical development (Fig. 8.1). In retrospect, this progressive development was valuable because, with each new model generation, assumptions in the old model were reevaluated, the necessity of the new model was justified, and optimization of the final product was pursued. The objective of H20TRANS was to evaluate the ecosystem-level significance of stomatal control mechanisms measured in the field on Douglas fir (Waring and Running, 1976).

After measuring stomatal closure under vapor pressure deficits of 30 mbar, rather than asking the reductionist question "By what mechanism does this closure occur?", the question became "What difference does this stomatal closure have in the seasonal hydrologic balance of the tree?" Later, after spending a considerable amount of time following diurnal stomatal conductance and tree hydraulic resistances (Running, 1980a,b), the predictability of diurnal transpiration rates seemed sufficiently high that developing an average *daily* calculation, and removing much of the internal flow dynamics to produce DAYTRANS, was an appropriate simplification for larger-scale questions (Running, 1984a).

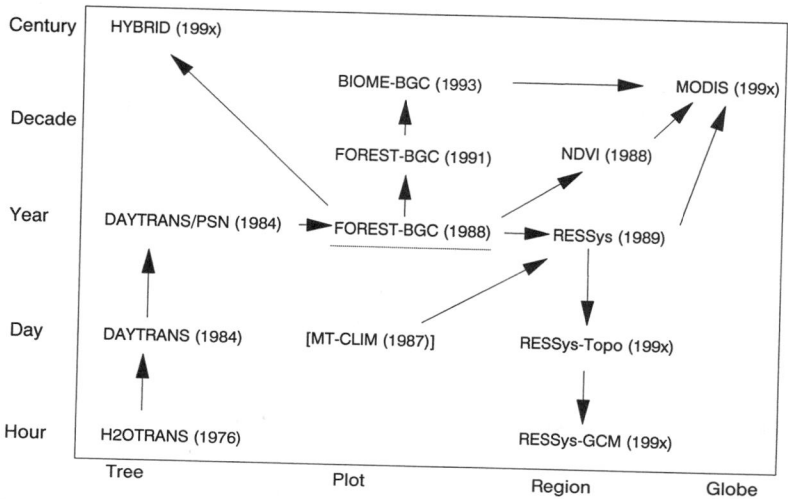

Figure 8.1 Evolution of BIOME-BGC and RESSys, showing model names and initial publication dates. Each new model shows a progression from its precursor in either temporal scale (*y*-axis) or spatial scale (*x*-axis). Models are H20TRANS (Waring and Running, 1976), DAYTRANS (Running, 1984a), DAYTRANS/PSN (Running, 1984b), FOREST-BGC (Running and Coughlan, 1988; Running and Gower, 1991), MT-CLIM (Running *et al.*, 1987), BIOME-BGC (this chapter), RESSys (Running *et al.*, 1989), and NDVI (Running and Nemani, 1988). All models designated 199X are in progress. MODIS is the Moderate Resolution Imaging Spectrometer which will be part of the NASA Earth Observing System.

Hunt *et al.* (1991b) suggested that similar simplifications are appropriate for modeling water relationships of vegetation within Global Climate Models and General Circulation Models (GCMs).

An initial focus on the tree hydrologic balance alone precluded exploration of interesting ecological questions concerning water use efficiency and net primary production (NPP). To meet these broad objectives, first a canopy photosynthesis routine was added to produce DAYTRANS/PSN (Running, 1984b). Because photosynthesis is only one component of NPP, it was then necessary to build an entire carbon cycle, with maintenance and growth respiration, carbon allocation, litterfall, and decomposition in the first generation of FOREST-BGC (Running and Coughlan, 1988). Adding the carbon cycle forced a reevaluation of the relevant model timestep, with the result that FOREST-BGC was built with a dual timestep, daily for the hydrologic balance and NPP and annually for the carbon cycle and, later, the nitrogen cycle.

Modeling a single tree has never been as personally interesting as understanding forest patterns across the landscape. An early hypothesis was that forest development in cold arid Montana was controlled substantially more by abiotic influences than by nutrient cycling or competition dynamics. This hypothesis is best illustrated by the striking difference in forest communities on opposing north–south slopes. Often, the south slopes are an arid grassland–*Pinus ponderosa* savannah, whereas the opposing north slope forest a few hundred meters away may be subalpine *Pinus contorta, Abies lasiocarpa,* and *Picea engelmannii.* Consequently, modeling the *site* was just as important as modeling the tree.

To describe the abiotic site conditions, a prerequisite to regional modeling was the ability to extrapolate meteorological driving variables from normal measuring sites (usually in valleys) to different slopes, aspects, and elevations. MT-CLIM (Mountain Microclimate Simulator; Running *et al.,* 1987) was developed to provide, conceptually, a continuous field of daily meteorological data across the landscape. MT-CLIM then was integrated with FOREST-BGC and other models to produce RESSys (Running *et al.,* 1989).

Because modeling trees one by one across the landscape is computationally impractical, a more generalized abstract means of defining a forest in spatial terms was needed. Whereas FOREST-BGC is based on 1-ha ground area, effectively this model is a one-dimensional calculation, so the horizontal representation is limited by homogeneity of the surface. As long as the model variables (meteorological, site, and vegetation) are representative, the spatial domain of the model is not limited. The most obvious assumption driven by this logic is the definition of the canopy as a "green sponge" with a thickness equivalent to the leaf area index (LAI). This logic ignores all the complexities of leaf-age class distribution, angular distribution, or canopy geometry, morphological adaptations, and so on. Requiring only LAI to define a canopy provided a critical connection to remote sensing, which then allowed us to determine LAI across the landscape (Running *et al.,* 1989). This abstraction of a forest stand precluded the need for an extensive database of tree heights and diameters, but also limited the ability of the model to grow "real" trees.

How can these generalized, abstract calculations be validated across a landscape? Validation is necessary but is extremely difficult over multikilometer scales. Some spatially integrated measurements, such as watershed hydrologic discharge and forest growth inventory data, are possibilities for the hydrologic and carbon cycle, respectively. Applying these operational data to the output of simulation models often requires translation of the data with assumptions that reduce confidence in the final results. Aircraft and tower-flux measurement systems are one option, although they are complicated, expensive, and require expertise available in only a handful of laboratories worldwide. New methodologies for *measuring* ecological processes over large spatial scales are needed desperately.

B. New Applications of FOREST-BGC and RESSys

As illustrated in Fig. 8.1, FOREST-BGC became the platform for exploration of a variety of time and space scales. We first used FOREST-BGC to evaluate and interpret satellite Normalized Difference Vegetation Indices (NDVIs) by comparing seasonal simulated evapotranspiration and NPP against weekly composited NDVIs for seven sites across North America (Running and Nemani, 1988). Other applications have emerged from FOREST-BGC, for example, forest site quality definition (McLeod and Running, 1988; Korol *et al.*, 1991), air pollution effects on forests (Kremer, 1991), and prediction of effects of doubled atmospheric CO_2 (Running and Nemani, 1991). The next FOREST-BGC version (Running and Gower, 1991) added a dynamic nitrogen cycle and carbon partitioning, and is the platform for the BIOME-BGC logic presented in this chapter.

Several new directions are under current development (Fig. 8.1). FOREST-BGC cannot simulate growth of forest stands; conversely, JABOWA/FORET-type forest population dynamics models simulate stand growth (Shugart, 1984) without the mechanistic physiology or sensitivity to climate of FOREST-BGC. We now are implementing the use of the FOREST-BGC annual photosynthesis calculation to drive the growth function in a FORET-type model, HYBRID, developed by A. Friend (personal communication).

Our first generation RESSys could not route water across the landscape, a constraint that is being resolved with RESSys-Top (renamed RHESSys for Regional Hydro-Ecological Simulation System; L. Band and R. Nemani, University of Toronto). RESSys-Top simulates explicitly both the magnitude and the timing of stream discharge hydrographs, which allows us to use this regionally integrated data set for validation. Also, an hourly version (RESSys-GCM) is being developed to better interface with GCMs for future work with BATS (Biosphere–Atmosphere Transfer Scheme; Dickinson *et al.*, 1986).

The most important lesson from this brief history is that the "scaling" decisions were a continuous progressive optimization; each new version of the model answered a specific question. As we gained confidence in existing models, we asked new questions and developed new models to attack these questions. No single model was adequate for all questions, yet the common heritage from H20TRANS and FOREST-BGC initially provided confidence in the new models, based on the validations for each of the previous models.

III. BIOME-BGC Development

As we move toward global-scale biospheric modeling in the NASA Earth Observing System (EOS) program, we want to generalize the "coniferous forest" logic in FOREST-BGC to other biomes and develop a more general ecosystem process framework that we call BIOME-BGC. Two key principles that we consider essential for future global-scale extrapolations are incorporated in our model logic. First, we consider it essential that the models interface well with remote sensing drivers; we have designed FOREST-BGC to be particularly sensitive to LAI for defining the vegetation canopy. LAI is one of the most promising products from satellite remote sensing of vegetation (Peterson *et al.*, 1987; Peterson and Running, 1989; Spanner *et al.*, 1990). Second, our models are controlled strongly by climate, both because we believe that climate control is important and because daily climate data are among the most globally available data sets. Additionally, we want to build a foundation for future coupling of GCMs with biome-level classifications, such as those described in BATS (Dickinson *et al.*, 1986).

A. Physiology across Biomes

The primary goal of this model is the calculation of water, carbon, and nitrogen cycles through an ecosystem. Table 8.1 summarizes the input and output variables for the BIOME-BGC model. Detailed documentation of the original FOREST-BGC is reported by Running and Coughlan (1988); the new dynamic carbon and nitrogen cycling version is presented by Running and Gower (1991).

Implicit in translating FOREST-BGC to BIOME-BGC is the assumption that the treatment of a number of key physiological responses, for example, stomatal control, can be treated generically. Stomatal or leaf conductance is controlled by absorbed radiation, air temperature, humidity deficits, and soil or leaf water potential, using the functional forms in Running and Coughlan (1988). We assume these basic controls to be universal, with only scalar adjustment necessary. Transpiration is then determined using the Penman–Monteith equation. Effects of water stress sometimes are ignored in other models, but in our models are incorporated through leaf conductance.

Although the exact formulation of a photosynthesis calculation changes, almost all models are based on a similar type of asymptotic response to radiation and carbon dioxide. The control of maximum photosynthesis rate by leaf nitrogen concentration is an extremely critical connection between the carbon and nitrogen cycles (Field and Mooney, 1986). Temperature response of photosynthesis can be approximated by

Table 8.1 BIOME-BGC Driving, Site, Life-Form, and Output Variables with Their Units

Required daily inputs	
Day of year	
Air temperature, maximum	°C
Air temperature, minimum	°C
Precipitation (water equivalent)	cm
Calculated and optional daily inputs	
Daylength	sec
Total solar radiation	$kJ \cdot m^{-2} \cdot day^{-1}$
Total photosynthetically active radiation	$kJ \cdot m^{-2} \cdot day^{-1}$
Relative humidity	%
Soil temperature (20-cm depth)	°C
Atmospheric CO_2	ppm
Site variables	
Latitude	°
Slope and aspect	°
Elevation	m
Albedo	%
Nitrogen fixation and deposition	$kg \cdot ha^{-1}$
Soil water content at field capacity	$m^3 \cdot ha^{-1}$
Soil water content at critical water potential	$m^3 \cdot ha^{-1}$
Initial water: soil, snowpack	$m^3 \cdot ha^{-1}$
Lifeform variables	
Precipitation interception coefficient	$mm \cdot LAI^{-1}$
Light extinction coefficient	LAI^{-1}
Turnover coefficient: leaf, stem, root	$\% \cdot year^{-1}$
Specific leaf area	$m^2 \cdot kg^{-1}$
Initial carbon: leaf, stem, root, soil, litter	$kg \cdot ha^{-1}$
Initial nitrogen: leaf, stem, root, soil, litter	$kg \cdot ha^{-1}$
Daily outputs	
Transpiration	$m^3 \cdot ha^{-1}$
Evaporation	$m^3 \cdot ha^{-1}$
Runoff	$m^3 \cdot ha^{-1}$
Soil water content	$m^3 \cdot ha^{-1}$
Predawn leaf and soil water potential	MPa
Daily photosynthesis	$kg \cdot ha^{-1}$
Daily maintenance respiration	$kg \cdot ha^{-1}$
Annual outputs	
Carbon fluxes: leaf, stem, root, soil, litter	$kg \cdot ha^{-1}$
Total growth respiration	$kg \cdot ha^{-1}$
Soil respiration for decomposition	$kg \cdot ha^{-1}$
Nitrogen fluxes: leaf, stem, root, soil, litter	$kg \cdot ha^{-1}$
Nitrogen mineralized	$kg \cdot ha^{-1}$
Nitrogen loss	$kg \cdot ha^{-1}$

an inverse parabolic function, whereas maintenance respiration generally is considered to be an exponential function with a Q_{10} of 2.

The general concept that growth respiration is a function of molecular energetics, independent of species and growth rate, is important for general ecosystem models (Penning deVries *et al.*, 1974). Similar unifying logic is needed to represent maintenance respiration; general representation of the respiring biomass is difficult since the percentage of living cells in plant tissues ranges from 0 to 100%. Tissue nitrogen concentration may allow a unifying measure of maintenance respiration activity (Ryan, 1991).

We define decomposition and nitrogen mineralization as controlled by soil temperature and moisture, as in Meentemeyer (1984), and represent litter decomposability with litter lignin concentration, a widely accepted parameter. Nitrogen retranslocation before litterfall of leaves and roots is an important nitrogen cycle component and is defined as 50%.

Leaf/root carbon allocation assumes that as water or nitrogen becomes less available, more fractional allocation of carbon is invested in the roots. However, the grasses have no permanent stem carbon allocation. Both deciduous forests and grasslands had 100% leaf carbon turnover annually, whereas the turnover for conifer forests was 33%.

B. Parameter Changes Required for BIOME-BGC

This chapter is our first step in testing how easily FOREST-BGC can simulate other ecosystems by parameterizing the model for either a deciduous broadleaf forest or a grassland. To enforce an initial minimalist philosophy, we chose not to define any new state variables, but only to change existing constant parameters. Although we suspect that, in our final BIOME-BGC version, some structure and state variables will be changed, we first want to see how realistically we can simulate these other biome types.

Table 8.2 presents the array of BIOME-BGC parameters changed from a coniferous forest in FOREST-BGC to a deciduous forest and grassland. These parameters were chosen from a large array of published data to be appropriate for ecosystems in an arid, cold, western climate. Major sources of data for grasslands are from William Parton, David Schimel, Dennis Ojima, and colleagues from the Natural Resource Ecology Laboratory at Colorado State University (e.g., Schimel *et al.*, 1991). Similarly, we used literature values from a large number of laboratories for *Populus* spp. and *Betula* spp. for deciduous broadleaves (e.g., Jurik *et al.*, 1988; Pregitzer *et al.*, 1990). All other model parameters were held constant across the three biomes.

Of great importance for an exercise in testing general ecosystem theory is knowing the parameters that we did *not* change, relying on parameters

Table 8.2 BIOME-BGC Physiological Parameters Changed from FOREST-BGC (Conifer List) for Broadleaf and Grass Simulations[a]

Parameter	Conifer	Broadleaf	Grass
Maximum leaf area index	10	6	3
Specific leaf area ($m^2 \cdot kg^{-1}$ dry mass)	5	17	5
Specific leaf area ($m^2 \cdot kg^{-1}$ carbon)	25	75	25
Leaf ON date (yearday)	0	120	120
Leaf OFF date (yearday)	365	300	240
Maximum stomatal conductance ($mm \cdot sec^{-1}$)	1.6	2.5	5.0
Boundary layer conductance ($mm \cdot sec^{-1}$)	100	100	10
Maximum photosynthetic rate ($\mu mol \cdot m^{-2} \cdot sec^{-1}$)	5	5	10
Critical leaf water potential (MPa)	-2.0	-2.0	-3.5
Leaf maintenance respiration ($g \cdot kg^{-1} \cdot day^{-1}$)	0.2	0.4	0.4
Stem maintenance respiration ($g \cdot kg^{-1} \cdot day^{-1}$)	0.2	0.2	0.3
Root maintenance respiration ($g \cdot kg^{-1} \cdot day^{-1}$)	0.4	1.1	0.6
Leaf turnover (% $\cdot year^{-1}$)	33	100	100
Stem turnover (% $\cdot year^{-1}$)	2	2	99
Root turnover (% $\cdot year^{-1}$)	80	80	40
Leaf lignin concentration (%)	25	18	17

[a] All other FOREST-BGC parameters remained constant for these simulations.

from western conifers to represent the deciduous forest and the grassland. For example, the slope of the function reducing stomatal conductance with increasing humidity deficit remained constant at 0.05 mm · sec^{-1} · ($\mu g \cdot g^{-1}$)$^{-1}$, effectively closing stomata at a 20-mbar vapor pressure deficit. The photosynthetic light compensation point of the canopy was held at 432 kJ · m^{-2} · day^{-1}. The photosynthetic temperature optimum was 20°C for all simulations. A Q_{10} of 2.0 was used for the temperature sensitivity of respiration, although the respiration rate coefficients were changed (Table 8.2).

All decomposition and nitrogen mineralization calculations were identical for the three ecosystems, except for the leaf lignin concentrations that define the distribution of leaf litter into the fast litter pool or the slower soil pool (Running and Gower, 1991). The range of allowed leaf nitrogen concentrations computed was kept between 0.0132 and 0.044 kg N · kg^{-1} C or 0.6 and 2.0% dry matter.

Because the carbon allocation logic in the new FOREST-BGC (Running and Gower, 1991) is an integral part of the model, it was not possible to stop the stem carbon allocation for the grass simulation. Instead, we set stem turnover to 99% (Table 8.2) so that stem carbon and, hence, respiration load would not accumulate. However, if one considers this "stem" carbon allocation for grasses to be annual seed production, it may be similar in magnitude and dynamics to expected seed carbon allocation.

We used a western Montana climate (Missoula, 1984) with annual precipitation of 33.7 cm and annual insolation of 4232 MJ $\cdot$ m^{-2} for the simulations. The climate is transitional from dry conifer forest to grasslands, so it is appropriate for both biome types. This area has few native deciduous trees (mainly in riparian zones), so we expected deciduous trees to gain less carbon in these simulations. A 20-year period was simulated by replicating the climate file, allowing for partial equilibration of LAI and the carbon and nitrogen cycle pools.

C. BIOME-BGC Simulation Results

The simulation results at Year 20, shown in Table 8.3, seem reasonable. More generally, the model stayed computationally stable and the carbon and nitrogen cycles remained reasonably balanced despite the free ranging nature of the dynamics in the model. For example, if the litterfall or decomposition rates differed significantly, insufficient nitrogen would be available via mineralization, causing leaf carbon production to decrease,

Table 8.3   BIOME-BGC Simulation Results at Year 20

	Conifer	Broadleaf	Grass
Carbon			
Gross primary productivity (GPP; Mg $\cdot$ ha^{-1} $\cdot$ year^{-1})	7.9	7.1	8.0
Maintenance respiration (Rm; Mg $\cdot$ ha^{-1} $\cdot$ year^{-1})	2.4	2.5	1.2
Growth respiration (Rg; Mg $\cdot$ ha^{-1} $\cdot$ year^{-1})	1.8	1.5	2.3
Net primary productivity (NPP; Mg $\cdot$ ha^{-1} $\cdot$ year^{-1})	3.7	3.1	4.5
Leaf area index (LAI; all sided)	3.2	2.8	1.9
Leaf to root ratio	0.32	0.38	0.36
Nitrogen			
Leaf nitrogen (kg $\cdot$ ha^{-1})	39	17	32
Leaf [N] (% DW)	1.4	2.0	1.9
Litter nitrogen (kg $\cdot$ ha^{-1})	134	135	210
Water			
Evaporation (E; cm $\cdot$ year^{-1})	5.9	5.9	2.9
Transpiration (T; cm $\cdot$ year^{-1})	24.5	29.6	30.7
Efficiencies			
Absorbed photosynthetic active radiation (APAR; MJ $\cdot$ m^{-2} $\cdot$ year^{-1})	692	496	290
Water use efficiency			
GPP/T; (g $\cdot$ kg^{-1})	2.2	1.6	2.5
NPP/T (g $\cdot$ kg^{-1})	1.5	1.0	1.5
Dry matter yield			
ε_c (NPP/APAR; g $\cdot$ MJ^{-1})	0.53	0.63	1.55
Transpiration efficiency			
ε_w (T/APAR; kg $\cdot$ MJ^{-1})	0.35	0.60	1.06

resulting in a negative NPP value. This did not happen, despite changing leaf turnover rates from 33 to 100%.

The carbon balance results of primary production and respiration components were of appropriate magnitude for each of these biome types in a dry, cold, continental climate. The LAI simulated by the model seemed low; a conifer forest should maintain a value near 4–5. However, it is significant that, for each biome simulation, the LAI equilibrated within a reasonable range. The Year 20 leaf and litter nitrogen pools also seemed reasonable, as did the leaf nitrogen concentration range (Table 8.3).

Differences in seasonal activity, as illustrated in Fig. 8.2, for cumulative net canopy photosynthesis are apparent. Although the grassland simulation had a smaller LAI and supported leaves from Yearday 120 to 240, the higher stomatal conductance and maximum photosynthetic rates of these leaves (Table 8.2) resulted in higher cumulative net carbon gain (Fig. 8.2). Also, lower maintenance respiration costs of 1.2–1.3 Mg · ha^{-1} · year^{-1} (Table 8.3) compared with those of the two forest biome types resulted in the grassland simulation having the highest NPP at Year 20. The broadleaf forest simulation showed noticeably lower water use efficiency by both computations (Table 8.3) which, if true, could be a significant competitive disadvantage for that life-form in a water-limited environment.

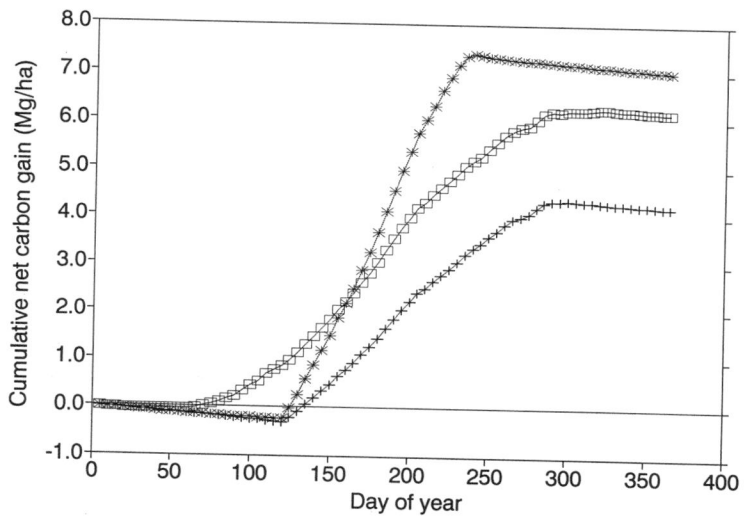

Figure 8.2 Cumulative net carbon gain over a year for a grassland (∗), coniferous forest (□), and deciduous broadleaf forest (+) using BIOME-BGC and the same arid cold climate: 1984 Missoula, Montana. See Table 8.2 for parameters redefined for the different life-forms.

IV. Global-Scale Application Using BIOME-BGC

A. Remote Sensing of Net Primary Production

Monteith (1977) provided a powerful example of how simplified models allow extrapolation across scale by estimating NPP solely from absorbed photosynthetically active radiation [APAR (MJ · ha^{-1})] rather than by using a carbon budget approach as in BIOME-BGC. Annual NPP (Mg · ha^{-1} · year^{-1}) is related to APAR by

$$NPP = \varepsilon_c \int APAR \, dt = \varepsilon_c \Sigma \, APAR, \tag{1}$$

where t is time in days and Σ APAR is the yearly sum of daily APAR (Monteith, 1977; Jarvis and Leverenz, 1984; Russell *et al.*, 1989). The dry matter yield [ε_c(g · MJ^{-1})] often is called the light use efficiency, PAR conversion efficiency, or dry matter:radiation quotient. It is confusing that ε_c is expressed per APAR, per intercepted PAR, per absorbed solar radiation, or per intercepted solar radiation (Prince, 1991).

Based on a photosynthetic quantum requirement of about 20 mol photons · mol^{-1} CO$_2$ (including photorespiration), there is a maximum ε_c of about 5.5 g · MJ^{-1}, which is reduced by uneven illumination, nutrient deficiency, and stomatal closure (Jarvis and Leverenz, 1984; Russell *et al.*, 1989; Prince, 1991). Further, ε_c is net after maintenance and growth respiration carbon losses are accounted for. APAR is primarily a function of LAI, so ε_c and Σ APAR can be considered analogous to the growth analysis terms net assimilation rate and leaf area duration, respectively (Warren-Wilson, 1981).

Kumar and Monteith (1981) and Asrar *et al.* (1984) showed that remotely sensed vegetation indices such as NDVI are proportional to the ratio of APAR to incident PAR; thus, NDVI can be used to determine NPP over large areas (Goward *et al.*, 1985; Tucker *et al.*, 1985). Equation (1) becomes

$$NPP = \varepsilon_c \Sigma (NDVI \cdot PAR) \tag{2}$$

(Prince, 1991). On regional scales, transpiration can be represented largely as a function of net radiation (both solar and thermal), whereas stomatal conductance is a function only of PAR. Since PAR is about half the solar radiation (McCree, 1981) and net radiation is mostly absorbed solar radiation, we can calculate a regional transpirational flux [T (kg · ha^{-1} · year^{-1})] as

$$T = \varepsilon_w \Sigma (NDVI \cdot PAR), \tag{3}$$

where ε_w is the transpirational efficiency (kg · MJ^{-1}). We do recognize that this approach was developed for annual unstressed croplands and that some measure of stress response is necessary for applications to natural vegetation. Both Eqs. (2) and (3) can have a satellite-derived surface resistance added explicitly for the reduction of photosynthesis and transpiration by stomatal closure resulting from water stress (Nemani and Running, 1989; Running, 1990).

B. Effect of Life-Form on ε_c

Simulations of NPP, T, and APAR with BIOME-BGC allow us to use a mechanistic model to explore the range and variability, resulting from different physiology and climate, of ε_c and ε_w among life-forms. Variations in measured ε_c among plants of a single life-form are demonstrated clearly (Fig. 8.3). Could these variations for a single life-form result primarily from climate? More important, the ranges of measured ε_c overlap considerably among the life-forms (Fig. 8.3). Are there significant differences of ε_c among life-forms resulting from different physiological

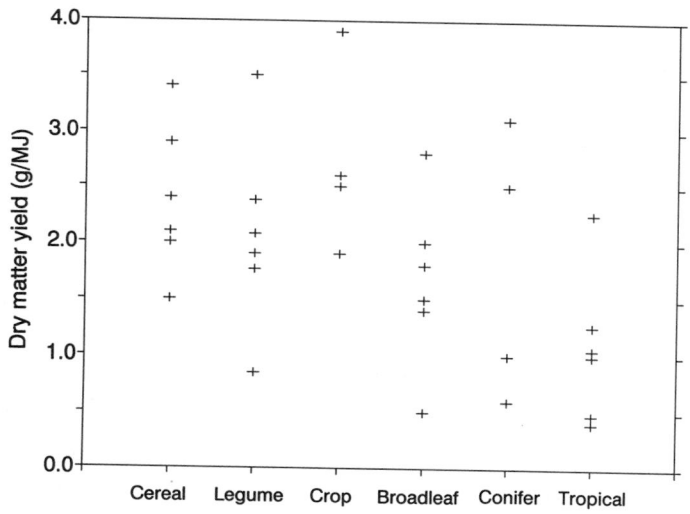

Figure 8.3 Comparison of measured dry matter yield (ε_c) among life-form types. Much of the data for woody life-forms are based on aboveground net production; we multiplied these data by two to account for belowground production. The data are from Prince (1991) for cereals (C$_3$ photosynthetic pathway only), legumes, and other agricultural crops; Rauner (1976), Cannell *et al.* (1987, 1988), and Linder (1985) for the deciduous broadleaves; Gholz *et al.* (1991), Grace *et al.* (1987), and Linder (1985) for conifers; and Luxmore and Saldarriage (1989) and Jordan (1971) for tropical forests.

characteristics (e.g., Table 8.2), or can a single average ε_c be used globally for all biomes?

Initial results from BIOME-BGC show that the differences in ε_c and ε_w between the conifer and deciduous trees in the same climate were small; ε_c values for both tree life-forms were much smaller than ε_c for the grass life-form (Table 8.3). These results compare favorably with the low end of measured values from other studies, using the cereals for comparison with grass (Fig. 8.3). The simulated differences of ε_c among the three life-forms result primarily from the higher net carbon gain of grass (Fig. 8.2), the length of time the leaves were green and, thus, absorbing PAR (Tables 8.2, and 8.3), and the large mass of respiring woody stems for the conifer and deciduous broadleaf (Table 8.3). Therefore, consistent differences in ε_c (and ε_w) among life-forms are predicted from the BIOME-BGC simulations when other factors are held constant.

C. Effect of Climate on ε_c

We hypothesized that simulations over long time periods may be useful to determine how ε_c and ε_w may vary with a changing climate. Between the cool, moist 1940s and the hot, dry 1930s in Superior, Montana, the yearly average of maximum temperature differed by about 2°C and annual precipitation by about 200 mm (Hunt *et al.*, 1991a), changes similar in magnitude to GCM projections for a doubled CO_2 climate in middle latitudes (Mitchell, 1989).

We also obtained annual increments of stem wood volume over the same time period for an entire stand of *Pinus ponderosa* near the Superior, Montana, weather station (Martin, 1987; McLeod and Running, 1988; Hunt *et al.*, 1991a). The simulated increments using FOREST-BGC explained 47% of the variation in annual increments of wood volume caused by climatic variability. Combining data from both the previous and the current year, we could explain 65% of the variation in annual growth increments (Hunt *et al.*, 1991a).

Using this FOREST-BGC validation with measured annual stem growth, we simulated the presumed effect of year-to-year climatic variations on ε_c. Figure 8.4 shows that simulated ε_c for this stand of *Pinus ponderosa* was correlated with annual precipitation. Also, simulated ε_c was correlated negatively with average yearly temperature (data not shown). The range in simulated ε_c for this conifer stand from 0.76 to 1.13 g · MJ^{-1} was larger than the simulated difference between the conifer and deciduous broadleaf life-forms using the same climate (Table 8.3). Therefore, a large range of possible ε_c and ε_w values is possible for each life-form.

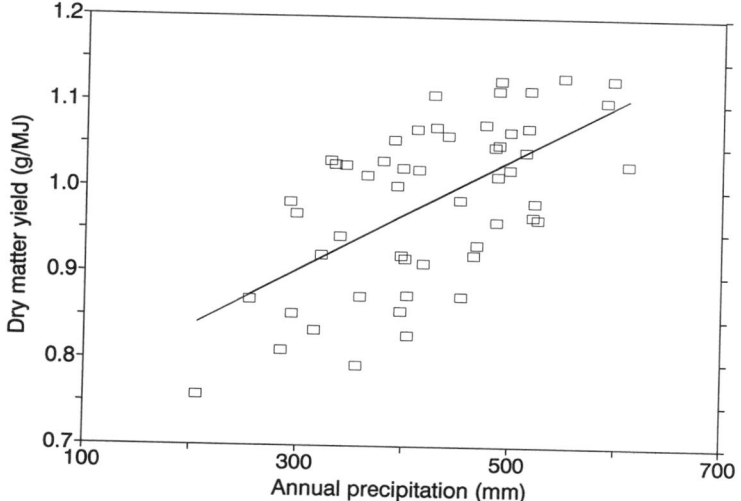

Figure 8.4 The relationship of dry matter yield (ε_c) of a coniferous forest with annual precipitation. Daily meteorological data from 1930 to 1982 were obtained from the National Weather Service station in Superior, Montana. The linear regression equation (line) between ε_c and annual precipitation was highly significant ($R^2 = 0.39$, $P < 0.0001$). Clearly, ε_c varies greatly within a life-form type as a result of year-to-year climatic variability.

V. Conclusions

In a general way, we see a clear hierarchy of the variables that most influence ecosystem processes. Climate and LAI alone can represent much of the activity of ecosystems (Running and Coughlan, 1988). Fundamental life-form variables of leaf longevity and turnover rate, specific leaf area, and maintenance respiration load from permanent biomass seem of next greatest importance. Of less importance are differences in stomatal control. These conclusions could be dismissed as circular logic since this is precisely the hierarchy incorporated into the model structure itself. However, the important test is how well the model simulates inter-biome variability of important ecosystem-level processes.

The scaling power of mechanistic models is in calibrating simple ecosystem models for use on larger spatial and temporal scales. The variations in ε_c with climate and life-form may be predicted from finer resolution models such as BIOME-BGC, providing robustness to the NDVI/APAR model that only an extremely large data set could otherwise provide. Conversely, it is not practical to run mechanistic models such as BIOME-BGC on a global scale, so using a simple satellite NDVI/APAR model allows us to estimate global vegetation NPP with some of the confidence that BIOME-GBC brings, but at a fraction of the computation and effort. Obviously, no single model or scale of resolution is adequate for all scientific questions, but a hierarchy of models can provide mechanistic confidence as we go up in scale.

Acknowledgments

This work was funded by NASA Grants NAGW-252 and NAS5-30920, and National Science Foundation Grant BSR-8817965. We thank Richard Waring, Sam Goward, and Steve Prince for discussions.

References

Asrar, G., Fuchs, M., Kanemasu, E. T., and Hatfield, J. L. (1984). Estimating absorbed photosynthetic radiation and leaf area index from spectral reflectance in wheat. *Agron. J.* **76,** 300–306.

Cannell, M. G. R., Milne, R., Sheppard, L. J., and Unsworth, M. H. (1987). Radiation interception and productivity of willow. *J. Appl. Ecol.* **24,** 261–278.

Cannell, M. G. R., Sheppard, L. J., and Milne, R. (1988). Light use efficiency and woody biomass production of poplar and willow. *Forestry* **61,** 125–136.

Dickinson, R. E., Henderson-Sellers, A., Kennedy, P. J., and Wilson, M. F. (1986). "Biosphere-Atmosphere Transfer Scheme (BATS) for the NCAR Community Climate Model." NCAR Technical Note NCAR/TN-275 + STR. National Center for Atmospheric Research, Boulder, Colorado.

Field, C., and Mooney, H. A. 1986. The photosynthesis-nitrogen relationship in wild plants. *In* "On The Economy of Plant Form and Function" (T. J. Givnish, ed.), pp. 25–55. Cambridge University Press, Cambridge.

Gholz, H. L., Vogel, S. A., Cropper, W. P., Jr., McKelvey, K., Ewel, K. C., Teskey, R. O., and Curran, P. J. (1991). Dynamics of canopy structure and light interception in *Pinus elliottii* stands of north Florida. *Ecol. Monogr.* **61,** 33–51.

Goward, S. G., Tucker, C. J., and Dye, D. G. (1985). North American vegetation patterns observed with the NOAA-7 advanced very high resolution radiometer. *Vegetatio* **64,** 3–14.

Grace, J. C., Jarvis, P. G., and Norman, J. M. (1987). Modelling the interception of solar radiant energy in intensively managed stands. *N. Z. J. For. Sci.* **17,** 193–209.

Hunt, E. R., Jr., Martin, F. C., and Running, S. W. (1991a). Simulating the effects of climatic variation on stem carbon accumulation of a ponderosa pine stand: Comparison to annual growth increment data. *Tree Physiol.* **9,** 161–171.

Hunt, E. R., Jr., Running, S. W., and Federer, C. A. (1991b). Extrapolating plant water resistances and capacitances to regional scales. *Agric. For. Meteorol.* **54,** 169–195.

Jarvis, P. G., and Leverenz, J. W. (1984). Productivity of temperate, deciduous and evergreen forests. *In* "Encyclopedia of Plant Physiology" (O. L. Lange, P. S. Nobel, C. B. Osmond, and H. Ziegler, eds.), Vol. 12D, pp. 233–280. Springer-Verlag, Berlin.

Jordan, C. F. (1971). Productivity of a tropical forest and its relation to a world pattern of energy storage. *J. Ecol.* **59,** 127–142.

Jurik, T. W., Weber, J. A., and Gates, D. M. (1988). Effects of temperature and light on photosynthesis of dominant species of a northern hardwood forest. *Bot. Gaz.* **149**, 203–208.

Korol, R. L., Running, S. W., Milner, K. S., and Hunt, E. R., Jr. (1991). Testing a mechanistic carbon balance model against observed tree growth. *Can. J. For. Res.* **21**, 1098–1105.

Kremer, R. G. (1991). Simulating ecosystem processes to estimate forest productivity under the influence of sulfur dioxide pollution. *Ecol. Model.* **54**, 111–126.

Kumar, M., and Monteith, J. L. (1981). Remote sensing of crop growth *In* "Plants and the Daylight Spectrum" (H. Smith, ed.), pp. 133–144. Academic Press, London.

Linder, S. (1985). Potential and actual production in Australian forest stands. *In* "Research for Forest Management" (J. J. Landsberg and W. Parsons (eds.), pp. 11–35. CSIRO, Melbourne.

Luxmore, R. J., and Saldarriage, J. G. (1989). PAR conversion efficiencies of a tropical rain forest. *Ann. Sci. For.* **46**,(Suppl.), 523–525.

Martin, F. C. (1987). "Evaluation of Site Quality Measures and Growth Curves for Even-Aged Ponderosa Pine under Different Levels of Stand Density and Site Productivity in Western Montana. Ph.D. Thesis, University of Montana, Missoula.

McCree, K. J. (1981). Photosynthetically active radiation. *In* "Encyclopedia of Plant Physiology" (O. L. Lange, P. S. Nobel, C. B. Osmond, and H. Ziegler, eds.), Vol. 12A, pp. 41–55. Springer-Verlag, Berlin.

McLeod, S. D., and Running, S. W. (1988). Comparing site quality indices and productivity in ponderosa pine stands of western Montana. *Can. J. For. Res.* **18**, 346–352.

Meentemeyer, V. (1984). The geography of organic decomposition rates. *Ann. Assoc. Am. Geographers* **74**, 511–560.

Mitchell, J. F. B. (1989). The "greenhouse" effect and climate change. *Rev. Geophys.* **27**, 115–139.

Monteith, J. L. (1977). Climate and the efficiency of crop production in Britain. *Phil. Trans. R. Soc. London B* **281**, 277–294.

Nemani, R. R., and Running, S. W. (1989). Estimation of regional surface resistance to evapotranspiration from NDVI and Thermal-IR AVHRR data. *J. Appl. Meteorol.* **28**, 276–284.

Penning de Vries, F. W. T., Brunsting, A. H. M., and Van Laar, H. H. (1974). Products, requirements and efficiency of biosynthesis: A quantitative approach. *J. Theoret. Biol.* **45**, 339–377.

Peterson, D. L., and Running, S. W. 1989. Applications in forest science and management. *In* "Theory and Applications of Optical Remote Sensing" (G. Asrar, ed.), pp. 429–473. Wiley, New York.

Peterson, D. L., Spanner, M. A., Running, S. W., and Teuber, K. B. (1987). Relationship of Thematic Mapper simulator data to leaf area index of temperate coniferous forests. *Remote Sens. Environ.* **22**, 323–341.

Pregitzer, K. S., Dickmann, D. I., Hendrick, R., and Nguyen, P. V. (1990). Whole-tree carbon and nitrogen partitioning in young hybrid poplars. *Tree Physiol.* **7**, 79–93.

Prince, S. D. (1991). A model of regional primary production for use with coarse-resolution satellite data. *Int. J. Remote Sens.* **12**, 1313–1330.

Rauner, J. L. (1976). Deciduous forests. *In* "Vegetation and the Atmosphere" (J. L. Montieth, ed.), Vol. 2, pp. 241–264. Academic Press, London.

Running, S. W. (1980a). Environmental and physiological control of water flux through *Pinus contorta. Can. J. For. Res.* **10**, 82–91.

Running, S. W. (1980b). Relating plant capacitance to the water relations of *Pinus contorta. For. Ecol. Mgmt.* **2**, 237–252.

Running, S. W. (1984a). "Documentation and Preliminary Validation of H20TRANS and DAYTRANS, Two Models Predicting Transpiration and Water Stress in Western Coniferous Forests. Rocky Mountain Forest and Range Experiment Station Research Paper RM-252. Fort Collins, Colorado.

Running, S. W. (1984b). Microclimate control of forest productivity: Analysis by computer simulation of annual photosynthesis/transpiration balance in different environments. *Agric. For. Meteorol.* **32,** 267–288.

Running, S. W. (1990). Estimating terrestrial primary productivity by combining remote sensing and ecosystem simulation. *In* "Remote Sensing of Biosphere Functioning" (R. J. Hobbs and H. A. Mooney (eds.), pp. 65–86.) Springer-Verlag, New York.

Running, S. W., and Coughlan, J. C. (1988). A general model of forest ecosystem processes for regional applications. I. Hydrologic balance, canopy gas exchange and primary production processes. *Ecol. Model.* **42,** 125–154.

Running, S. W., and Gower, S. T. (1991). FOREST-BGC, a general model of forest ecosystem processes for regional applications. II. Dynamic carbon allocation and nitrogen budgets. *Tree Physiol.* **9,** 147–160.

Running, S. W., and Nemani, R. R. (1988). Relating seasonal patterns of the AVHRR vegetation index to simulated photosynthesis and transpiration of forests in different environments. *Remote Sens. Environ.* **24,** 347–367.

Running, S. W., and Nemani, R. R. 1991. Regional hydrologic and carbon balance responses of forests resulting from potential climate change. *Climatic Change* **19,** 349–368.

Running, S. W., Nemani, R. R., and Hungerford, R. D. (1987). Extrapolation of synoptic meteorologica data in mountainous terrain and its use for simulating forest evapotranspiration and photosynthesis. *Can. J. For. Res.* **17,** 472–483.

Running, S. W., Nemani, R. R., Peterson, D. L., Band, L. E., Potts, D. F., Pierce, L. L., and Spanner, M. A. (1989). Mapping regional forest evapotranspiration and photosynthesis by coupling satellite data with ecosystem simulation. *Ecology* **70,** 1090–1101.

Russell, G., Jarvis, P. G., and Montieth, J. L. (1989). Absorption of radiation by canopies and stand growth. *In* "Plant Canopies: Their Growth, Form and Function" (G. Russell, B. Marshall, and P. G. Jarvis, eds.), pp. 21–39. Cambridge University Press, Cambridge.

Ryan, M. G. (1991). Effects of climate change on plant respiration. *Ecol. Appl.* **1,** 157–167.

Schimel, D. S., Kittel, T. G. F., Knapp, A. K., Seastedt, T. R., Parton, W. J., and Brown, V. B. (1991). Physiological interactions along resource gradients in a tallgrass prairie. *Ecology* **72,** 672–684.

Shugart, H. H. (1984). "A Theory of Forest Dynamics." Springer-Verlag, New York.

Spanner, M. A., Pierce, L. L., Peterson, D. L., and Running, S. W. (1990). Remote sensing of temperate coniferous forest leaf area index: The influence of canopy closure, understory vegetation and background reflectance. *Int. J. Remote Sens.* **11,** 95–111.

Tucker, C. J., Vanpraet, C. L., Sharman, M. J., and Van Ittersum G. (1985). Satellite remote sensing of total herbaceous biomass production in the Senegalese Sahel: 1989-1984. *Remote Sens. Environ.* **17,** 233–249.

Waring, R. H., and Running, S. W. (1976). Water uptake, storage and transpiration by conifers: A physiological model. *In* "Ecological Studies: Analysis and Synthesis" (O. L. Lange, L. Kappen, and E.-D. Schulze, eds.), Vol. 19, pp. 189–202. Springer-Verlag, Berlin.

Warren-Wilson, J. (1981). Analysis of growth, photosynthesis and light interception for single plants and stands. *Ann. Bot.* **48,** 507–512.

9

How Ecophysiologists Can Help Scale from Leaves to Landscapes

Richard H. Waring

I. Role of Ecophysiologists

At its core, ecophysiology is the study of how biochemistry and biophysics affect the competitive status of plants. With predictions for rapid changes in global climate and in atmospheric chemistry, we urgently need a better understanding of how biochemistry, biophysics, and plant responses interact on landscape and global scales. We need to understand the ways in which environmental stresses combine to affect acquisition, assimilation, and allocation to make any general predictions of the future performance of terrestrial vegetation. In turn, knowledge of plant performance and environmental interactions has implications in predicting the future composition of the atmosphere. Ecophysiologists should play an important role in this research. To be successful, we first face the challenge of incorporating aspects of our field of knowledge with those of other fields of study.

In scaling, we must provide underlying ecophysiological principles that can account for important changes observed at higher levels of organization (Passioura, 1979). We must proceed with caution, however, and listen carefully to researchers from other disciplines, because not all principles scale directly. For example, our knowledge of stomatal movement on the leaf level does not explain transpiration at the canopy or landscape level. Boundary layer conductance and advection become important considerations as one scales from leaves upward (Jarvis and McNaughton, 1986; McNaughton and Jarvis, 1991).

Despite this caveat, our discipline has much to offer. Many of our contributions will be at the ecosystem level, some at the watershed, and a few on regional and global scales. The most successful endeavors on these scales will provide new insights both to the core of ecophysiology and to allied fields operating on larger scales. In this chapter, I identify some of the areas in which ecophysiologists have made and can continue to make important contributions to higher levels of organization.

II. Promising Research Areas

A. Assessing the Availability and Acquisition of Resources

One way of scaling from the single plant to the plant community level is by developing process models of how plants interact and compete for resources. Both the relative growth of different life-forms and the spatial extraction of resources—light, water, and nutrients—have been considered (Sharpe *et al.*, 1985). More often, however, models of competition involve a single resource, that is, water for desert succulents (Hunt *et al.*, 1987), light for forest trees (Botkin *et al.*, 1972), or nutrients for weedy species (Tilman, 1984).

More detailed physiological knowledge about the potential of various species to alter their growth rates, allocation, and reproductive patterns will be necessary if we are to assess critically potential changes in community composition under a projected change in climate, but general principles can be used to estimate photosynthesis, decomposition, and mineral cycling in an ecosystem, all of which set limits on the levels of resources available to all plants in a community (Running and Coughlan, 1988; Aber *et al.*, 1991).

Ecophysiologists have developed improved quantitative measures of the efficiency with which plants obtain water, nutrients, and other resources. These measures serve ecologists by defining environmental gradients more functionally. For example, water potentials of plants and nutrient flux through the sap stream help interpret available water and nutrients over what can be assessed by measurements of soil properties alone (Richie, 1979; Stark *et al.*, 1989).

Valuable concepts have been taken from other fields and applied to ecophysiology. For example, from economics we gained optimization techniques and recognized that multiple environmental stresses might induce plants to expend surplus products such as carbohydrates for more scarce ones such as nitrogen, water, or light (Bloom *et al.*, 1985; Chapin *et al.*, 1987). Climatically driven models that consider multiple constraints on photosynthesis are able to predict growth of trees much more precisely than the statistical approaches normally applied by dendrochronologists (Hunt *et al.*, 1991).

To test these and other integrative concepts on a large scale, long-term field experiments now complement growth room and open-top chamber experiments in verifying models (Linder, 1987; McMurtrie *et al.,* 1990). One principle emerging from long-term experiments is the recognition that the partitioning patterns of growth may be altered substantially under chronic pollution, fertilization, or irrigation. Another is that changes in community composition are likely when permanent shifts in the availability of resources occur (Bazzaz *et al.,* 1987).

Long-term experiments shed new insights on episodic events also. In many cases, a single application of fertilizer, a hard frost, infrequent drought, or insect attack has only a very temporary impact on growth allocation and primary productivity (Wickman, 1980; Miller, 1981).

B. Identifying the Origin of Resources

Originally, ecophysiologists did not concern themselves with where plants obtained their source of water, CO_2, or nutrients. It was sufficient to define the environment operationally, recognizing only those resources that impinged on a plant in a demonstrable way, whether through a mechanical force causing bending, light interception affecting photosynthesis, or ion uptake altering nutrition (Mason and Langenheim, 1957).

Concern for the spread of toxic compounds provided an incentive for identifying the origin of poisons; from this followed an interest in the origin of critical resources. As a result of this broader perspective, we have discovered that some plants pump water and other resources from deep soil horizons and that these resources are shared with more shallowly rooted individuals (Richards and Caldwell, 1987). In other cases, where salt and fresh water mix, the differential uptake of saline water by salt-tolerant and intolerant plants has been documented isotopically (Sternberg and Swart, 1987). Similarly, the relative contributions of groundwater and surface water to transpiring plants can be assessed (White *et al.,* 1985). The enzymatic activity by which some symbiotic plants fix nitrogen from the atmosphere is evaluated now also by careful analysis of ^{15}N and ^{14}N in tissue and in soil (Shearer and Kohl, 1986).

The broader perspective gained by studying the sources of material led to the more integrative measures of stress. For example, the incorporation of stable isotopes in tissue produced under chronic water stress bears witness in the composition of leaves and other organs (DeLucia *et al.,* 1988).

C. Identifying the Fate of Resources

Originally, the fate of materials that passed through plants was of little interest to ecophysiologists, but environmental questions arising from the transfer of toxic compounds through food chains broadened our scope. Following the discovery that the natural abundances of ^{13}C and

^{12}C differ in plant tissue, depending on the photosynthetic pathway (Bender, 1968; Smith and Epstein, 1971), ecophysiologists could assess the composition of vegetation more functionally. We and other scientists soon found wider applications of this knowledge. Anthropologists determined quantitative changes in the food resources of ancient civilizations (van der Merwe, 1982); animal ecologists similarly determined wildlife feeding patterns (Tieszen and Boutton, 1989); climatologists and ecosystem scientists used the principles derived from basic physiology (Farquhar *et al.*, 1982) to look for isotopic signals of change in soils (Balesdent *et al.*, 1988), and in the atmosphere (Keeling *et al.*, 1979) to test their models.

Other elements with at least two forms of stable isotopes, for example, sulfur and nitrogen, help ecophysiologists measure the uptake of pollutants by various life-forms (Winner *et al.*, 1978; Peterson and Fry, 1987). This line of investigation promises to grow and extend to include the biosphere (Rundel *et al.*, 1989; Schlesinger, 1990).

D. Animal–Host Plant Interactions

Ultimately, plant survival may depend on the form in which assimilates are allocated to protect against or attract herbivores. Ecophysiologists have helped explain important animal–plant interactions by combining our knowledge of plant biochemistry and growth processes with that of animal physiologists and population biologists who understand the behavior of selected herbivores (White, 1984; Coley *et al.*, 1985; Waring *et al.*, 1985; Christiansen *et al.*, 1987; Mattson *et al.*, 1988; Waring *et al.*, 1991). The field of animal–plant interactions is growing, providing insights useful for protecting vegetation and for sustaining populations of dependent animals.

One finding from long-term experiments is that specific amino acids such as arginine may accumulate in some species when critical nutrients are not supplied in balance (Nasholm and Ericsson, 1990). Sustained damage from insect and pathogen outbreaks have been correlated with the accumulation of such amino acids (Turner and Lambert, 1986). Balanced growth and nutrition is seen from a new perspective when predicting the susceptibility or resistance of plants to herbivores or pathogens (Entry *et al.*, 1991).

III. Landscape Ecology

Satellite images of the surface of the earth are a powerful tool for envisioning changes in the operation of plants across landscapes. At present, most of the remote sensing literature deals with the classification of landscapes (Iverson *et al.*, 1989), but remotely sensed information collected continuously may allow assessment of process rates by observing changes in temperature, soil water status, radiation levels, and the photosynthetic capacity of vegetation (Sellers, 1987). Satellite-derived regional estimates of leaf area index have been coupled with mechanistic ecosystem models and climatic extrapolations to estimate daily and annual rates of photosynthesis, transpiration, and net primary production over an area of 20,000 km^2 in Montana (Running *et al.*, 1989).

Potentially, much more physiological and environmental information may be derived through remote sensing (Ustin *et al.* 1991; Nicholson *et al.*, 1990; Carson *et al.*, 1991; Engman, 1991; Seguin *et al.*, 1991). Narrow-band spectroscopy may provide information about subtle shifts in xanthophyll pigments associated with changes in photosynthesis diurnally (Gamon *et al.*, 1990). The insights of ecophysiologists are required to evaluate the ability of remote sensing techniques to discriminate biochemical, structural, and environmental features under laboratory and field conditions (Ustin *et al.* 1991; Gamon *et al.*, 1990; Yoder and Daley, 1990; Carlson *et al.*, 1991). Attempts to follow global changes in the phenology, photosynthesis, and transpiration of vegetation at daily and monthly resolution make remote sensing a promising field for joint endeavors.

IV. Challenges for the Future

The four major branches of ecophysiology and the field of landscape ecology offer opportunities for interesting and productive liaisons with allied fields. Also, at the biochemical and biophysical core of ecophysiology, new opportunities present themselves. We need to know much more about how various enzymatic and other reactions affect isotopic fractionation, not only of carbon but of nitrogen and perhaps of other groups of stable isotopes. We must seek insights that permit better generalizations of the process of respiration (Ryan, 1991). We need to understand what causes shifts in the allocation of resources within plants so that we may assess the impact of new combinations of environment on plant growth and distribution.

More effort is required in familiarizing ourselves with various kinds of quantitative process models. Mechanistic models can help identify the relative importance of variables and encourage experiments that allow confirmation of underlying assumptions. When detailed process models are confirmed, they may serve to evaluate the accuracy of less refined, more general models.

Broadening the disciplinary composition of workshops is one of the most efficient means of making progress. This workshop on scaling from leaves to landscape has contributed in a notable way toward bringing disciplines together.

Acknowledgments

I gratefully acknowledge the invitation of Jim Ehleringer and Chris Field to participate in the workshop. John Gamon provided detailed notes on the discussion that took place in a session I chaired on the subject of this paper. Those notes proved valuable in developing the text. Finally, I wish to recognize Barbara Yoder, John Runyon, and Chris Field who make constructive suggestions on earlier drafts of this paper.

References

Aber, J. D., Melillo, J. M., Nadelhoffer, K. J., Pastor, J., and Boone, R. D. (1991). Factors controlling nitrogen cycling and nitrogen saturation in northern temperate forest ecosystems. *Ecol. Appl.* **1,** 303–315.

Balesdent, J., Wagner, G. H., and Mariotti, A. (1988). Soil organic matter turnover in long-term field experiments as revealed by carbon-13 natural abundance. *Soil Sci. Soc. Am. J.* **52,** 118–124.

Bazzaz, F. A., Chiariello, N. R., Coley, P. D., and Pitelka, L. F. (1987). Allocating resources to reproduction and defense. *BioScience* **37,** 58–67.

Bender, M. M. (1968). Mass spectrometric studies of carbon-13 variations in corn and other grasses. *Am. J. Sci. Radiocarbon Suppl.* **10,** 468–472.

Bloom, A. J., Chapin, F. S., III, and Mooney, H. A. (1985). Resource limitation in plants.. an economic analogy. *Annu. Rev. Ecol. Syst.* **16,** 363–392.

Botkin, D. B., Janak, J. F., and Wallis, J. R. (1972). Some ecological consequences of a computer model of forest growth. *J. Ecol.* **60,** 849–872.

Carlson, T. N., Belles, J. E., and Gillies, R. R. (1991). Transient water stress in a vegetation canopy: Simulation and measurements. *Remote Sens. Environ.* **35,** 175–186.

Chapin, E. S., III, Bloom, A. J., Field, C. B., and Waring, R. H. (1987). Plant responses to multiple environmental factors. *BioScience* **37,** 49–57.

Christiansen, E., Waring, R. H., and Berryman, A. A. (1987). Resistance of conifers to bark beetle attack: Searching for general relationships. *For. Ecol. Mgmt.* **22,** 89–106.

Coley, P. D., Bryant, J. P, and Chapin, F. S., III (1985). Resource availability and plant antiherbivore defense. *Science* **230,** 895–899.

DeLucia, E. H., Schlesinger, W. H., and Billings, W. D. (1988). Water relations and the maintenance of Sierran confiers on hydrothermally altered rock. *Ecology* **69,** 303–311.

Engman, E. T. (1991). Application of microwave remote sensing of soil moisture for water resources and agriculture. *Remote Sens. Environ.* **35,** 213–226.

Entry, J. A., Cromack, K., Jr., Hansen, E., and Waring, R. (1991). Responses of western coniferous seedlings to infection by *Armillaria ostoyae* under limited light and nitrogen. *Physiol. Biochem.* **81,** 89–94.

Farguhar, G. D., O'Leary, M. H., and Berry, J. A. (1982). The relationship between carbon isotope discrimination and the intercellular carbon dioxide concentration in leaves. *Aust. J. Plant Physiol.* **9,** 121–138.

Gamon, J. A., Field, C. B., Bilger, W., Bjorkman, O., Fredeen, A. L., and Penuelas, J. (1990). Remote sensing of the xanthophyll cycle and chlorophyll fluorescence in sunflower leaves and canopies. *Oecologia* **85,** 1–7.

Hunt, E. R., Jr., and Nobel, P. S. (1987). A two-dimensional model for water uptake by desert succulents: Implications of root distribution. *Ann. Bot.* **59**, 559–569.

Hunt, E. R., Jr., Martin, F. C., and Running, S. W. (1991). Simulating the effects of climatic variation on stem carbon accumulation of a ponderosa pine stand: comparison to annual growth increment data. *Tree Physiol.* **9**, 161–171.

Iverson, L. R., Graham, R. L., and Cook, E. A. (1989). Applications of satellite remote sensing to forested ecosystems. *Landscape Ecol.* **3**, 131–143.

Jarvis, P. G., and McNaughton, K. G. (1986). Stomatal control of transpiration: Scaling up from leaf to region. *Adv. Ecol. Res.* **15**, 1–49.

Keeling, C. D., Mook, W. G., and Tans, P. P. (1979). Recent trends in $^{13}C/^{12}C$ ratio of atmospheric carbon dioxide. *Nature* **277**, 121–123.

Linder, S. (1987). Responses to water and nutrients in coniferous ecosystems. *In* "Potentials and Limitations of Ecosystem Analysis (E.-D. Schulze and H. Zwolfer, eds.), pp. 180–202. Springer-Verlag, New York.

Mason, H. L., and Langenheim, J. H. (1957). Language analysis and the concept of environment. *Ecology* **38**, 325–330.

Mattson, W. J., Levieux, J., and Bernard-Dagan, C. (eds.) (1988). "Mechanisms of Woody Plant Defenses against Insects; Search for Pattern." Springer-Verlag, New York.

McMurtrie, R. E., Rook, D. A., and Kelliher, F. M. (1990). Modelling the yield of *Pinus radiata* on a site limited by water and nitrogen. *For. Ecol. Mgmt.* **30**, 381–413.

McNaughton, K. G., and Jarvis, P. G. (1991). Effects of spatial scale on stomatal control of transpiration. *Agric. For. Meteor.* **54**, 279–301.

Miller, H. G. (1981). Forest fertilization: Some guiding concepts. *Forestry* **54**, 157–167.

Nasholm, T., and Ericsson, A. (1990). Seasonal changes in amino acids, protein and total nitrogen of fertilized Scots pine trees. *Tree Physiol.* **6**, 267–281.

Nicholson, S. E., Davenport, M. L., and Malo, A. R. (1990). A comparison of the vegetation response to rainfall in the Sahel and East Africa using Normalized Difference Vegetation Index from NOAA AVHRR. *Climatic Change* **17**, 209–242.

Passioura, J. B. (1979). Accountability, philosophy and plant physiology. *Search* **20**, 347–350.

Peterson, B. J., and Fry, B. (1987). Stable isotopes in ecosystem studies. *Annu. Rev. Ecol. Syst.* **18**, 293–320.

Richards, J. H., and Caldwell, M. M. (1987). Hydraulic lift: substantial nocturnal water transport between soil layers by *Artemisia tridentata* roots. *Oecologia* **73**, 486–489.

Richie, G. A. (1979). The pressure chamber as an instrument for ecological research. *Adv. Ecol. Res.* **9**, 165–254.

Rundel, P. W., Ehleringer, J. R., and Nagy, K. A. (eds.) (1989). "Stable Isotopes in Ecological Research." Springer-Verlag, New York.

Running, S. W., and Coughlan, J. C. (1988). A general model of forest ecosystem processes for regional applications. I. Hydrologic balance, canopy gas exchange and primary production processes. *Ecol. Model.* **442**, 125–154.

Ryan, M. G. (1991). Effects of climate change on plant respiration. *Ecol. Appl.* **1**, 157–167.

Schlesinger, W. H. (1990). "Biogeochemistry: An Analysis of Global Change." Academic Press, San Diego.

Seguin, B., Lagouarde, J.-P., and Savane, M. (1991). The assessment of regional crop water conditions from meteorological satellite thermal infrared data. *Remote Sensing Environ.* **35**, 141–148.

Sellers, P. J. (1987). Canopy reflectance, photosynthesis, and transpiration. II. The role of biophysics in the linearity of their interdependence *Remote Sensing Environ.* **21**, 143–183.

Sharpe, P. J. H., Walker, J., Penridge, L. K. and Wu, H.-I. (1985). A physiological based continuous-time Markov approach to plant growth modelling in semi-arid woodlands. *Ecol. Model.* **29**, 189–213.

Shearer, G., and Kohl, D. H. (1986). N^2-fixation in field settings: estimations based on natural ^{15}N abundance. *Aust. J. Plant Physiol.* **13,** 699–756.

Smith, B. N., and Epstein, S. (1971). Two categories of ^{13}C/^{12}C ratios for higher plants. *Plant Physiol.* **47,** 380–384.

Stark, N., Essig, D., and Baker, S. (1989). Nutrient concentrations in *Pinus ponderosa* and *Pseudotsuga menziesii* xylem sap from acid and alkaline soils. *Soil Sci.* **148,** 124–131.

Sternberg, L. L., and Swart, P. K. (1987). Utilization of fresh water and ocean water by coastal plants of southern Florida. *Ecology* **68,** 1898–1905.

Tieszen, L. L., and Boutton, T. W. (1989). Stable carbon isotopes in terrestrial ecosystem research. *In* "Stable Isotopes in Ecological Research" (P. W. Rundel, J. R. Ehleringer, and K. A. Nagy, eds.), pp. 167–195. Springer-Verlag, N.Y.

Tilman, G. D. (1984). Plant dominance along an experimental nutrient gradient. *Ecology* **65,** 1445–1453.

Turner, J., and Lambert, M. M. (1986). Nutrition and nutritional relationships of *Pinus radiata. Annu. Rev. Ecol. Syst.* **17,** 325–350.

Ustin, S. L., Wessman, C. A., Curtiss, B., Kasischeke, E., Way, J., and Vanderbilt, V. C. (1991). Opportunities for using the EOS imaging spectrometers and synthetic aperture radar in ecological models. *Ecology* **72,** 1934–1945.

van der Merwe, N. J. (1982). Carbon isotopes, photosynthesis, and archaeology. *Amer. Sci.* **70,** 596–606.

Waring, R. H., McDonald, A. J. S., Larsson, S., Ericsson, T., Wiren, A., Arwidsson, E., Ericsson, A., and Lohammar, T. (1985). Differences in chemical composition of plants grown at constant relative growth rates with stable mineral nutrition. *Oecologia* **66,** 157–160.

Waring, R. H., Savage, T., Cromack, K., Jr., and Rose, C. (1991). Thinning and nitrogen fertilization in a grand fir stand infected with western spruce budworm. IV. An ecosystem management perspective. *For. Sci.* **38,** 275–286.

White, J. W. C., Cook, E. R., Lawrence, J. R., and Broecker, W. S. (1985). The D/H ratios of sap in trees: Implications for water sources and tree ring D/H ratios. *Geochim. Cosmochim. Acta* **49,** 237–246.

White, T. C. R. (1984). The abundance of invertebrate herbivores in relation to the availability of nitrogen in stressed food plants. *Oecologia* **63,** 90–105.

Wickman, B. E. (1980). Increased growth of white fir after a Douglas-fir tussock moth outbread. *J. For.* **78,** 31–33.

Winner, W. E., Bewley, J. D., Krouse, H. R., and Brown, H. M. (1978). Stable sulfur isotope analysis of SO$_2$ pollution impact on vegetation. *Oecologia* **36,** 351–361.

Yoder, B. J., and Daley, L. S. (1990). Development of a visible spectroscopic method for determining chlorophyll a and b in vivo in leaf samples. *Spectroscopy* **5,** 44–50.

III

Global Constraints and Regional Processes

Physiological processes at the leaf or whole-plant level ultimately are integrated at a regional or global level. Top-down models predominate in this area, but mechanistic input from bottom-up approaches are necessary for understanding what defines the limits and interactions at these levels of scale. The top-down approach in the chapter by Vitousek develops a general approach to using global-scale information for constraining smaller scale interpretations. In the next chapter, Tans discusses a strategy for combining global and regional data to localize and quantify carbon sources and sinks further. Jarvis and Dewar review the global carbon cycle in the context of constraints on the magnitude and location of carbon sources and sinks. In the final chapter of this section, Caldwell, Matson, Wessman, and Gamon present a summary of discussions at the workshop that were related to future development of bottom-up and top-down approaches and offer a series of guidelines to be considered when addressing issues of scale.

10

Global Dynamics and Ecosystem Processes: Scaling Up or Scaling Down?

Peter M. Vitousek

I. Introduction

Integrating information from physiological ecology to the ecosystem level or from ecosystem ecology to regional or global levels should be comparatively straightforward. The central concerns at each of these levels are the identification and characterization of the pathways of transfer and transformation of energy and material, and of the mechanisms regulating those pathways. Research at each level is based on a metabolic approach that focuses on inputs and outputs of energy and resources; conservation of material is an important constraint at all levels. Finally, most of the same resources (carbon dioxide, water, nitrogen, other nutrients) are of interest on all scales. Physiological, ecsytem, and global studies, therefore, involve levels of biological organization that ultimately must fit together naturally, if perhaps not easily. The same is not necessarily true of all efforts to integrate across levels of biological organization. Differences in the resources of interest make linkage from population to ecosystem or global ecology qualitatively more difficult.

Great progress has been made in linking physiological and ecosystem ecology in the past decade, to the benefit of both fields. In this paper, I review briefly ways in which that integration has come about. I suggest that, although our goal (and, to some extent, accomplishment) has been to base our understanding of ecosystem phenomena on physiological mechanisms, our path to that goal has been directed primarily by ecosystem-level measurements and experiments. I then discuss how physiological and ecosystem ecology (now often as an integrated unit) can be and are being integrated into global studies.

II. From Physiology to Ecosystem

One way in which physiological and ecosystem ecology have been brought together has been to consider organisms as open systems that take up resources from the environment, use or store them, and eventually give up products while optimizing net carbon gain (or some other measure of fitness) for a given set of resource availability, environment, and organism characteristics. The products given off by organisms need not be in the same form as the original resources, but they must be transformations of them. Ecosystems then are viewed as partially closed systems: effectively closed in some resources (nitrogen, other nutrients) and open in others (energy, carbon, water) (Fig. 10.1). Within an ecosystem, the "closed" resources are depletable, that is, quantities taken up by one plant are subtracted from the amount available to others for as long as they are retained within the organism.

Although all plants use the same basic resources in fundamentally similar ways, plants often differ in their rates of resource use or in the relative amounts of different resources that they use. These differences can be determined genetically or they can arise phenotypically as a result of differences in environments or resource availability. When considered pairwise, these differences in resource use often are described as efficiencies or utility functions, for example, water use efficiency (carbon dioxide taken up per unit of water lost) or nitrogen use efficiency (some permutation of carbon dioxide taken up per unit of nitrogen taken up) (Hirose, 1975; Chapin, 1980; Berendse and Aerts, 1987). More generally, similarities and differences in resource use among organisms can be

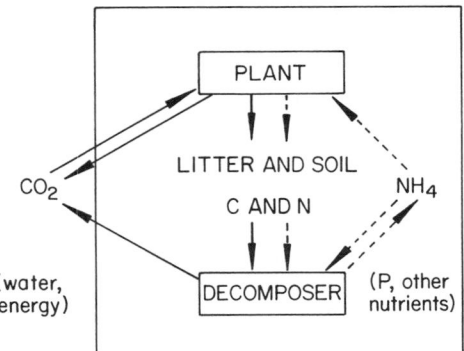

Figure 10.1 A conceptual model illustrating how physiological mechanisms interact with ecosystem dynamics. Organisms (here plants and decomposers) are viewed as open systems that exchange limited resources within a partially closed ecosystem.

evaluated in terms of the basic stoichiometry of life processes (Reiners, 1986).

Differences among species in the rate and ratio of resource use can be important at the population and community levels of organization, since they can allow the coexistence of species at equilibrium and can underlie specialization in different environments or at different stages of succession (Tilman, 1988). Differences in the ratio of resources that plants obtain or release are also important at the ecosystem level, because they imply differences in the ratio of elements in plant tissues or their major products (leaf, stem, and root litter), which, in turn, represent the major input of energy and resources to animals and decomposers.

For simplicity, I will concentrate on decomposers, but similar reasoning can be used with respect to animals. A functional ecosystem could be made up of producers, decomposers, and a finite (closed) amount of certain resources (especially nutrient elements). Such a simple system nevertheless is capable of generating extremely strong positive or negative feedbacks between trophic levels that remain important in more complex and realistic ecosystems. For example, suppose that nitrogen is deficient in a forest ecosystem. Trees are known to respond by retaining nitrogen in biomass through patterns of carbon allocation, tissue longevity, and nitrogen reabsorption. Consequently, the litter that is produced by such nitrogen-deficient trees has relatively high carbon/nitrogen and low nitrogen/phosphorus ratios (Vitousek, 1982); it also tends to be enriched in recalcitrant carbon compounds (Coley *et al.*, 1985). Decomposers that use such tissue act slowly; they also retain nitrogen in their tissue, and may take up inorganic nitrogen from the soil. The nitrogen demand of such microoganisms further reduces nitrogen availability to plants, sustaining a positive feedback to increased nitrogen limitation of primary production (Vitousek and Howarth, 1991).

Most modern views of ecosystem ecology make no special claim that ecosytems are fundamental units that cannot be understood in terms of the properties of the organisms, inorganic constituents, and interactions that they contain. However, ecosystem ecologists do maintain—indeed, have established—that any mechanistic understanding of ecosystems must emphasize trophic interactions and the feedbacks they engender. In a sense, these feedbacks are ecosystem ecology; they control the large-scale energy and nutrient dynamics of ecosystems and cannot be derived from the study of a single species, or even trophic level, in isolation, although the feedbacks are driven by these organisms. Resource-based feedbacks of the sort outlined earlier are an explicit part of most modern ecosystem models (cf. Pastor and Post, 1986; Parton *et al.*, 1988a).

Despite the essentially reductionist nature of modern ecosystem ecology, it is not obvious that a research strategy based on physiological ecology and aimed at a bottom-up understanding of ecosystems could have made substantial progress without ecosystem-level measurements and experiments. Certainly our present understanding has not developed in that way. For example, the watershed ecosystem approach (Likens *et al.*, 1977; Bormann and Likens, 1979) yielded ecosystem-level input–output budgets for a variety of elements. Ecosystem-level experiments showed when and how those budgets were disrupted when aspects of the system were altered (Bormann and Likens, 1979). Much of our progress in understanding ecosystem function has been achieved by starting with these integrated ecosystem-level signals and tracing them within the system to determine what (in terms of physiological or geochemical mechanisms) controls them.

The plant–decomposer positive feedback described earlier is a good illustration. It was identified as a result of studies designed to determine why some ecosystems lose large amounts of nitrogen after clearcutting whereas others lose little. Research began with measurements of nutrient loss in disturbed ecosystems through ecosystem-level experiments designed to determine where (in the nitrogen cycle) losses were prevented and continued with mechanistic studies of the processes that were identified as potential regulators of ecosystem-level losses (Vitousek *et al.*, 1982).

III. From Ecosystem to Global Scale

A similar conceptual framework can be used to integrate from our understanding of ecosystems to regional and global phenomena. Ecosystems can be considered relatively open systems embedded in a relatively closed global system (Fig. 10.2). Unlike ecosystems with respect to physiology, the global system truly is closed in all resources except energy.

In this section, I will concentrate on spatial rather than temporal scaling. Most efforts to integrate from ecosystems to regional and global scales involve determining ways to scale up, that is, to move from reasonably well-known plots and processes to dynamics on coarser spatial scales (Rosswall *et al.*, 1988). At the simplest level, scaling up can involve determining the flux of resources or products at a point, and then multiplying that flux by the area of that ecosystem type in a region or globally. The type of ecosystem can be defined as crudely as a biome or as finely as land classification and computer capacity allow. More sophisticated steps include determining fluxes as functions of gradients of the environmental factors that control those fluxes (Matson *et al.*, 1989) and calculating fluxes from geographically referenced process-level ecosystem models (Parton *et al.*, 1988b; Burke *et al.*, 1990). There is also substantial

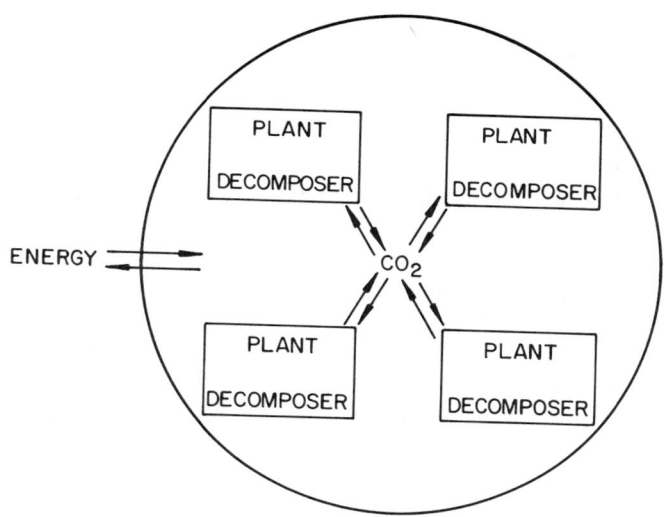

Figure 10.2 A conceptual model illustrating how ecosystem-level mechanisms interact on a global scale. Here, ecosystems are viewed as relatively open systems that exchange material within a closed global system.

discussion (but no real information) on how to incorporate nonadditive effects resulting from the distribution and connectivity of ecosystems on landscapes. All these approaches depend increasingly on remote sensing of ecosystem characteristics as a basis for classification or modeling; several groups are reaching toward the development of process-level models that can be parameterized and driven by information derived from remote sensing (cf. Running, 1990).

This scaling-up approach is essential to the development of global ecology. Ultimately, it provides the only way to incorporate the ecosystem-level mechanisms that will allow us to deal with the effects of atmospheric, climatic, or ecosystem change. However, just as ecosystem ecology has advanced in large part through the use of ecosystem-level measurements and experiments, the science of global ecology is likely to develop most efficiently if it is driven by regional and global measurements designed to answer globally significant research questions. On one level, this statement is unexceptionable; clearly only through regional and global scale measurements can we check our calculations based on scaling up. Mesoscale measurement campaigns such as FIFE (First International Satellite Land Surface Climatology Program Field Experiment) (Sellers *et al.*, 1988) are therefore of interest to ecologists as well as meteorologists. Perhaps more controversially, I suggest that for many globally significant research questions, our understanding will develop most rapidly if we scale down from global measurements to the ecosystem level.

IV. Global Measurements to Ecosystem Mechanisms

The best developed examples of scaling from global measurements to ecosystem dynamics have been developed for carbon dioxide, the most carefully studied biological resource or product in the atmosphere. Fung *et al.* (1987) measured latitudinal variations in the amplitude and timing of the striking seasonal variation in carbon dioxide concentrations, and used them to calculate latitudinal variations in the monthly net uptake or release of carbon dioxide by terrestrial ecosystems. There are no coarse-scale monthly measurements of net ecosystem production (net carbon exchange) against which these global measurements and calculations could be compared, but Fung *et al.* (1987) developed a combination of coarse-scale models and remote sensing to estimate carbon exchange on a scale comparable to that of the global measurements. They used information on carbon dioxide flux from soils (as a function of biome and temperature) to calculate carbon dioxide release from ecosystems, and assumed that the timing of carbon dioxide uptake could be calculated from the seasonal timing of absorption of photosynthetically active radiation globally, which in turn could be determined using the satellite-based Advanced Very High Resolution Radiometer. They then compared the measured latitudinal variation in the amplitude of the annual cycle of carbon dioxide concentrations with their calculations of net carbon exchange from terrestrial ecosystems with excellent results. This study was, of course, an initial effort but it represents a clear example of a case in which global measurements are far more solid than the ecosystem-level measurements with which we wish to compare them. Hence, the global data can serve as constraints on ecosystem-level measurements. Work by Tans *et al.* (1990; Chapter 11) demonstrates that global measurements of carbon dioxide also can serve as guides to determining which ecosystems appear to be responsible for the "missing carbon sink"; thus, these ecosystems should receive intensive study. In fact, I suggest that these global measurements are more likely to focus research on the critical processes regulating carbon storage in terrestrial and marine ecosystems than would any number of efforts to scale up from ecosystem measurements.

The same approach is applicable to a wide range of trace gases. Of these, I will emphasize nitrous oxide, but first will discuss methane briefly. Methane is of interest because it is increasing rapidly in the troposphere and because there are very strong latitudinal gradients and seasonal variations in methane concentrations. This variation has been used to determine methane source distributions (Fung *et al.*, 1991) in a manner analogous to that used to detect carbon dioxide. For this analysis, a necessary starting point was the calculation of an approximate global source budget based on ecosystem-level measurements. This estimate then was refined using constraints imposed by the spatial and seasonal distributions in methane concentrations, and the variations in methane sources that these distributions imply. For methane in particular, the major sources differ strongly in carbon isotopic composition (of both ^{13}C and ^{14}C) (Cicerone and Oremland, 1988), and spatial and temporal gradients in the isotopic composition of methane constrain the determination of global methane sources fairly closely. One result of these global measurements has been to emphasize the importance of a methane source that is relatively enriched in ^{13}C and modern in ^{14}C, a description that helps identify biomass combustion as a significant global source of methane.

Nitrous oxide is more difficult to work with on a global scale, in large part because its low fluxes (relative to atmospheric concentrations) preclude the development of strong latitudinal or seasonal variations in nitrous oxide concentrations. Nevertheless, careful long-term observations have demonstrated that there are small but repeatable latitudinal and interhemispheric gradients in nitrous oxide concentrations (Cicerone, 1989; Prinn *et al.*, 1990). Calculations of source distributions based on these gradients yield the conclusion that the tropics are a major source for nitrous oxide under unperturbed background conditions, and that tropical regions are also a significant source of the ongoing global increase in tropospheric nitrous oxide (Prinn *et al.*, 1990). The initial runs of the Prinn group model used ecosystem-level observations that fluxes of nitrous oxide from most tropical forests are greater than those from most temperate forests (Keller *et al.*, 1986; Matson and Vitousek, 1987), in that the source budget for tropical forests used was calculated by extrapolating from ecosystem-level information and mechanisms (Matson and Vitousek, 1990). The global analysis then can be used to constrain and refine the field-based estimates.

The global measurements of nitrous oxide also support observations of elevated nitrous oxide flux after tropical land conversion (Luizao *et al.*, 1989). However, the magnitude of human-caused increases in flux that have been observed in the field to date is far less than that required by the global analysis (Prinn *et al.*, 1990). The results of the global analysis are, therefore, directing field research to sites, land-use practices, and mechanisms that might now be causing elevated nitrous oxide flux in tropical regions.

V. Conclusions

In this chapter, I have argued strongly that scaling down from regional or global measurements and models to the ecosystem level provides the most effective way to begin to integrate ecosystem and global processes at the global scale, just as ecosystem measurements and models have driven the integration of physiological and ecosystem ecology. However, it should be recognized that this approach represents a limited (if powerful) tool for global studies. It is often the most effective way to determine current occurrences on a global or regional scale; it does not explain why particular events occur or how they will change with elevated carbon dioxide, land-use change, climate change, or other factors. A knowledge of current events is useful in determining why changes occur, but predictive capability can come only from a painstaking mechanism-based understanding of global phenomena in terms of ecosystem and physiological mechanisms. In other words, interaction between the top-down and bottom-up approaches can yield better results more rapidly than can bottom-up analyses alone; the interaction can yield an understanding that is wholly inaccessible to top-down analyses.

Acknowledgment

Preparation of this manuscript was supported by NASA Grant NAGW-1068.

References

Berendse, F., and Aerts, R. (1987). Nitrogen-use-efficiency: A biologically meaningful definition? *Funct. Ecol.* **1**, 293–296.

Bormann, F. H., and Likens, G. E. (1979). "Pattern and Process in a Forested Ecosystem." Springer-Verlag, New York.

Burke, I. C., Schimel, D. S., Yonker, C. M., Parton, W. J., Joyce, L. A., and Lauenroth, W. K. (1990). Regional modelling of grassland biogeochemistry using GIS. *Landsc. Ecol.* **4**, 45–54.

Chapin, F. S., III. (1980). The mineral nutrition of wild plants. *Annu. Rev. Ecol. Syst.* **11**, 233–260.

Cicerone, R. J. (1989). Analysis of sources and sinks of atmospheric nitrous oxide (N_2O). *J. Geophys. Res.* **94**, 18,265–18,272.

Cicerone, R. J., and Oremland, R. (1988). Biogeochemical aspects of atmospheric methane. *Global Biogeochem. Cycles* **2**, 299–328.

Coley, P. D., Bryant, J. P., and Chapin, F. S., III (1985). Resource availability and plant anti-herbivore defense. *Science* **230**, 895–899.

Fung, I., Tucker, C. J., and Prentice, K. C. (1987). Application of AVHRR vegetation index to study atmospheric-biosphere exchange of CO_2. *J. Geophys. Res.* **92**, 2999–3016.

Fung, I., John, J., Lerner, J., Matthews, E., Prather, M., Steele, L. P., and Fraser, P. J. (1991). Global budgets of atmospheric methane: Results from a three-dimensional global model synthesis. *J. Geophys. Res.* **96**, 13,033–13,065.

Hirose, T. (1975). Relations between turnover rate, resource utility and structure of some plant populations: A study of the matter budgets. *J. Fac. Sci. Univ. Tokyo* **11**, 355–407.

Keller, M., Kaplan, W. A., and Wofsy, S. C. (1986). Emissions of N_2O, CH_4, and CO_2 from tropical soils. *J. Geophys. Res.* **91**, 11,791–11,802.

Likens, G. E., Bormann, F. H., Pierce, R. S., Eaton, J. S., and Johnson, N. M. (1977). "Biogeochemistry of a Forested Ecosystem." Springer-Verlag, New York.

Luizao, F. J., Matson, P. A., Livingston, G. L., Luizao, R., and Vitousek, P. M. (1989). Nitrous oxide flux following tropical land clearing. *Global Biogeochem. Cycles* **3**, 281–285.

Matson, P. A., and Vitousek, P. M. (1987). Cross-system comparison of soil nitrogen transformations and nitrous oxide fluxes in tropical forests. *Global Biogeochem. Cycles* **1**, 163–170.

Matson, P. A., and Vitousek, P. M. (1990). Ecosystem approach to a global nitrous oxide budget. *Bioscience* **40**, 667–672.

Matson, P. A., Vitousek, P. M., and Schimel, D. S. (1989). Regional extrapolation of trace gas flux based on soils and ecosystems. *In* "Exchange of Trace Gases between Terrestrial Ecosystems and the Atmosphere" (M. O. Andreae and D. S. Schimel, eds.), pp. 97–108. Wiley, Chichester.

Parton, W. J., Cole, C. V., Stewart, J. W. B., Ojima, D. S., and Schimel, D. S. (1988a). Dynamics of C, N, P, and S in soils: A model. *Biogeochemistry* **5**, 109–131.

Parton, W. J., Mosier, A. R., and Schimel, D. S. (1988b). Rates and pathways of nitrous oxide production in a shortgrass steppe. *Biogeochemistry* **6**, 45–58.

Pastor, J., and Post, W. M. (1986). Influence of climate, soil moisture, and succession on forest carbon and nitrogen cycles. *Biogeochemistry* **2**, 3–27.

Prinn, R., Cunnold, D., Rasmussen, R., Simmonds, P., Alyea, F., Crawford, A., Fraser, P., and Rosen, R. (1990). Atmospheric emissions and trends of nitrous oxide deduced from 10 years of ALE-GAGE data. *J. Geophys. Res.* **95**, 18,369–18,385.

Reiners, W. A. (1986). Complementary models for ecosystems. *Am. Nat.* **127**, 59–73.

Rosswall, T. H., Woodmansee, R. G., and Risser, P. G. (eds.) (1988). "Scales and Global Change." Wiley, Chichester.

Running, S. W. (1990). Estimating terrestrial primary productivity by combining remote sensing and ecosystem simulation. *In* "Remote Sensing of Biosphere Functioning" (R. J. Hobbs and H. A. Mooney, eds.), pp. 65–86. Springer-Verlag, New York.

Sellers, P. J., Hall, F. B., Asrar, G., Strebel, D. E., and Murphy, R. E. (1988). The first ISLSCP field experiment (FIFE). *Bull. Am. Meteorol. Soc.* **69**, 22–27.

Tans, P. P., Fung, I. Y., and Takahashi, T. (1990). Observational constraints on the global atmospheric CO_2 budget. *Science* **247**, 1431–1438.

Tilman, D. (1988). "Plant Strategies and the Dynamics and Structure of Plant Communities." Princeton University Press, Princeton, New Jersey.

Vitousek, P. M. (1982). Nutrient cycling and nutrient use efficiency. *Am. Nat.* **119**, 553–572.

Vitousek, P. M., and Howarth, R. W. (1991). Nitrogen limitation on land and in the sea: How can it occur? *Biogeochemistry* **13**, 87–115.

Vitousek, P. M., Gosz, J. R., Grier, C. C., Melillo, J. M., and Reiners, W. A. (1982). A comparative analysis of potential nitrification and nitrate mobility in forest ecosystems. *Ecol. Monogr.* **52**, 155–177.

11

Observational Strategy for Assessing the Role of Terrestrial Ecosystems in the Global Carbon Cycle: Scaling Down to Regional Levels

Pieter P. Tans

I. Introduction

Very accurate measurements of the increase of the atmospheric carbon dioxide (CO_2) concentration have been made for several decades. The records gathered at the Mauna Loa Observatory since 1958 (Pales and Keeling, 1965; Komhyr *et al.*, 1989) document a textbook example of how humans have been changing the global environment, yet our knowledge of the global carbon budget is still surprisingly uncertain. From the Mauna Loa measurements we know the "bottom line" by how much CO_2 in the atmosphere increases every year. Using United Nations economic statistics, we also can account with reasonable accuracy the amount of CO_2 that is being emitted by the burning of fossil fuels. Beyond that, our accounting stops. Estimates of the oceanic uptake of CO_2 are not based on direct observations of inorganic carbon in the oceans, but on models (Broecker *et al.*, 1979). These models, in which ocean circulation and mixing play important roles, are calibrated with observations of ^{14}C, tritium, and other tracers. The terrestrial biosphere, which contains about 800 Pg C (1 Pg equals 10^{15} g or 10^9 metric tons), has been estimated to lose up to 6 Pg C · year^{-1} to the atmosphere (Woodwell, 1978), as well as to increase its storage of carbon by about 1 Pg C · year^{-1} (Bacastow and Keeling, 1973). The carbon storage in soils has been estimated to be large, on the order of 1500 Pg C (Schlesinger, 1977), but with a generally slow turnover rate. Ongoing changes have been difficult to estimate on a global scale (Schlesinger, 1984) but seem to suggest a current loss rate of about 0.8 Pg C · year^{-1}, mainly because of land-use practices.

We know from the Mauna Loa record that the increase in atmospheric CO_2 (3 Pg C · year^{-1} during the last decade) accounts for about half the CO_2 emitted by fossil fuel burning (5.5 pg C · year^{-1}). Because the destruction of tropical forests and the loss of organic matter from soils are adding another 1–2 Pg C · year^{-1} to the atmosphere, a sink for carbon of about 3.5–4.5 Pg C · year^{-1} must be found. Oceanographers generally believe that the oceans cannot absorb carbon at such a fast rate (Broecker *et al.*, 1979). The oceans are estimated to be capable of taking up 2 Pg C · year^{-1} at most, suggesting that there must be considerable uptake of CO_2 by terrestrial ecosystems.

The waters in the deep oceans will hold carbon for many centuries before they return to the surface where they can equilibrate with the atmosphere. On the other hand, the storage of carbon in forests or soils appears to be much less long lived and more subject to human intervention and climatic change. Therefore, as long as we are unable to sort out the proportions of carbon taken up by the oceans and terrestrial ecosystems, our capability to forecast increases in atmospheric CO_2 many decades ahead, for given scenarios of fossil fuel burning and land management, is poor.

We would like to learn, first, where the CO_2 is going currently and, second, why the natural systems are responding the way they are. Finally, we also would like to develop an understanding of the response of the natural systems to the changes in global climate that are expected to occur. A credible answer to the first of these questions would be enormously helpful to the formulation of answers to the second and third. This chapter explores how we could use atmospheric observations to narrow the uncertainties about the large-scale regional sources and sinks of CO_2 significantly. It will not address the additional observations that will still be needed to gain more insight into the physiological and ecological mechanisms driving the storage of carbon.

II. Atmospheric Concentration Gradients and Transport Modeling

The combined effect of all sources and sinks of CO_2 is reflected in the atmospheric concentration patterns of this gas. If a collection of ecosystems acts as an important sink of CO_2 on a regional scale, the regionally averaged CO_2 concentration in the atmospheric boundary layer should be lower than in an adjacent region where net sources dominate (Fig. 11.1). This simple mass balance principle can be forged into a tool for locating major source and sink areas and to estimate their magnitudes. Two main elements are required to make the approach successful: precise observations of the concentration of CO_2 and numerical models of atmospheric transport. The mass balance approach works because, except for the oxidation of CO to CO_2, CO_2 is a conserved species in the atmosphere. The amount of CO oxidized annually has been estimated to be as large as 0.9 ± 0.3 Pg C (Seiler and Conrad, 1987), but its spatial pattern and seasonal dependence are fairly well known.

Until now, the "inverse method" has been applied successfully only on the largest of scales, roughly speaking to the difference in concentration between the Northern and Southern hemispheres or between half-hemispheres (Fung *et al.*, 1983; Pearman and Hyson, 1980; Enting and Mansbridge, 1989; Keeling *et al.*, 1989; Tans *et al.*, 1989, 1990). In the atmosphere, the correct balance between sources and sinks is required on these broad scales to result in the latitudinal concentration distribution that is observed. The basic idea of the inverse method is that atmospheric transport models are used to simulate the concentration patterns resulting from hypothetical sets of surface sources and sinks. Only a limited set of large-scale distributions of sources and sinks is consistent with the observed latitudinal concentration gradient. This set can be further constrained by estimates of fossil fuel combustion, deforestation, and air–sea exchange, thus leading to quantitative estimates of "unknown" sources and sinks. Probably the most important result of these studies thus far is the recognition that there must be a very large sink for CO_2 in the Northern hemisphere, either in the northern oceans or in terrestrial ecosystems.

To estimate the air–sea transfer of CO_2, one needs not only the partial pressure difference between the air and the water (ΔpCO_2), but also a relationship between the transfer velocity and wind speed. Data on ΔpCO_2 of the Northern hemisphere surface ocean waters led Tans *et al.* (1990) to the conclusion that the oceans are not absorbing much CO_2. These investigators used two different parameterizations, representing probable extremes, of the dependence of the transfer velocity on wind speed. This result suggested that the major sink must be on land. However, the latitudinal gradient of the $^{13}C/^{12}C$ ratios of atmospheric CO_2 provides contrary evidence (Fig. 11.2). Fossil fuels, like all material derived from C-3 photosynthesis, are depleted in ^{13}C relative to the atmosphere. Most fossil fuels are burned in temperate latitudes of the Northern hemisphere, so the $^{13}C/^{12}C$ ratio of atmospheric CO_2 should be lower by about 2.5‰ in the Arctic than in Antarctica if fossil fuels are

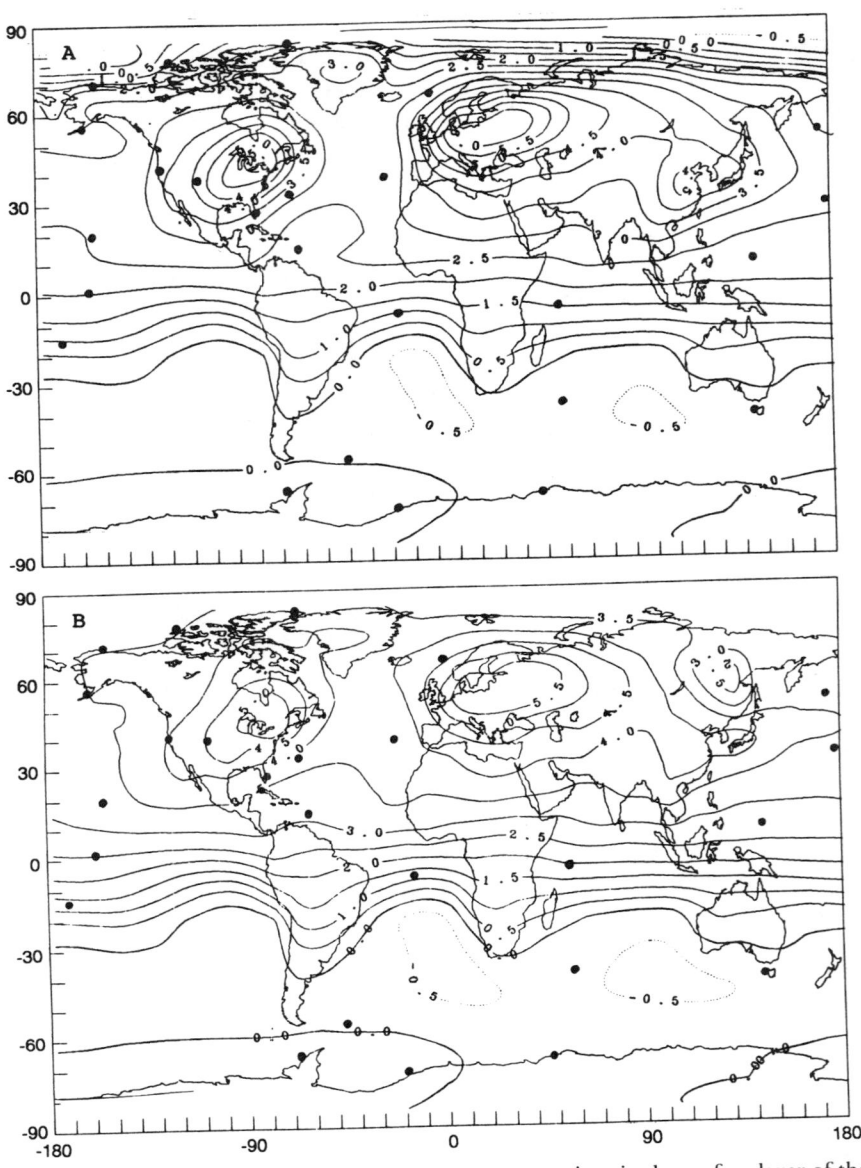

Figure 11.1 Simulated annual average CO_2 concentrations in the surface layer of the NASA/GISS General Circulation Model. All concentrations are relative to the South Pole. Location of CMDL flask sampling sites is indicated with dots. (A) The sources consist of fossil fuel combusition (5.3 Pg C · year^{-1}), tropical deforestation (0.3 Pg C · year^{-1}), equatorial waters (1.0 Pg C · year^{-1}), and the Antarctic Ocean south of 50°S (0.5 Pg C · year^{-1}), whereas uptake occurs in the Arctic Ocean (−0.7 Pg C · year^{-1}), the North Pacific and North Atlantic central gyres (−1.0 Pg C · year^{-1} each), and the combined Southern Ocean gyres between 15° and 50°S (−1.4 Pg C · year^{-1}). This pattern of sources

the only influence on the isotopic ratio. A second influence is air–sea exchange. The isotopic equilibrium between gaseous and dissolved CO_2 is temperature dependent, and colder waters will tend to lower $^{13}C/$ ^{12}C in atmospheric CO_2. The sum of these two effects alone fits the observations fairly well (Fig. 11.2), suggesting that there cannot be much CO_2 uptake by terrestrial ecosystems (Keeling *et al.*, 1989). Terrestrial photosynthetic uptake discriminates against ^{13}C, increasing the $^{13}C/^{12}C$ ratio of the CO_2 left behind in the atmosphere. If this third influence was taken into account to the extent required by the ΔpCO_2 data, the resulting isotopic gradient between the Aratic and Antarctica would be less then 0.05‰. These isotopic gradients were all calculated by a two-dimensional (latitude, height) numerical transport model (Tans *et al.*, 1989), rather than the three-dimensional model used for CO_2 (Tans *et al.*, 1990).

There are presently two major shortcomings in down-scaling this approach from the global scale to broad regional areas, such as the central midwestern or southeastern United States. The atmospheric data have, until now, been collected primarily over oceanic areas because the CO_2 concentrations of oceanic air masses are much less variable than those over land, allowing the use of relatively inexpensive methods, namely, the collection and measurement of weekly air samples in flasks. However, that strategy has left the allocation of CO_2 sinks between the oceans and the land uncertain, because no concentration gradients have been measured between the oceans and the land in the same latitude zones. A second major shortcoming is the treatment by the global models of the flow around topography and mixing of air in the vertical dimension. Development of limited-domain, high-resolution climate models embedded in lower-resolution global models (Dickinson *et al.*, 1989) paves the way for data-model comparison on regional scales. Nevertheless, not nearly enough data in the vertical dimension are currently available to constrain the models sufficiently for use in deciphering regional sources or sinks.

---

and sinks corresponds to Scenario 1 of Tans *et al.* (1990). (B) Sources are fossil fuels (5.3 Pg C · year^{-1}), tropical deforestation (1.0 Pg C · year^{-1}), boreal forests (0.4 Pg C · year^{-1}), equatorial waters (1.3 Pg C · year^{-1}), and the Antarctic Ocean (0.5 Pg C · year^{-1}), whereas the sinks are global CO_2 fertilization of terrestrial ecosystems (-1.0 Pg C · year^{-1}), uptake by terrestrial ecosystems at temperate latitudes (-2.3 Pg C · year^{-1}), the Arctic Ocean (-0.23 Pg C · year^{-1} total), the North Pacific and North Atlantic gyres (-0.36 Pg C · year^{-1} total), and the Southern Ocean gyres (-1.6 Pg C · year^{-1}). This pattern corresponds to Scenario 7 of Tans *et al.* (1990). (Figures courtesy of Inez Fung, NASA Goddard Space Flight Center, Institute for Space Studies, New York.)

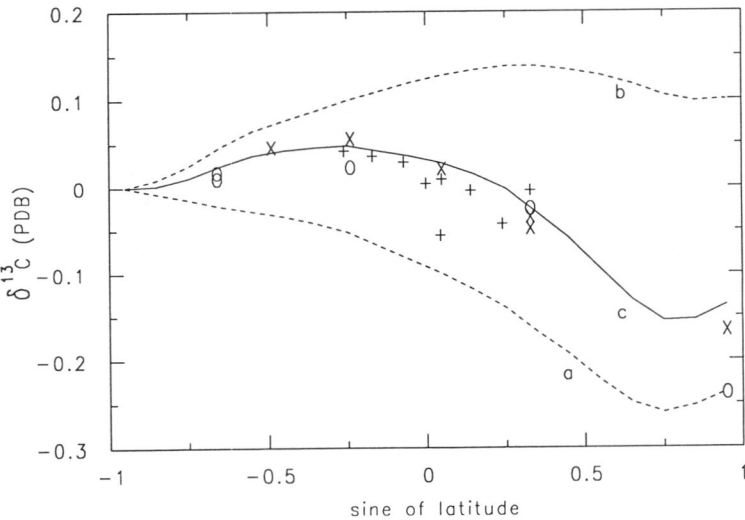

Figure 11.2 Annual mean gradient of the $^{13}C/^{12}C$ ratio of atmospheric CO_2 as a function of latitude. Data and calculated values are plotted as differences from the South Pole values. Data from the FGGE cruises in the mid-Pacific (+; Keeling *et al.*, 1989); data from land stations (x; Keeling *et al.*, 1989); data from land stations (O; Francey *et al.*, 1990). Calculated gradients with a two-dimensional atmospheric transport model (---; Tans *et al.*, 1989) for fossil fuel combustion only (a) and for isotopic exchange with dissolved CO_2 only (b). The air–sea exchange rate is empirical (Tans *et al.*, 1990), using monthly mean wind speeds; the surface ocean $^{13}CO_2$ data are from Kroopnick (1985). Curve c (—) represents the sum of a and b. The calculations appear to fit the data reasonably well without invoking any uptake of CO_2 by terrestrial ecosystems at mid- to high northern latitudes.

At this time, the level of regional detail to which this approach can be extended is not clear. Costs will be approximately proportional to the number of measurement sites. At some point, the need for detailed auxiliary data (e.g., land use, vegetation, meteorological data) or a need for much higher resolution transport modeling may overwhelm available resources. Extrapolation with the help of satellite information will be necessary.

III. General Requirements for Measurements

Observations of the large-scale concentration patterns of CO_2 over continental areas should be started now. This priority is not diminished by the fact that the present generation of numerical transport models still needs considerable improvement before they can be used with confidence to extract information about regional scale sources and sinks from the observations. The reason is simply that we will not get a second chance to monitor the events of the current year, whereas the transport models are applied always to data that have been collected already. Further, the data collected, especially vertical profiles, will contribute to the improvement of the models. Finally, we must accumulate time series of sources and sinks to contribute to our study of feedback mechanisms between climate change and carbon storage.

Regardless of the experimental method chosen, there are three basic requirements. First, the calibration should be tied very firmly to that of the rest of the global network, to a precision of no less than 0.1 part per million (ppm). Second, the measurements must be made year-round. Third, all results should be confirmed by independent methods to the maximum extent feasible.

The object is to obtain data that are directly comparable to other global data, since a great deal of information is contained in relatively small concentration differences. The differences on the largest spatial scales are on the order of a few ppm. The measurements should be calibrated against the World Meteorological Organization (WMO) mole fraction scale, presently maintained by the Scripps Institution of Oceanography. Almost all the CO_2 monitoring data gathered worldwide are being reported with reference to this scale. In the case of *in situ* instruments in the field, the measurements not only should be calibrated by intermittent injections of reference gases into the analyzer, but also should be compared regularly to whole air flask samples analyzed in the same way as other samples in a global network.

An important element missing from high-intensity measurement campaigns designed to study exchange of trace gases between ecosystems and the atmosphere is often the time factor. Because of cost and other limitations, such campaigns typically last a few weeks. However, if we want to know the annual balance sheet, we need measurements over periods of net ecosystem uptake and over periods of net loss to the atmosphere, as well as in the transitions between them. Since it is difficult to predict these times precisely, some optimal combination of continuous monitoring and more intensive campaigns of short duration is needed.

IV. Methods for Monitoring the Carbon Cycle on the Continents

In this section, I will suggest a few possible independent approaches to monitoring the carbon budget on large spatial scales on the continents. Although the technology is available for all these applications, most have not yet been attempted to a significant extent.

A. High Precision Concentration Measurements

High precision concentration measurements on very high towers represent a relatively inexpensive technique with great potential payoff. Measurements obtained exclusively near the ground have little value alone. For example, a modest amount of plant respiration could cause a very large CO_2 concentration increase under a nocturnal inversion layer. Since we are trying to develop a mass balance, it is important to know how much of the air mass above the ground is affected. Therefore, the towers should be as high as possible, and situated on relatively flat terrain, to be in air flow patterns that are not unnecessarily complicated. Several dozen antennas exist in the United States with a height of about 2000 ft each; their use for this purpose should be investigated. Concentration measurements should be performed continuously at a number of levels on these towers, in addition to monitoring of wind speed, wind direction, temperature, and humidity at each level. Such a study should yield enough information to extract average CO_2 concentrations in the lowest few hundred meters of the atmosphere likely to be representative of large areas. Such concentration data then can be used directly in the global/regional transport models; differences from concentrations measured over the oceans will hold important clues about sources and sinks on large spatial scales. The tower data also are likely to be very useful for model improvement, especially in the treatment of boundary layer transfer processes.

At the same time, we would obtain a continuous record of the column burden of CO_2 along the height of the tower. Changes in this column burden over time are due partly to surface sources in the vicinity of the tower and partly to advection. Therefore, changes in the column burden would constitute a flux measurement of sorts, although the "footprint" of such a flux measurement would be defined poorly since the concentrations at greater elevations on the tower are influenced by sources and sinks that are progressively farther away.

The addition of other species to the measurements would strengthen their usefulness, narrowing the range of possible interpretations. ^{222}Radon is probably the most crucial species to consider. It originates in soils and its 3.8-day half-life makes it an excellent tracer for mixing processes in the boundary layer and throughout the lower atmosphere (e.g., Trumbore *et al.*, 1990). It is an excellent candidate for tracking the contribution of soil respiration and litter decomposition to the overall carbon budget. Another reason for adding radon to the species measured is that its penetration to higher levels in the atmosphere provides a further constraint on the way numerical models handle vertical mixing, quite independent of the constraint provided by CO_2. In the latter case, the models will have to match the attenuation and the phase shift of the seasonal CO_2 cycle as a function of altitude (e.g., Tanaka *et al.*, 1987). Because of its short half-life, radon could provide a sensitive test of the relative importance of intensive convective mixing events of short duration, whereas CO_2 would be influenced more by slow and gradual mixing over large scales.

If more gaseous and particulate tracers can be measured, there will be low-frequency (hours, days) correlations between them and the meteorological parameters. For instance, at our background atmospheric monitoring site in Point Barrow, Alaska, there are episodes with wind from the "clean air sector" with all the characteristics of "background" air in which the concentrations of CO_2, CH_4, and soot are highly correlated. We interpret these as large-scale pollution plumes that have traveled over great distances (Hansen *et al.*, 1989). In this case, the ratios of the concentration "anomalies" hold information about the relative source strengths of CO_2 and CH_4 for large geographical areas.

The comparison between models and measured concentrations thus far has been limited to long-term averages. The tower measurements should yield this information as a function of the wind sector, which would allow a more detailed comparison with modeled concentrations because the model data also could be provided as a function of wind direction. The comparison could be supported by back-trajectories calculated on the basis of observed wind fields, a standard procedure for the existing atmospheric background observatories (Harris and Kahl, 1990).

Solar radiation, humidity, precipitation, and soil moisture must be monitored in the vicinity of every tower or obtained by other means because of their obvious importance in photosynthesis and respiration. To serve as ground truth for satellite spectral measurements, the visible and near-infrared radiation upwelling from below should be measured from the top of the tower with high spectral resolution instruments similar to those planned to be placed on satellites in the near future. Chlorophyll (Knipling, 1970) and other pigments (Gamon *et al.*, 1990), as well as plant water status and atmospheric column water vapor (Gao and Goetz, 1990), can be recovered from their spectral signatures. This approach may supply part of the foundation for extrapolation of the carbon budget to other parts of the globe based on satellite observations.

B. Flux Measurement through Eddy Correlation

A completely different technique is flux measurement through eddy correlation with fast-response instrumentation. The only site, until now, at which this technique has been applied continuously for more than a year is the Harvard Forest in Massachusetts. The method is nonintrusive and measures the flux directly, but has a small "footprint" compared to the scales that are being addressed here. It is on the order of a kilometer, depending on the distance of the sensors above the canopy. Aircraft can be employed occasionally to measure fluxes by eddy correlation methods over much greater areas (e.g., Desjardins *et al.*, 1982). In that case, high-precision concentration data should be obtained by the aircraft at the same time to assess the spatial representativeness of the tower measurements, both in the horizontal dimension in the boundary layer and in the vertical dimension extending well above the height of the tower.

C. Weekly Vertical Profile Determination

Regular (weekly) vertical profiles extending at least 5 km into the free troposphere at a series of fixed continental sites would define a climatology of the seasonal cycle of CO_2 that should place severe constraints both on the way vertical transport is treated in the models and on the sources and sinks driving the modeled annual cycle. If the profiles are obtained by flask sampling, a number of stable species other than CO_2 (and isotopic ratios) can be measured easily. Above the boundary layer, weekly sampling is probably sufficient to establish a seasonal cycle with confidence, as is being done currently for ground-level oceanic air masses. In both cases, there generally is considerable mixing time between sampling and the last period of contact of the air mass with strong surface sources or sinks. Also, in this case, the addition of [222]radon to the measurements would enhance their usefulness significantly for the reasons just outlined.

D. Studies of Land Use, Nutrient Budget, and Carbon Allocation

Studies of land use, nutrient budgets (which also could be supported by tower or flux studies), carbon allocation, and species composition of the vegetation in the general vicinity of a tower site will yield another form of independent evidence for carbon loss or storage in an area. This type of evidence would be much closer to the second and third objectives mentioned in the introduction, that is, determining why the observed changes are taking place.

V. Summary

An experimental strategy has been proposed to monitor the net carbon budget of terrestrial ecosystems on very broad regional scales in continental areas. Precise measurements of atmospheric concentration differences on large spatial scales can be combined with numerical models of atmospheric transport to yield estimates of surface sources and sinks. This technique already has been successful on global scales; it is feasible to scale down one step further to regional levels.

Acknowledgments

The author thanks Chris Field and Inez Fung for fruitful discussions.

References

Bacastow, R., and Keeling, C. D. (1973). Atmospheric carbon dioxide and radiocarbon in the natural carbon cycle. II. Changes from AD 1700 to 2070 as deduced from a geochemical model. *In* "Carbon and the Biosphere" (G. M. Woodwell and E. V. Pecan, eds.), pp. 86–135. Proc. 24th Brookhaven Symposium in Biology, Upton, New York, May 1972. Available as CONF-720510 from National Technical Information Service, U.S. Department of Commerce, Springfield, Virginia.

Broecker, W. S., Takahashi, T., Simpson, H. J., and Peng, T.-H. (1979). Fate of fossil fuel carbon dioxide and the global carbon budget. *Science* **206**, 409–418.

Desjardins, R. L., Brach, E. J., Alvo, P., and Schuepp, P. H. (1982). Aircraft monitoring of surface carbon dioxide exchange. *Science* **216**, 733–735.

Dickinson, R. E., Errico, R. M., Giorgi, F., and Bates, G. T. (1989). A regional climate model for the western United States. *Climatic Change* **15**, 383–422.

Enting, I. G., and Mansbridge, J. V. (1989). Seasonal sources and sinks of atmospheric CO_2. Direct inversion of filtered date. *Tellus* **41B**, 111–126.

Francey, R. J., Robbins, F. J., Allison, C. E., and Richards, N. G. (1990). The CSIRO global survey of CO_2 stable isotopes. *In* "Baseline 88" (S. R. Wilson and G. P. Ayers, eds.), pp. 16–27. Bureau of Meteorology, Melbourne, Australia.

Fung, I., Prentice, K., Matthews, E., Lerner, J., and Russell, G. (1983). Three-dimensional tracer model study of atmospheric CO_2: Response to seasonal exchanges with the terrestrial biosphere. *J. Geophys. Res.* **88**, 1281–1294.

Gamon, J. A., Field, C. B., Bilger, W., Bjorkman, A., Fredeen, A. L., and Penuelas, J. (1990). Remote sensing of the xanthophyll cycle and chlorophyll fluorescence in sunflower leaves and canopies. *Oecologia* **85**, 1–7.

Gao, B.-C., and Goetz, A. F. H. (1990). Column atmospheric water vapor and vegetation liquid water retrievals from airborne imaging spectrometer data. *J. Geophys. Res.* **95**, 3549–3564.

Hansen, A. D. A., Conway, T. J., Steele, L. P., Bodhaine, B. A., Thoning, K. W., Tans, P., and Novakov, T. (1989). Correlations among combustion effluent species at Barrow, Alaska: Aerosol black carbon, carbon dioxide and methane. *J. Atmos. Chem.* **9**, 283–300.

Harris, J. M., and Kahl, J. D. (1990). A descriptive atmospheric transport climatology for the Mauna Loa Observatory, using clustered trajectories. *J. Geophys. Res.* **95**, 13,651–13,667.

Keeling, C. D., Bacastow, R. B., Carter, A. F., Piper, S. C., Whorf, T. P., Heimann, M., Mook, W. G., and Roeloffzen, H. (1989). A three-dimensional model of atmospheric CO_2 transport based on observed winds. 1. Analysis of observational data. *In* "Aspects of Climate Variability in the Pacific and the Western Americas" (D. H. Peterson, ed.), Amer. Geophys. Union, Washington D.C.

Knipling, E. B. (1970). Physical and physiological basis for the reflectance of visible and near-infrared radiation from vegetation. *Remote Sens. Environ.* **1**, 155–159.

Komhyr, W. D., Harris, T. B., Waterman, L. S., Chin, J. F., and Thoning, K. W. (1989). Atmospheric carbon dioxide at Mauna Loa Observatory. 1. NOAA GMCC measurements with a non-dispersive infrared analyzer, 1974-1985. *J. Geophys. Res.* **94**, 8533–8547.

Kroopnick, P. M. (1985). The distribution of ^{13}C of ΣCO_2 in the world oceans. *Deep Sea Res.* **32**, 57–84.

Pales, J., and Keeling, C. D. (1965). The concentration of atmospheric carbon dioxide in Hawaii. *J. Geophys. Res.* **70** 6053–6076.

Pearman, G. I., and Hyson, P. (1980). Activities of the global biosphere as reflected in atmospheric CO_2 records. *J. Geophys. Res.* **85,** 4468–4474.

Schlesinger, W. H. (1977). Carbon balance in terrestrial detritus. *Annu. Rev. Ecol. Syst.* **8,** 51–81.

Schlesinger, W. H. (1984). Soil organic matter: A source of atmospheric CO_2. *In* "The Role of Terrestrial Vegetation in the Global Carbon Cycle" (G. M. Woodwell, ed.), Chap. 4, pp. 111–127. Wiley, New York.

Seiler, W. and Conrad, R. (1987). Contribution of tropical ecosystems to the global budgets of trace gases, especially CH_4, H_2, CO and N_2O. *In* "Geophysiology of Amazonia" (R. E. Dickinson, edr.), Chap. 9, pp. 133–162. Wiley, New York.

Tanaka, M., Nakasawa, T., and Aoki, S. (1987). Time and space variations of tropospheric carbon dioxide over Japan. *Tellus* **39B,** 3–12.

Tans, P. P., Conway, T. J., and Nakazawa, T. (1989). Latitudinal distribution of the sources and sinks of carbon dioxide derived from surface observations and an atmospheric transport model. *J. Geophys. Res.* **94,** 5151–5172.

Tans, P. P., Fung, I. Y., and Takahashi, T. (1990). Observational constraints on the global atmospheric CO_2 budget. *Science* **247,** 1431–1438.

Trumbore, S. E., Keller, M., Wofsy, S. C., and Da Costa, J. M. (1990). Measurements of soil and canopy exchange rates in the Amazon forest using ^{222}Rn. *J. Geophys. Res.* **95,** 16,865–16,873.

Woodwell, G. M. (1978). The carbon dioxide question. *Sci. Am.* **238,** 34–43.

12

Forests in the Global Carbon Balance: From Stand to Region

Paul G. Jarvis and Roddy C. Dewar

I. Introduction

In the global carbon cycle, carbon is transported bidirectionally between the atmosphere and terrestrial ecosystems (vegetation and soils) and the oceans. The amounts of carbon estimated to be stored in the main compartments (atmosphere, terrestrial biota, soils, and oceans) are large. The annual fluxes between the compartments are also large, although net fluxes are much smaller (Fig. 12.1).

This cycle has been perturbed substantially by humans over the last 130 years through large changes in land use and through the burning of fossil fuels. These changes continue today at an accelerating rate. The resultant anthropogenically induced fluxes of carbon dioxide are very small in relation to the gross fluxes that occur naturally, but are large enough to modify the net fluxes, leading to an increase in the CO_2 content of the atmosphere at a current rate of 3.2 gigatonnes (Pg) of carbon per year (3.2×10^{15} g/year).

This increase in CO_2 content of the atmosphere has been the largest single contributor to the enhanced "greenhouse effect," cumulatively, about 55% of the total effect up to the present. It is, therefore, vital to know whether the atmospheric CO_2 content is likely to continue to increase proportionately with increases in the consumption of fossil fuels.

The current annual increase in CO_2 content of the atmosphere accounts for a little more than half the known current release of CO_2 through the burning of fossil fuels (5.7 Pg carbon per year). The fate of the remainder of this CO_2 released, in addition to any released in land-

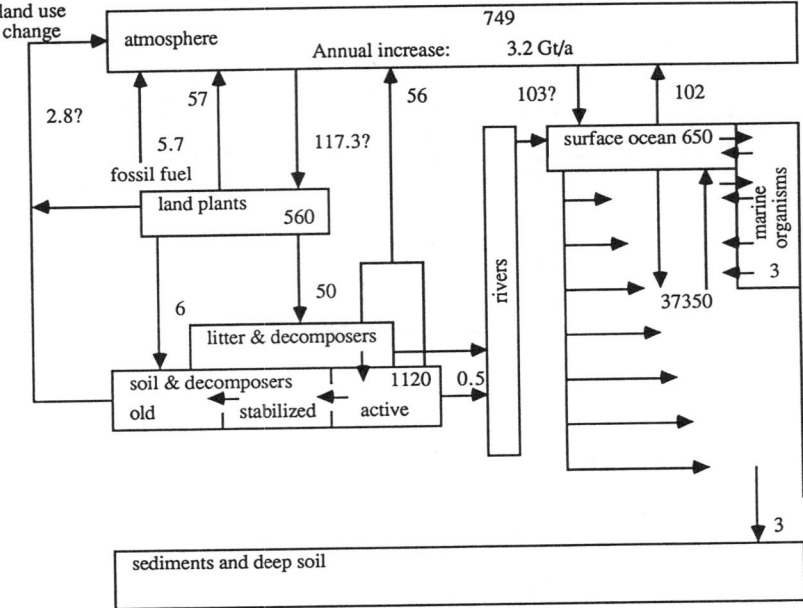

Figure 12.1 A diagram to show 1990 conceptions of the sizes of the carbon reservoirs (Pg) and of the carbon fluxes (Pg/year) in the global cycle. Adapted from King *et al.* (1987). The fluxes resulting from land use change (2.8 Pg/year), gross primary productivity of terrestrial plants (117.3 Pg/year), and ocean uptake (103 Pg/year) have been assigned artificially precise values consistent with Table 12.1 As the question mark indicates, these fluxes are, in fact, highly uncertain.

use changes, is poorly known. The airborne fraction of future CO_2 releases must depend on the continuing capacity of the sinks for CO_2 to take up a substantial part of the CO_2 released. To be able to evaluate this ability and, indeed, possibly manage the sinks to diminish the "greenhouse effect," we first must know the size and locations of the major sinks for CO_2 at the present time and, second, must interpret the nature of these sinks and predict their likely role in response to future changes in CO_2, land use, and climate.

There are, however, considerable uncertainties in our knowledge of the present magnitude and spatial distribution of the sinks for the anthropogenically produced CO_2, other than the atmospheric sink. These uncertainties arise from lack of atmospheric CO_2 data of adequate spatial resolution, so the sinks are poorly defined. Also, there is inadequate understanding of the physiological processes governing the responses of plants and soils to, for example, the fertilization effect of the rise in atmospheric CO_2 concentration, so the nature of the sinks is uncertain. The major sources arising from land-use changes are also poorly known because of inadequate survey data.

A tentative global carbon balance sheet is given in Table 12.1. It must be emphasized that the only quantities known with any degree of certainty are the annual fossil fuel emission and the increase in atmospheric content of carbon. The likely errors attached to the other terms are of the order of $\pm 100\%$.

II. Carbon Balance Concept

A. Anthropogenic Perturbation

Table 12.1 presents the global carbon balance in terms of anthropogenically induced perturbations to the natural carbon cycle. In this section, we discuss the carbon balance concept in more general terms to highlight problems in the definition of major sources and sinks for CO_2. The concept of a global carbon balance can be expressed as a statement of mass conservation,

$$\frac{\delta C_a}{\delta t} = \text{sources} - \text{sinks},$$

(1)

where $\delta C_a / \delta t$ (Pg/year) is the rate of increase in the atmospheric CO_2 content, expressed as a mass of carbon. We assume that, in the middle of the last century, prior to major land-use changes and the exploitation of fossil fuels, atmospheric CO_2 (C_a) was in a steady state, that is, $\delta C_a / \delta t \approx 0$ and the wide range of natural sources and sinks cancelled. In this assessment, we are concerned primarily with identifying enhanced sinks that are a consequence of the recent anthropogically induced sources, in terms of which Eq. (1) may be written

$$\frac{\delta C_a}{\delta t} = E + \Delta E - D - \Delta D.$$

(2)

The sources, E (emissions), formerly were cancelled by the sinks, D (deposition), prior to the start of the current increase in atmospheric CO_2 concentration (i.e., $E = D$); ΔE and ΔD represent perturbations to the steady state that may, for example, be the result of fossil fuel production and deforestation, on the one hand, and CO_2 and nitrogen fertilization, on the other. Therefore, Eq. (2) may be expressed entirely in terms of perturbations

$$\frac{\delta C_a}{\delta t} = \Delta E - \Delta D.$$

(3)

Table 12.1 Projected Annual Global Balance Sheet for Enhanced CO_2 Sources and Sinks in 1990

Sources	Carbon dioxide (Gt/year)	Reference	Sinks	Carbon dioxide (Gt/year)	Reference
Fossil fuels	5.7	Houghton et al. (1990)	Atmosphere	3.2	Houghton et al. (1990)
Tropical deforestation	2.1?[a]	Hammond (1990), Houghton (1991a)	Oceans	1.0?	Tans et al. (1990)
CO and CH_4 from burning vegetation and soil changes	0.7?	Enting and Mansbridge (1991)	Temperate and boreal forests	1.8?	Enting and Mansbridge (1991)
			Tropical forests and grasslands	2.5?	Enting and Mansbridge (1991)
Total sources	8.5?		Total sinks	8.5?	

[a] Values followed by ? are uncertain.

In practical terms, it is not always easy to judge whether a source or sink is enhanced currently relative to its strength 130 years ago. We are not, however, concerned with the large sources and sinks that were features of the unperturbed carbon cycle and are still functioning, as far as we can tell, much the same way as before.

B. Consideration of Temporal and Spatial Scale

The concept of a steady state, as just expressed, and, hence, the definition of significant perturbations, is dependent on the period of time that is considered. For example, over periods of a few months $\delta C_a/\delta t \neq 0$, even in the absence of major anthropogenic perturbations, because of the influence of the seasons on the assimilation and dissimilation of CO_2. On the other hand, over periods of 200 years $\delta C_a/\delta t$ may effectively be zero as far as deforestation is concerned, because the CO_2 released in a short burst of forest removal at the present time may be reassimilated subsequently by the following secondary forest over such a period.

In examining the role of the terrestrial biosphere in the global carbon balance, a year is a convenient period to consider because, in an unperturbed world, vegetation and soils are in a steady state on this time scale, assimilating and respiring CO_2 in approximately equal amounts over the year (i.e., $\delta C_v/\delta t = 0$, where C_v is the carbon content of the vegetation and soil). Most agricultural crops are CO_2 neutral in this way. In some cases, however, where the harvested product is put into storage (e.g., grain) or consumed over a longer period of time, it may be more desirable to think of a steady state over 3 or 4 years, or, in the case of shifting agriculture in the tropics, over 10 years.

Spatial scale also must be considered. In steady-state tropical forest, for example, trees die and gaps appear within which regeneration occurs. Over a large enough area, however, the standing crop of carbon remains effectively constant with time, that is, $\delta C_v/\delta t = 0$. Spatial resolution of the global carbon balance, as currently set by atmospheric and oceanic circulation models, is of the order of 1° latitude and longitude. Spatially separated sources and sinks also create problems in the identification of perturbations. For example, is the CH_4 produced in the tundra and oxidized to CO and CO_2 in the troposphere and on the surfaces of soils a new source, or has it been much the same for the past 130 years? Similarly, are changes in the ocean sink likely to have occurred in the last 25 years?

The importance of the productivity of tropical grasslands and the large amount of carbon stored in them, and the significance of the burning of these grasslands by humans has been stressed (e.g., Hall and Scurlock, 1991). In the course of a year, the grasses grow up, seed, die down, and decay in what is essentially a CO_2 neutral system. The rates of growth and oxidation of the biomass are unimportant, if the stocks of carbon are not changing. We are concerned with whether large changes occur in the amounts of carbon stored in the soil on a timescale of a year or more, as a result of CO_2 fertilization on the one hand or burning on the other. Only if CO_2 fertilization enhances the accumulation of organic matter in the soil or burning results in enhanced oxidation of carbon in the soil will there be change in the global flux (Crutzen and Andreae, 1990). In contrast, large-scale deforestation, with its associated burning, represents a very evident transfer of carbon to the atmosphere that will take several hundred years to replace, and is consequently an appreciable source of carbon on the relevant time scale. More difficult to evaluate is the long-term significance of short-term, small-scale perturbations such as wildfire disturbance in the Canadian boreal forest. Perturbations of this type give rise to episodic sources and sinks of carbon on the order of 0.1 Pg/year (Kurz *et al.*, 1991) on a natural time scale of tens of years, but the future source or sink status of these forests is highly sensitive to human- or climate-induced changes in the frequency of fires.

C. Changes in Land Use and Productivity

These considerations of source and sink enhancement and temporal and spatial scale are relevant in defining the sources and sinks that we need to include in establishing the present day balance sheet for anthropogenically produced CO_2. In considering the nature of these sources and sinks on land, we may distinguish two cases (Houghton, 1991b). The amount of carbon held in terrestrial ecosystems may change as a result of (1) changes in vegetation area resulting from human impacts on land use and (2) changes in the environment that affect the amount of carbon per unit area of land on which no land-use change has occurred (e.g., changes in climate, CO_2 fertilization). The former changes are easier to determine than the latter, because of the large differences in stocks of carbon between forests and the agricultural systems that replace them, and because land-use changes are, to some extent, documented. In contrast, changes in carbon per unit area are likely to be small relative to the stocks of carbon in vegetation and soils. If integrated over a large enough area, however, such changes may affect the net CO_2 exchange significantly. For example, a 1% change in global terrestrial carbon stocks would yield a global carbon flux of 17 Pg/year yet these stocks are thought to be known globally within only ±20% (Houghton *et al.*, 1985).

III. Methodology for Determining Enhanced Sources and Sinks

Four approaches currently are being applied to determine the spatial distribution of the major sources and sinks of atmospheric CO_2 of the present day CO_2 budget:

- prediction of surface sources and sinks on the basis of atmospheric CO_2 signals obtained from flask networks, through the inversion of an atmospheric transport model (inverse modeling)
- prediction of atmospheric CO_2 signals on the basis of known or hypothesized surface sources and sinks through use of an atmospheric transport model in the forward sense (forward modeling)
- direct measurement of net CO_2 exchange, usually coupled with process-based modeling to extrapolate over time (flux measurement)
- direct measurement and estimation of changes over time in the mass of stored carbon in particular terrestrial ecosystems (stock taking)

Inverse and forward modeling might be regarded as methods for generating specific hypotheses that would be tested directly by flux measurements and stock taking. Process-based modeling also plays a role in interpreting the nature of the sources and sinks and of predicting the likely CO_2 fluxes in a changed climate. The following four sections consist of a brief elaboration of these methods.

A. Inverse Modeling

There are three major networks for sampling the composition of the atmosphere at intervals of 1–3 weeks by collecting samples into flasks that are transported for analysis to a central point. These networks are the CSIRO/NERDDC network centered on the CSIRO Division of Atmospheric Research at Aspendale, Victoria, Australia; the SIO/UG network centered on the Scripps Institute of Oceanography at La Jolla, California; and the NOAA/GMCC network centered on Boulder, Colorado.

Between 20 and 30 sampling stations constitute each network. One advantage of a network of sampling stations is that all analyses for that network are made on the same machines so that systematic calibration and other instrument errors are minimized. As a result of flask exchange arrangements, some sampling stations are common to more than one network; this enables comparisons to be made between analyses made in the different networks.

The sampling stations were selected originally to give information on average background concentrations of gases, unaffected by local sources and sinks as much as possible. Thus, the stations tend to be located where they will receive air masses that have travelled for some considerable distances over the oceans. In the most recent transport models, 10 or 20 latitudinal zones are defined and some of the zones may not contain a sampling station. For the purposes of defining the latitudinal distribution of sources and sinks, these samples are now being regarded as representative zonal mean samples of air. It is clear that this assumption may not always be realistic and that more adequate sampling may be required in a zone.

A number of trace gases are analyzed at each center (e.g., Wilson and Ayers, 1990). Those relevant to this discussion are CO_2, ^{13}C in CO_2, ^{18}O in CO_2, O_2, CO, and CH_4.

Consistent records of good quality data from an adequate number of sampling stations over a sufficient time period exist only for CO_2. The differences among networks for CO_2 are small and data from two networks have been pooled in some inversions. Only the CO_2 data are currently suitable for inverse modeling, but inverse modeling probably will be possible with the other gases within the next 10 years.

Of the four main features of the concentration of each gas, all or some contribute to the inversion: an intraannual seasonal oscillation, an interhemispheric zonal variation, an interannual variation, and an annual trend.

To effect the inversion, these CO_2 data are filtered by fitting splines to them and then are combined with an atmospheric transport model to predict the surface distribution of sources and sinks. In the early days, one-dimensional models were used. Now two-dimensional models (latitude and altitude or pressure) have been used (Enting and Mansbridge, 1989, 1991; Tans *et al.*, 1989). It is anticipated that three-dimensional models including longitude will be used soon, but it seems likely that there will be insufficient flask sampling stations for this to yield a better definition of sources and sinks. The inversion can be validated using gases for which the sources and sinks are known, such as ^{85}Kr, CFCs, and radon.

The ^{13}C data have shown conclusively that the seasonal oscillation is the result of the seasonal cycle of assimilation and dissimilation of CO_2 by vegetation. However, the data for ^{13}C, O_2, and ^{18}O are still too preliminary for useful inverse modeling (Francey and Tans, 1987; Keeling, personal communication, 1991; R. J. Francey, personal communication, 1991).

B. Forward Modeling

An alternative approach is to run transport models in a forward sense to test surface source/sink hypotheses from a comparison of the simulated and observed atmospheric CO_2 distributions. For example, Fung *et al.*

(1983, 1987) assumed a distribution of vegetation types with a specified phenology and magnitude of assimilatory and dissimilatory processes and compared seasonal amplitudes of the predicted CO_2 concentrations at the flask sampling sites with measured amplitudes. Excellent agreement was obtained, giving support to the assumed source and sink distributions and to the transport model. Tans *et al.* (1990) used the three-dimensional tracer transport model of Fung *et al.* (1983), derived from the general circulation model (GCM) developed at GISS, to simulate the global distribution of atmospheric CO_2 in response to several surface source/sink hypotheses. Observational constraints from measured values of the north–south gradient in atmospheric CO_2 concentrations and of CO_2 partial pressure (pCO_2) in northern and equatorial oceans during the period from 1981 to 1987 were applied to aid selection of acceptable source/sink hypotheses.

Inverse and forward modeling methods (Fig. 12.2) each have their limitations (Enting and Newsam, 1990). A major limitation of the former is that small errors in the atmospheric CO_2 signal and in the transport model parameters can be magnified into arbitrarily large errors in the predicted source/sink distribution. Inversion of more realistic atmospheric transfer models (e.g., three-dimensional models instead of two-dimensional models) leads to less severe error amplification, in which source resolution is limited by availability, if not variability, in the data.

A related limitation of forward modeling is that it provides only a weak test of the source hypotheses. The problem is that errors in the sources are suppressed because of atmospheric mixing by a factor that is the reciprocal of the amplification factor by which inverse methods magnify errors when deriving sources from concentrations. Because spatial and temporal variations in pCO_2 are obtained less easily than comparable atmospheric CO_2 data, inverse modeling currently provides the best option for improving the definition of surface sources and sinks and is likely to continue to do so for some time as data and models improve.

C. Flux Measurements

The net CO_2 exchange of grasslands and forests can be measured by flux-gradient methods and by eddy covariance methods. These techniques involve placing instruments above vegetation and measuring the net flux of CO_2 across the plane (Fig. 12.3). In principle, hourly mean fluxes can be measured continuously for periods of a year or more. In practice, severe constraints are set by the instruments and logistics so data usually are obtained for only about one-fourth of that time. An appropriate strategy is, therefore, to develop a process-based model of the CO_2 flux of the vegetation in relation to environmental variables that will enable the net flux to be integrated for a year or more. The flux measurements

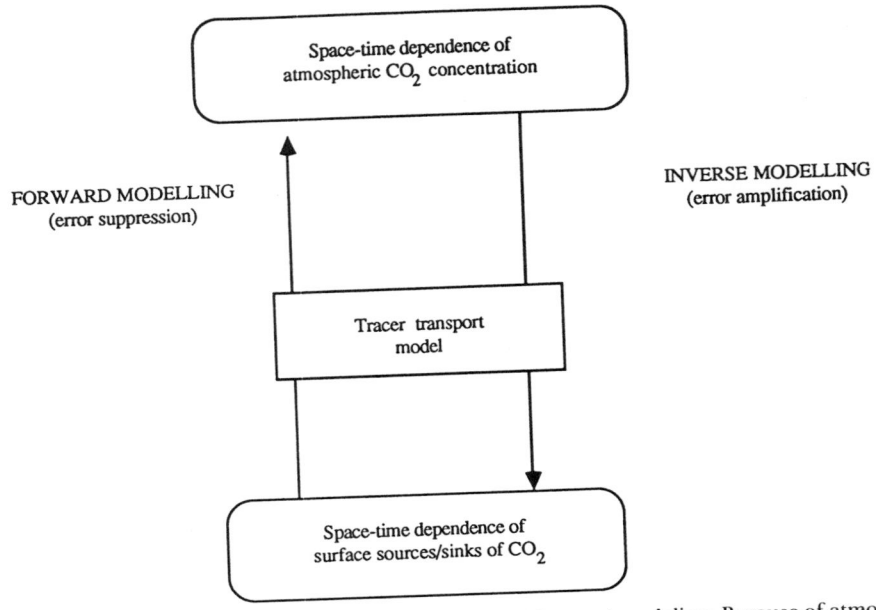

Figure 12.2   The difference between inverse and forward modeling. Because of atmospheric mixing in the tracer transport model, forward modeling suppresses errors in the data on CO_2 sources and sinks, so the predicted atmospheric CO_2 distribution is relatively insensitive to such errors. For the same reason, inverse modeling predicts CO_2 sources and sinks that are sensitive to small errors in atmospheric CO_2 data (see Enting and Newsam, 1990).

then are regarded as a test of the model. The latter approach has the added advantage that, if the model is parameterized appropriately, predictions can be made of the likely CO_2 fluxes in a changed climate.

In a region on the scale of a GCM grid square, the vegetation consists of a mosaic of components with different surface properties and CO_2 fluxes. Three approaches are used commonly to obtain estimates of CO_2 fluxes at this scale: (1) flux measurements can be made intermittently with aircraft-mounted instrumentation; (2) remotely sensed data can be used to define the components of the vegetation mosaic, and radiative data such as the simple ratio,[1] which essentially depends on the amount of chlorophyll present, can be correlated directly with the CO_2 flux; (3) fluxes measured or estimated on the stand scale can be scaled up to the regional scale.

To effect spatial integration over a region, an appropriate model of

[1] The ratio of surface reflectance for near infrared (0.75–0.88 μm) and visible (0.63–0.68 μm) wavelengths.

the planetary convective boundary layer is needed (e.g., McNaughton and Spriggs, 1986; King, 1991; McNaughton and Jarvis, 1991), as are the distribution of vegetation types and parameterizations of their CO_2 fluxes. The so-called "scaling-up" problem is an area of active research at this time.

D. Stock Taking

Direct measurements of the change in mass of stored carbon over a period of time are practical in even-aged crops and forest plantations but are generally impractical in heterogeneous areas of natural vegetation; the sampling problem is usually too great. Nonetheless, this approach may prove useful in certain well-defined types of vegetation, such as agricultural crops and tropical grasslands (Hall and Scurlock, 1991). There are

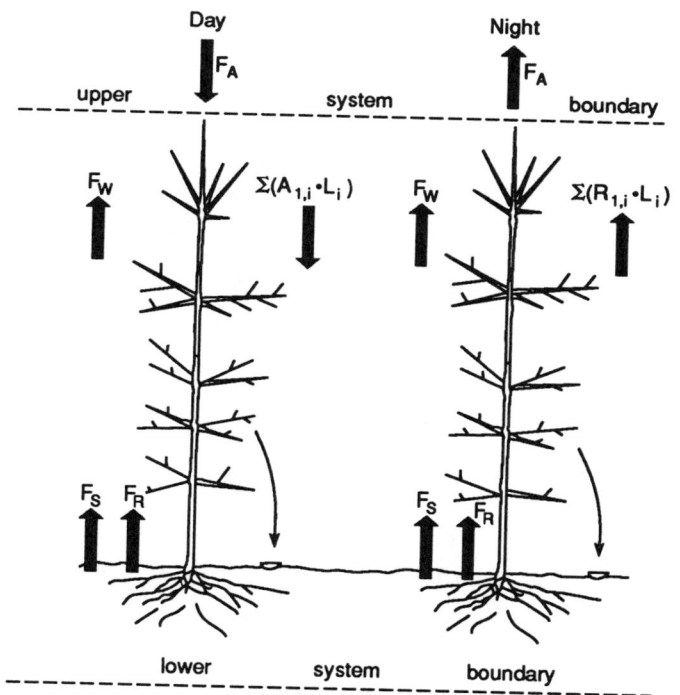

Figure 12.3   The limits of the system of interest. There is no CO_2 flux across the lower boundary and an influx or efflux across the upper boundary during the day or the night (F_A). Within the system is an assimilatory influx to all the leaves during the day, $\Sigma(A_{1,i} \cdot L_i)$, and an efflux from the leaves at night, $\Sigma(R_{1,i} \cdot L_i)$. There are also CO_2 effluxes in day and night from the respiring wood in stems, branches, and twigs (F_W), from the respiring roots (F_R), and from microbial respiration in the soil (F_S).

very evident difficulties in estimating the changes in carbon stored in soils (e.g., Prentice and Fung, 1990; Schlesinger, 1990). Tree ring analysis may provide a means of estimating mass changes in individual trees, but quantitatively scaling this up to heterogeneous forests raises problems.

IV. Current Enhanced Sources

A. Fossil Fuels

The current magnitude of the fossil fuel source of 5.7 Pg/year is reasonably accurately known from both the production and consumption of fossil fuel and cement (Rotty, 1987a). The spatial distribution of the source has been represented on a 5° grid by Marland *et al.* (1985); the temporal distribution has been represented by Rotty (1987b).

B. Burning Vegetation

Carbon monoxide (CO) originating from burning both fossil fuels and vegetation is oxidized in the atmosphere and on soil surfaces to CO_2. In addition, some methane (CH_4) is oxidized in the atmosphere to CO and contributes to this source. The magnitude and distribution of recent enhanced burning of native tropical grasslands and the consequent oxidation of soil organic carbon may contribute to the global CO_2 flux (Hall and Scurlock, 1991), but its magnitude is poorly known (D. O. Hall, personal communication, 1991). Enting and Mansbridge (1991) summarized previous estimates of the magnitude of this CO_2 source and take a middle value of 0.77 Pg/year for the non-fossil component. (The fossil fuel component is included already in the fossil fuel CO_2 source.)

The tropospheric and surface distributions of CO are given by Crutzen and Gidel (1983); the surface distribution of CH_4 is given by Matthews and Fung (1987; see Enting and Mansbridge, 1991).

C. Tropical Deforestation

The magnitude and distribution of tropical deforestation and other changes in land use in the tropics are much less well known (Hammond, 1990). Estimates by Houghton *et al.* (1987) of the net release of carbon to the atmosphere from deforestation in the tropics in 1980 range between 0.9 and 2.5 Pg. These estimates are based on rates of deforestation and carbon stocks of forests (FAO/UNEP, 1981) combined with a simple model of the history of carbon fluxes associated with different uses of the deforested land.

Since 1980, increases in the rates of tropical deforestation (Myers, 1989; Hammond, 1990) suggest that the net flux from these changes is currently nearer the higher end of the 1980 range. Houghton (1991a) estimated the net release of carbon from the tropics in 1989 as between 1.1 and 3.6 Pg with a more likely range of 1.5 to 3.0 Pg. This uncertainty results from inadequate information about (1) the area deforested per year, (2) carbon stocks within forests (including reductions resulting from thinning or degradation), and (3) the sources and sinks created by new land use (e.g., permanent or temporary clearance) (Brown and Lugo, 1984).

Uncertainties over these quantities are briefly elaborated next, following Houghton (1991a).

1. The FAO/UNEP study (1981) projected an average rate of deforestation in Brazil of 1.9 million ha/year for the period 1980 to 1985 for both closed and open forests. In contrast, Setzer and Pereira (1988) found the area in 1987 to be 8 million ha (3.2 million of which was closed forest) based on the number of fires and area of smoke recorded by satellite. On the other hand, Myers (1989) estimated the area to be 5 million ha for closed forests only, based on surveys. Another estimate, also based on satellite data, suggests an area of 2.1 million ha (Fearnside *et al.*, 1990). Thus, estimates for closed forests alone vary by a factor of two or more between low and high limits.

   Large differences exist between estimates of deforestation in other parts of the world, particularly in Asia. Globally, in 1980 the rate of tropical deforestation was about 11.3 million ha/year (FAO/UNEP, 1981). Almost 10 years later, Myers (1989) estimated an annual rate of tropical deforestation of about 21.5 million ha, an increase of 90%, but the preliminary update of FAO data gives an estimate of no more than 16.8 million ha (Lanly *et al.*, 1991). Both the past and recent estimates of deforestation were based on forest surveys, not on satellite data, and their accuracy is unknown. Given the possible unreliability of government returns, the use of satellite data offers the best method of reducing the uncertainty of estimates of changes in the area of world forests, although a number of technical difficulties exist at present, as evidenced by the difference between the Setzer and Pereira (1988) and Fearnside *et al.* (1990) estimates for the Amazon basin.

2. The estimates of carbon stocks within forests vary by almost 100% (Brown and Lugo, 1984). Three sources of uncertainty are (1) conversion factors from merchantable wood volumes to total carbon stocks, (2) the small areas (<30 ha) where carbon stocks have been sampled directly, and (3) the inclusion of thinned or degraded stands in the survey for wood volume. Greater accuracy will be achieved when disturbed and undisturbed stands can be identified,

and when the stocks of the forests actually cleared, not the average stocks, are obtained.

3. The practice of shifting cultivation, in which short periods of cropping alternate with long periods of fallow, is common throughout the tropics, and leads to a temporary release of carbon to the atmosphere followed by refixation during regrowth of secondary forest. Major uncertainties lie in the fraction of deforestation for shifting cultivation and, more importantly, in the length of fallow periods and the extent of secondary forest regrowth. Again, satellite imagery would determine whether, and how rapidly, deforested lands return to forests (see Cross, 1990).

D. Temperate and Boreal Deforestation

Deforestation in the temperate or boreal forests of Western Europe and North America largely came to an end after World War II (1945) and in many countries has been replaced by afforestation. Although there is continuing pressure from timber interests in the Pacific Northwest to harvest much of the remaining old-growth forest and replace it with plantations, with a consequent substantial loss of carbon to the atmosphere from the sites (Harmon *et al.*, 1990), the areas involved are small. Where afforestation has been practiced widely, as in the United Kingdom, carbon is being sequestered by extensive new young forests with a doubling time of about 20 years (Jarvis, 1992), but the area involved is small on a global scale. The extend of deforestation, afforestation, and forest growth across the temperate/boreal zone, especially Russia, must be known but is not readily available. Consequently, we cannot determine whether there is a current temperate/boreal forest CO_2 source resulting from deforestation of any significance, but it seems unlikely.

V. Current Enhanced Sinks

A. Increase in the Atmospheric Concentration

The trend in the mean annual atmospheric CO_2 concentration is obtained by demodulating the seasonal oscillations at a site over a series of years. Although Conway *et al.* (1988) arrived at a median rate of increase of atmospheric CO_2 of 1.23 ppmv/year for the flask sampling locations in the NOAA/GMCC network from 1981 to 1984, there was a wide range from 0.86 to 1.65 ppmv/year. A figure of ~1.5 ppmv/year for the period from 1980 to 1985 was widely accepted, but that estimate has risen to ~1.8 ppmv/year for 1990 (Houghton *et al.*, 1990). An increase of 1.5 ppmv/year corresponds to a rate of increase of atmospheric carbon of 3.185 Pg/year (1 ppmv $\cong$ 2.123 Pg).

B. Controversial Ocean Sink and the Tans Hypothesis

The magnitude and, to a lesser extent, the distribution of the ocean sink is a matter of controversy. A large variation exists in the estimates of the global net uptake of CO_2 by the oceans using models that take into account air–sea gas exchange, aqueous carbonate chemistry, and transport from the surface to deep ocean layers (Table 12.2). The three-dimensional ocean GCMs, however, tend to predict lower oceanic uptake than the earlier box models, and there appears to be some convergence toward the smaller estimates based on direct observations of oceanic CO_2 partial pressures (pCO_2) (Takahashi, 1989; Tans et al., 1990). Three-dimensional ocean circulation models, which take account of surface water biological activity and can be coupled to atmospheric circulation models, are required to study realistically the feedbacks between climate change and oceanic uptake of CO_2 (M. Fasham, personal communication, 1990), although such models need to be constrained by more data than are currently available.

Tans et al. (1990) put forward evidence to suggest that the oceans are currently taking up no more than 0.5–1.0 Pg carbon per year. Their estimates were based on forward modeling atmospheric CO_2 concentrations using a three-dimensional atmospheric transport model (Fung et al., 1983) and estimates of fluxes of CO_2 across the atmosphere–ocean boundary that were derived from actual measurements of (pCO_2) coupled with assumptions about the transport coefficients across the atmo-

Table 12.2 Projected Enhanced CO_2 Fluxes into the Ocean for the 1980s Based on Different Methods[a]

Method	Reference	CO_2 flux (Gt/year)
Box-diffusion model	Siegenthaler (1983)	2.4
Outcrop-diffusion model	Siegenthaler (1983)	3.6[b]
Ocean GCM	Maier-Reimer and Hasselmann (1987)	2.0[c]
Ocean GCM	Maier-Reimer (1990)	1.2[d]
Ocean GCM	Sarmiento et al. (1992)	1.9
pCO_2 data	Takahashi (1989)	1.6
pCO_2 data and 3-D tracer transport model	Tans et al. (1990)	< 1.0

[a] Table after Wigley (1991).

[b] Probably an overestimate because of assumed instantaneous exchange between high-latitude surface waters and the deep ocean.

[c] From inversion of the global atmospheric CO_2 concentration (Wigley, 1991) using this ocean general circulation model (GCM).

[d] From a revised version of the Maier-Reimer and Hasselmann (1987) model, cited by Watson et al. (1990).

sphere–ocean boundary (see Section III,B). They concluded that (1) the global oceanic carbon sink is no larger than 1 Pg/year and (2) terrestrial ecosystems in the northern temperate zone are a net sink of between 2 and 4 Pg/year.

The Tans *et al.* hypothesis is of considerable significance to our current interest in the role of the terrestrial biota. If the net oceanic uptake of carbon is less than 1 Pg/year, the atmospheric carbon budget must be balanced by a new global uptake of carbon of at least 3 Pg/year into the terrestrial biota.

The Tans *et al.* (1990) hypothesis is supported in part by two independent inversions of the variations in atmospheric CO_2 concentration using two-dimensional atmosphere transport models (Enting and Mansbridge, 1989, 1991; Tans *et al.*, 1989) to yield the net surface sink as a function of latitude. Figure 12.4 shows a substantial sink in mid-latitudes of the southern hemisphere, which is attributable to the oceans.

On the other hand, the Tans *et al.* hypothesis is at variance with views developed from the circulation of bomb-created ^{14}C in the oceans and with preliminary attempts to invert the [13]CO_2 atmospheric signal (Roelof-fzen *et al.*, 1990). At present, the conflict with the ^{14}C data can be resolved only by assuming change with time in CO_2 uptake by the oceans or by ambiguities in the calibration of global carbon cycle models (Enting, 1990). As far as interpretation of the [13]CO_2 data is concerned, no firm

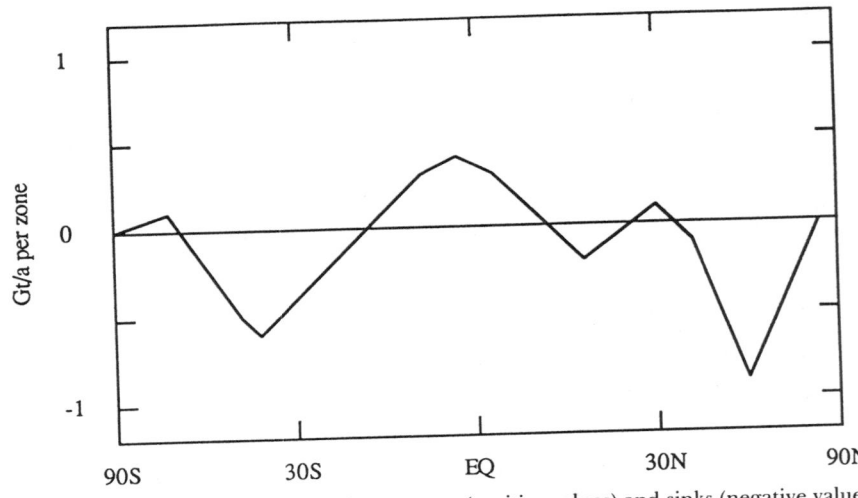

Figure 12.4 Nonfossil net surface sources (positive values) and sinks (negative values) of atmospheric carbon in Pg/year per zone of $18 \times 10^{12} m^2$, calculated taking the role of CO into account. Figure reprinted, with permission, from Enting and Mansbridge (1991).

conclusions can be drawn at this time. Longer runs (another 10 years) of data and additional sampling sites are needed, particularly in the zone 37° to 54°N, to resolve present discrepancies in the data.

Although inversion of the atmospheric CO_2 signal has inherent limitations (see, for example, Taylor, 1989), agreement between the recent inverse modeling of Enting and Mansbridge (1991) and the forward modeling of Tans *et al.* (1990) and Fung *et al.* (1987) is impressive. We therefore accept the conclusions of Tans *et al.* (1990) as a working hypothesis.

Acceptance of the Tans *et al.* (1990) hypothesis leaves a clearly defined "missing sink" in the global CO_2 balance at present that, we may conjecture, is found in the terrestrial biota. This conjecture is consistent with the analysis (Houghton, 1991b) of changes in global fluxes over the last 135 years, to be discussed in Section VI.

C. Terrestrial Biotic Sinks

Deconvolution of the atmospheric CO_2 signal through inversion of a two-dimensional atmospheric transport model provides a strong indication of the location and magnitude of these sinks. Figure 12.4 shows a strong mid-latitude northern hemisphere sink in the three zones between latitudes 37° and 64°N, peaking between 44° and 53°N. A similar strong mid-latitude northern hemisphere sink is implied by the results of Enting and Mansbridge (1989) and Tans *et al.* (1989) using similar methods, and by the three-dimensional forward modeling studies of Keeling *et al.* (1990). The area of ocean at these latitudes and the differences in (pCO_2) between ocean and atmosphere are too small for this sink to be wholly oceanic. Allowing for some uptake by the ocean, of less than 1 Pg/year (Tans *et al.*, 1990), we may therefore hypothesize from Enting and Mansbridge (1991) that the trees and soils of the temperate/boreal forest are assimilating CO_2 at a rate of up to 2 Pg/year, perhaps as a result of atmospheric CO_2 and nitrogen fertilization by both wet and dry deposition and reafforestation.

Figure 12.4 also indicates a small net annual mean source in the tropical equatorial zone between 12°S and 12°N of 0.615 Pg carbon per year (Enting and Mansbridge, 1991). This net source is the result of release from the ocean, release as a result of both deforestation and cultivation of tropical grasslands, and possible enhanced uptake by tropical forests and grasslands, again as a result of atmospheric fertilization or regrowth of forests on previously cleared land.

If we accept the Tans *et al.* (1990) figures of 0.80–1.32 Pg/year for ocean release of carbon in this zone as realistic, and take a middle value of 1.06 Pg/year, the terrestrial biota must be a net sink of at least 0.44 Pg/year (1.06–0.62). Since we have postulated already that deforestation is releasing carbon at a rate of ~2.1 Pg/year (Table 12.1), we are left with the conclusion that the tropical forests and grasslands of the world in the equatorial zone between 12°S and 12°N are taking up CO_2 from the atmosphere at a rate of about 2.5 Pg/year (2.1 ± 0.4) (Enting and Mansbridge, 1991), that is, there is also a major equatorial biotic sink.

Some direct evidence for support of this hypothesis is available. During the wet season of 1988, Fans *et al.* (1990) measured the net exchange of CO_2 above the Amazon forest near Manaus continuously for 12 days. From this small database they established a relationship between net CO_2 flux and solar radiation that enabled them to estimate the net CO_2 uptake over a period of 50 days in April and May. They concluded that the net uptake of CO_2 was 0.25 kg/(ha · hr) and that, if this amount were scaled up over the entire Amazon Basin for a year, the total CO_2 uptake would amount to 1.2 Pg/year. Taking into account the distribution of tropical forest and other vegetation elsewhere in this equatorial zone, a total uptake of carbon of 2.5 Pg/year for the equatorial zone is not conceivable. Nonetheless, the measurements made by Fan *et al.* (1990) cover only a part of the year and do not include the dry season, which is significant over much of the Amazon Basin and elsewhere in the tropics. Further, as these investigators point out, the fluxes measured were close to the detection limits of their equipment (eddy covariance using a closed path NDIR CO_2 analyzer). Clearly there is a requirement for a more comprehensive measurement program to establish whether tropical forests and grasslands are a significant sink for atmospheric CO_2 on a yearly basis.

VI. Historical Trend of the Global Terrestrial Sink

As discussed earlier, the net flux of CO_2 from terrestrial ecosystems to the atmosphere has two components: a flux resulting from changes in the area of different land types and a flux resulting from the response per unit area of ecosystems of particular type and extent to changes in CO_2, temperature, nitrogen deposition, and other environmental variables. Houghton (1991b) explored the difference between the net global, nonfossil flux of carbon from terrestrial ecosystems to the atmosphere, based on inversions of ice core data by Siegenthaler and Oeschger (1987),[2]

[2]Inversion of atmospheric CO_2 concentrations using global carbon models gives estimates of the net flux of carbon to or from the atmosphere. When the contribution of fossil fuels is subtracted from this net flux, the residual is the nonfossil flux and, when the oceans are taken into account, the net terrestrial flux of carbon to the atmosphere. Two estimates of this net terrestrial flux were determined by Siegenthaler and Oeschger (1987) using two different models of oceanic circulation; a box diffusion model and an outcrop model. The outcrop model simulates direct exchange of carbon between the atmosphere and deeper, colder waters and estimates absorption of more carbon than the box diffusion model. Hence, the outcrop model accommodates a larger biotic source.

and the most recent estimate of the flux arising from change in land use for the period 1850 to 1985 (Houghton and Skole, 1990). The difference between these two terrestrial fluxes may indicate a change in terrestrial carbon stocks per unit area in response to changes in CO_2, N, and temperature, but could also result from changes in the absorptive capacity of oceans. In Fig. 12.5, the pattern of this residual flux or "missing" flux of carbon over the last 135 years shows an increasing global sink for carbon from the start of the 20th century to a maximum of about 1.8 Pg/year in 1975, followed by a decrease to about 1.5 Pg/year by 1985. The estimate of the missing flux of carbon depends on which ocean model is used in the inversion of the ice core data (Siegenthaler and Oeschger, 1987) and on the range of estimates of change in tropical biomass used in deriving the land-use flux (Houghton, 1991a), but all cases show the same temporal pattern.

Although it is not possible to assign a cause of the apparent trend in the missing flux of carbon directly from this analysis, 85% of the variance over the 135-year period can be explained by variation in the atmospheric CO_2 concentration. Between 1940 and 1980, variations in global temperature and atmospheric CO_2 concentrations explained 89% of the variance in the missing flux, with physiologically reasonable regression coefficients.

The post-1975 decrease in the missing sink carbon is not inconsistent with a short-term warming-enhanced increase in respiration. However, if CO_2 fertilization leads to enhanced N uptake and sequestration in trees, litter quality and, hence, decomposition may decline in the medium term.

VII. Carbon Dioxide Fertilization

A. Hypothesis

The analysis of Tans *et al.* (1990) suggests a sink of 2–4 Pg/year in the northern temperate zone for the years 1981 to 1987. The global sink determined by Houghton (1991b) is at most 1.5 Pg/year, but could be larger in the temperate zone if offset by a net source elsewhere. It is important to recall that the estimate of land-use flux includes accumulation of carbon on land from regrowth of logged forests or from reforestation of abandoned agricultural lands. Therefore, the temporal pattern of the global biotic sink (Houghton 1991b) adds weight to the hypothesis that the present-day northern and equatorial terrestrial sinks, identified by spatially resolved inverse modeling (Tans *et al.*, 1990; Enting and Mansbridge, 1991) are, in part, a result of CO_2 fertilization.

In Sections IV and V we reviewed the status of knowledge about the present-day spatial distribution of sources and sinks of CO_2. We have

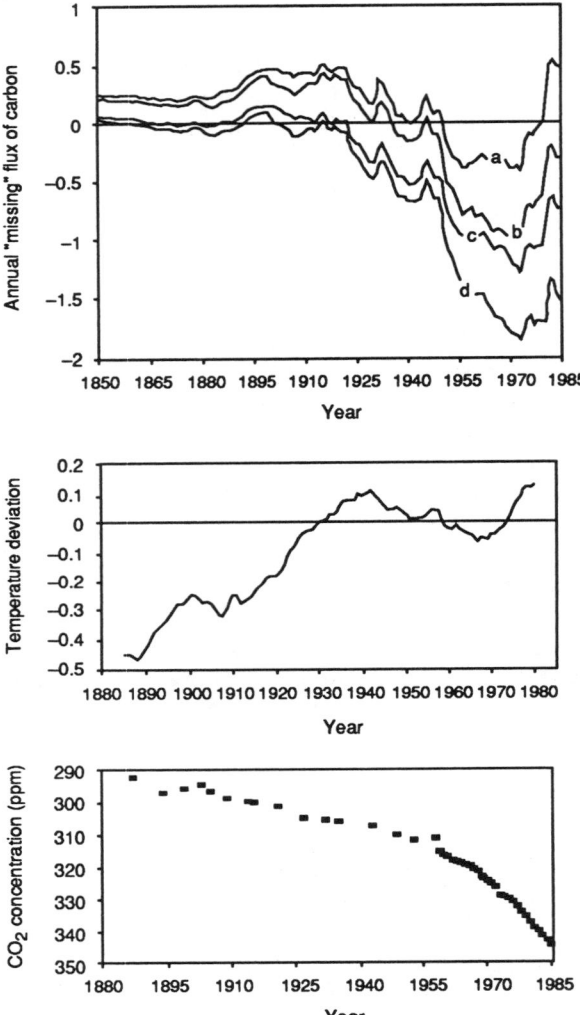

Figure 12.5 Top: The difference between the net biotic flux as determined from inversion of CO_2 concentrations (Siegenthaler and Oeschger, 1987) and the land-use flux. The curves are based on the following differences: (a) outcrop-low biomass; (b) outcrop-high biomass; (c) box diffusion-low biomass; (d) box diffusion-high biomass (most likely). Negative values indicate a biotic sink not accounted for by land-use change; positive values indicate a biotic source in addition to that from land-use change. Middle: Mean global surface temperature derivations from the 1951–1980 baseline (Hansen and Lebedeff, 1987). Bottom: Concentration of atmospheric CO_2 (from Siegenthaler and Oeschger, 1987), plotted with higher concentrations lower so the overall trend parallels the trend in the missing flux (*top*). Figure reprinted, with permission, from Houghton (1991b).

suggested already that enhanced growth of terrestrial ecosystems, stimulated by CO_2 fertilization and eutrophication, and the regrowth of deforested areas are potentially significant terrestrial sinks for CO_2. If CO_2 fertilization is operating in northern latitudes, there is no intuitive reason that the same mechanism is not operating in the tropics. Enting and Mansbridge (1991) point out that, as a proportion of biomass, their estimate of an equatorial sink is much smaller than the northern temperate sink of Tans *et al.* (1990); this result may reflect real differences between the physiological response of CO_2 fertilization in the two regions. However, afforestation may account for some of the CO_2 uptake in northern mid-latitudes also. We would emphasize that mechanistic interpretation of sinks is crucial for the projection of future increases in atmospheric CO_2 concentrations.

B. Evidence

There is a wealth of direct evidence, from many CO_2 enhancement experiments, of a positive response of plant growth and productivity to increased ambient CO_2 concentrations, at least in the short term (e.g., Acock and Allen, 1985; Eamus and Jarvis, 1989). On the basis that CO_2 fertilization may lead to a 0.5% increase in net primary productivity per 1% increase in atmospheric CO_2, Goudriaan (1990) used a compartment model of the global carbon cycle to conclude that CO_2 fertilization could sequester about 1.5 Pg carbon per year globally. Although much work needs to be done to determine quantitatively the long term effect of CO_2 fertilization on the amount of carbon stored in different ecosystems, indirect evidence for a significant CO_2 fertilization effect has been inferred from inversions of current and past atmospheric CO_2 concentrations with atmospheric transport and global carbon models. We now require stand-scale, process-based models of forest and grassland ecosystems to justify and explain these inferences.

C. Process-Based Terrestrial Ecosystem Models

Models that represent the biological processes underlying the responses of terrestrial ecosystems to the environment have been used to investigate CO_2 fertilization and its implications from a mechanistic standpoint. The GEM forest ecosystem model (Rastetter *et al.*, 1991a), which represents carbon and nitrogen cycling in trees and soils, has been run using historical CO_2 concentration, nitrogen deposition, and global temperature data. Preliminary results on net ecosystem CO_2 exchange (E. B. Rastetter, personal communication, 1991) appear to be qualitatively consistent with the trend in the 135-year record of the missing global carbon sink (Figure 12.5) constructed by Houghton (1991b). Using the Hurley grassland ecosystem model, Thornley (Thornley and Verberne, 1989; Thornley *et al.*, 1991) estimated that the post-1850 increase in CO_2 could lead to a 30% increase in the soil carbon density of nitrogen-fertilized pastures, because of CO_2-stimulated increase in the litter input. The implications from both models is that CO_2 fertilization may lead to enhanced sinks whose sizes depend on soil nutrient status and temperature also.

VIII. Moving Forward

The role of terrestrial ecosystems, and forests in particular, in the global carbon cycle is uncertain. The removal of this uncertainty is a prerequisite for predicting the future response of atmospheric CO_2 concentrations to continuing anthropogenic perturbations and the impact on terrestrial biota. There is a need for a close interaction between the development of atmospheric models, biotic models, and new measurement programs over a range of spatial scales to remove current uncertainties. We see this need for enhanced research activity in four main areas:

- improvement of the database on changes in land use and area of forests and grasslands in the temperate/boreal and tropical zones
- measurement of changes over time in carbon stocks in significant ecosystems, particularly boreal, temperate and tropical forests, tropical grasslands, and tundra
- measurement of net influxes/effluxes of carbon in the same significant ecosystems
- development of process-based models of CO_2 balance on site, biome, regional, and GCM scales to investigate sensitivity to CO_2 and nitrogen fertilization and changing temperature

These activities are elaborated in the following sections.

A. Database Improvement

In both the temperate/boreal and equatorial zones, a much better definition is required of changes in zonal land use and, within the zonal mosaic of vegetation types, of carbon sequestration. In both zones, we need to know the extent of different vegetation types and their current rates of net carbon sequestration in both plant and soil. In the temperate/boreal region, this would best be achieved by collaboration with the United States, Canada, and the Independent Commonwealth of Soviet States, especially the last, to obtain data on the distribution and recent changes in area and growth rates of the forests. In the equatorial zone, collaboration also will be needed to obtain reliable data on changes in land use, particularly with respect to the areal extent of deforestation, new plantations, and native grasslands. A GIS approach for these areas would be an effective way forward to define and evaluate land-use changes.

B. Changes in Carbon Stocks

Inversions of atmospheric CO_2 variations from the NOAA/GMCC and CSIRO/NERDDC networks have led to strong support for the earlier, more intuitive hypotheses that the temperate/boreal forests in the Northern hemisphere are a major sink for CO_2 (Section V,C). The same approach also has led to the hypothesis that the equatorial belt of tropical forests and grassland is a CO_2 sink of more than sufficient magnitude to cancel out the combined sources of oceanic upwelling and deforestation. In combination with database improvement, hypotheses with respect to particular ecosystems, such as tropical grassland, can be tested directly by measuring changes in soil and vegetation carbon stocks. This approach is, however, more difficult to follow in heterogeneous, diverse forest vegetation.

C. Carbon Dioxide Flux Measurements

In both the equatorial and temperate/boreal zones, there is a clear need to make direct measurements of the net CO_2 flux of the main vegetation types, particularly in forests, to obtain a net CO_2 balance over a full year. Simultaneous measurements of CO_2 and water vapor flux are now practical on a regular basis using new eddy covariance technology with both open and closed path systems of NDIR gas analysis. Where logistically practical, such measurements should be made over a full year or more, as at the Harvard Forest (S. C. Wofsy, personal communication). When it is logistically impractical to measure fluxes for periods of more than a few weeks at a time, the seasons should be sampled adequately. Such measurements provide the only information on how an active stand of vegetation is functioning on a time scale short enough to allow processes to be evaluated and reasonably mechanistic models to be made for interpolation, extrapolation, and prediction.

D. Coherent Suite of Models over a Range of Scales

Development of models over a range of scales is integral to forecasting the likely effects of global change on ecosystems and the feedback of ecosystems onto global change. We discuss next the need for the following coordinated enhancement of modeling activity:

- development of generic (i.e., globally relevant) process-based models of forest, grassland, and tundra ecosystems on the scale of individual sites or stands (stand scale modeling)
- spatial integration of site-level models (biome and regional scale modeling)
- process-based modeling of atmosphere–plant–soil interactions at the GCM grid scale (GCM scale modeling)
- inverse modeling on a global scale (global scale modeling)

1. Stand Scale Modeling Fundamental uncertainties exist in our understanding of the movement of carbon between atmosphere, vegetation, and soils on local scales. Process-based models are needed to (1) integrate current understanding of photosynthesis, maintenance respiration, allocation, growth, nutrient uptake, plant–soil water relations, and soil decomposition in relation to the environment and (2) generate testable hypotheses of whole-system responses to global change, including CO_2 and nitrogen fertilization and increasing temperature. Close interaction with manipulative laboratory experiments is required to refine and test process submodels for (1). Interaction with field experiments on globally relevant systems is required for (2). There are many existing forest and grassland models (ITE, HURLEY, MAESTRO, BIOMASS, FOREST-BGC, GEM, VEGIE, CENTURY, FORGRO). Some of these models describe important physiological processes in an empirical fashion and are of limited use in prediction beyond the data sets with which they are parameterized. Thus, we see a need for further development of current process-based models, in parallel with purpose-designed experiments, to improve understanding of relevant processes and to enable prediction of future responses to a changing climate.

2. Biome and Regional Scale Modeling On larger scales the averaging of surface properties involves unsolved problems. The great diversity of soil and vegetation types and growing conditions, on the larger scale, may cause differences in the atmospheric transport processes over different parts of the landscape. Spatial differences in roughness, albedo, and canopy conductance of the vegetation cause differences in the turbulent transport processes and may cause mesoscale circulations to develop within the convective planetary boundary layer (PBL).

In principle, the regional scale CO_2 flux can be obtained as the area-weighted sum of the fluxes from the different vegetation types in the region. Since the fluxes from the diverse vegetation types are likely to have been measured on only a few occasions, a model must be used to provide estimates of the fluxes as functions of weather and season.

King *et al.* (1989) have derived regional-scale exchanges of CO_2 by spatial extrapolation from smaller-scale ecosystem-specific models. Each ecosystem model is run by sampling from a distribution of input parameters that reflects the spatial heterogeneity of climate and soil type within each biome across the region. The CO_2 flux for each biome is the product of the mean of the model CO_2 flux and the area of that biome. Predicted regional fluxes are the sums of the biome fluxes. King *et al.* (1989) have applied this technique to predict the net seasonal CO_2 flux from coniferous forest and tundra ecosystems within the 64°N to 90°N latitude belt; results from the extrapolation compare favorably with independent estimates of seasonal CO_2 exchange and net primary production. In principle, this method can be applied at any scale and predictions tested at each stage using CO_2 flux measurements. The validity of the area-averaging scheme used can be tested by inverse modeling of large scale CO_2 fluxes measured with aircraft to give estimates of the fluxes from the component vegetation types and effective surface conductance. In practice, a major challenge is the assembly of databases necessary for defining the input distributions. Preliminary sensitivity analysis can be used to select a subset of parameters that has the greatest influence at the regional scale.

Although direct summation of fluxes measured from the range of vegetation types must yield a valid estimate of the regional flux, a different situation exists if the fluxes must be modeled because, for example, the measurements are confined to a few occasions for resource and logistic reasons. There is a major difference between model estimates of regional-scale fluxes and fluxes on the smaller scales. First, simulated CO_2 flux at a site using mean parameter values can differ significantly from the regional mean CO_2 flux calculated from a distribution of parameter values (King, 1991; Rastetter *et al.*, 1991b). This result is almost always the case for nonlinear systems, and illustrates that no single site can be identified as representative of a region when nonlinear responses between plant processes and environment are involved. Spatial scaling techniques must be used to identify the relevant small-scale complexity needed for prediction of large-scale responses; premature simplification can lead to misguided assessment of large-scale behavior (Luxmoore *et al.*, 1991).

Second, for patches of vegetation with a dimension of a few hundred meters, the driving variables for the exchanges of heat, water vapor, and CO_2 can be regarded as externally set and independent of the vegetation. However, at the larger regional scale, these variables can no longer be treated as independent because the heat, water vapor, and CO_2 exchanged at the ground surface are contained within the PBL and feedback to influence the fluxes at the surface. Photosynthesis depletes the CO_2 content and the fluxes of heat and water vapor from the surface warm and humidify the boundary layer. In addition, they drive its growth during the day by eroding the base of the capping inversion and entraining warmer, drier air that is richer in CO_2 from above (McNaughton, 1989). In moving from stand to regional scale, the small-scale systems become part of the large-scale system and new negative feedback pathways are introduced. At the regional scale, these feedbacks can be taken into account only in the context of a three-dimensional PBL model that takes mesoscale circulations into account. Such a soil–vegetation–atmosphere transfer scheme (SVATS) may be self-standing or part of a

GCM (Mascart *et al.*, 1991), creating a need for the development of area-averaging schemes for surface properties and fluxes within the framework of a PBL model.

3. GCM Scale Modeling   Long-term forecasts of global change made with GCMs must include the effects of dynamic feedbacks between climate and the terrestrial biosphere. On the one hand, the atmosphere drives vegetation and soil dynamics through changes in rainfall, temperature, solar radiation, humidity, CO_2 concentration, wind speed, and deposition of pollutants. On the other hand, vegetation and soils define the terrestrial boundary condition in atmospheric GCMs through modification of surface roughness, albedo, and exchanges of CO_2, CH_4, and CO, and thus modify substantially the partitioning of latent and sensible heat, so the inclusion of CO_2 exchange and carbon cycle models in GCMs is essential.

GCMs inherently run on fine time scales (hours). A major challenge in modeling climate–vegetation feedbacks is to incorporate processes over a wide range of time scales, from water, energy, and gas exchange on fine time scales (less than days) to successional changes in vegetation on long time scales (years to centuries) (Fung *et al.*, 1991).

A further challenge is to address disparities in spatial scale. Atmospheric GCMs currently operate on spatial scales on the order of 100–1000 km. Much higher spatial resolution is needed for terrestrial climate variables, most critically for precipitation (10–50 km). A reciprocal challenge exists in the development of process-based vegetation models at the GCM scale, when most information on mechanisms is limited to the scale of individual leaves, seedlings, and, rarely, whole plants. Regional scale modeling is a necessary link between stand-scale and GCM-scale modeling, whereby ecologically sound interactive vegetation models at the GCM scale are developed by integrating smaller scale processes.

Although premature simplification of interactive vegetation models could give misleading results on the GCM scale, there is a positive role to be played by "toy" models developed directly at the GCM scale that represent, in a simplified manner, the essential attributes of key components and processes (Parton *et al.*, 1991; Raich *et al.*, 1991). These models can be viewed as an important step in building more realistic ecological models of the terrestrial biosphere, by which specific hypotheses can be explored more readily. Such models would identify the most important ecophysiological processes that need to be modeled with finer resolution. This top-down approach complements the bottom-up approach of regional-scale modeling.

In developing an interactive vegetation model, there are two possible approaches to the classification of vegetation types. In the first, a taxonomic approach is adopted based on biome classification of present-day land cover (desert, temperate forest, savanna, grassland, tundra, tropical forest). In the second approach, vegetation and soil functional types are defined in terms of characteristics that determine their interaction with the atmosphere (evergreen, deciduous, annual, perennial). If changes in extent and type of biomes are to be considered, then, for ecosystems in which change is predominantly made by humans, the first approach may be more appropriate in a scenario-based description of land-use change, whereas the second approach is more relevant to climate-driven changes, under which present-day correlations between biome type and environment may no longer hold. Functional classification is also a more rigorous, process-based approach to parameterizing interactive models that describe the response of biomes of a given type and extent to changes in climate.

4. Global Scale Inverse Modeling   Inversion of atmospheric signals provides a powerful way forward for the definition of sources and sinks as the database and models improve. Three-dimensional transport models, with much greater resolution, are likely to be developed in the immediate future, but will be limited seriously in their application by inadequacies in the existing flask sampling networks.

IX. Conclusion

It is evident that a substantial, coherent, integrated program of measurement and modeling is needed over a range of scales to identify and explain the role of forests in the current perturbed global carbon balance, and to make predictions about the likely future progression of both atmospheric CO_2 and forest sustainability. The largest impediment to answering questions raised here by measurement and modeling over a range of scales is likely to be the availability of intellectual resources. We need more young people with the appropriate skills tackling these questions which means we need an immediate enhancement and redirection of educational resources to coordinated training in biotic and atmospheric processes on scales varying from the leaf to the globe. We need a new breed of scientist. At present, it is not clear from where it will come.

Acknowledgments

We thank M. J. Apps, D. A. Bennetts, W. R. Emanuel, M. R. J. Fasham, I. G. Enting, R. J. Francey, G. I. Pearman, D. O. Hall, M. H. Unsworth, J. H. M. Thornley, I. Y. Fung, R. A. Houghton, A. W. King, W. A. Kurz, R. J. Luxmoore, E. Matthews, R. V. O'Neill,

W. M. Post, K. C. Prentice, E. B. Rastetter, P. R. Rowntree, and G. R. Shaver for valuable discussions and access to manuscripts in advance of publication.

References

Acock, B., and Allen, L. H. (1985). Crop responses to elevated carbon dioxide concentration. *In* "Direct Effects of Increasing Carbon Dioxide on Vegetation" (B. R. Strain and J. D. Cure, eds.), pp. 53–97. U.S. Department of Energy, Washington, D.C.

Brown, S., and Lugo, A. E. (1984). Biomass of tropical forests: A new estimate based on forest volumes. *Science* **223**, 1290–1293.

Conway, T. J., Tans, P., Waterman, L. S., Thoning, K. W., Masarie, K. A., and Gammon, R. H. (1988). Atmospheric carbon dioxide measurements in the global troposphere, 1981-1984. *Tellus* **40B**, 81–115.

Cross, A. (1990). "Tropical Deforestation and Remote Sensing: The Use of NOAA/AVHRR Data over Amazonia." UNEP/GRID report to CEC(D G VII). Article B946/88, NERC, UK.

Crutzen, P. J., and Andreae, M. O. (1990). Biomass burning in the tropics: Impact on atmospheric chemistry and biogeochemical cycles. *Science* **250**, 1669–1678.

Crutzen, P. J., and Gidel, L. T. (1983). A two-dimensional photochemical model of the atmosphere. 2. The tropospheric budgets of the anthropogenic chlorocarbons, CO, CH_4, CH_3CL and the effect of various NO_x sources on tropospheric ozone. *J. Geophys. Res.* **88**, 6641–6661.

Eamus, D., and Jarvis, P. G. (1989). The direct effects of increase in the global atmospheric CO_2 concentration on natural and commercial trees of forests. *Adv. Ecol. Res.* **19**, 1–55.

Enting, I. G. (1990). Ambiguities in the calibration of carbon cycle models. *Inverse Problems* **6**, L39–L46.

Enting, I. G., and Mansbridge, J. V. (1989). Seasonal sources and sinks of atmospheric CO_2: Direct inversion of filtered data. *Tellus* **41B**, 111–126.

Enting, I. G., and Mansbridge, J. V. (1991). Latitudinal distribution of sources and sinks of CO_2: Results of an inversion study. *Tellus* **43B**, 156–170.

Enting, I. G., and Newsam, G. N. (1990). Atmospheric constituent inversion problems: Implications for baseline monitoring. *J. Atmos. Chem.* **11**, 69–87.

Fan, S.-M., Wofsy, S. C., Bakwin, P. S., and Jacob, D. J. (1990). Atmosphere-biosphere exchange of CO_2 and O_3 in the central Amazon forest. *J. Geophys. Res.* **95**, 16,851–16,864.

FAO/UNEP (1981). "Tropical Forest Resources Assessment Project." FAO, Rome.

Fearnside, P. H., Tardin, A. T., and Filho, L. J. M. (1990). "Deforestation Rate in Brazilian Amazonia." National Secretariat for Science and Technology, Brazilia.

Francey, R. J., and Tans, P. P. (1987). Latitudinal variation in oxygen -18 of atmospheric CO_2. *Nature* **327**, 495–497.

Fung, I. Y., Prentice, K., Matthews, E., Lerner, J., and Russell, G. (1983). Three-dimensional tracer model study of atmospheric CO_2: Response to seasonal changes with the terrestrial biosphere. *J. Geophys. Res.* **88C**, 1281–1294.

Fung, I. Y., Tucker, C. J., and Prentice, K. C. (1987). Application of advanced very high resolution radiometer vegetation index to study atmosphere–biosphere exchange of CO_2. *J. Geophys. Res.* **92**, 2999–3015.

Fung, I. Y., Moore, B., Prather, M., Running, S., and Tiessen, H. (1991). Linkages between terrestrial ecosystem and the atmosphere. *In* "1990 Global Change Institute on Earth System Modelling." Office for Interdisciplinary Earth Sciences, Boulder, Colorado.

Goudriaan, J. (1990). Atmospheric CO_2, global carbon fluxes and the biosphere. *In* "Theoretical Production Ecology: Reflections and Prospects" (R. Rabbinge, J. Goudriaan, H.

van Keulen, F. W. T. Penning de Vries, and H. H. van Laar, eds.), pp. 17–40. Simulation Monograph 34, Pudoc, Wageningen.

Hall, D. O., and Scurlock, J. M. O. (1991). Tropical grasslands and their role in the global carbon cycle. *In* "Facets of Modern Ecology" (G. Esser and D. Overdieck, eds.), pp. 659–677. Elsevier, Amsterdam.

Hammond, A. L. (ed.) (1990). "World Resources 1990-91." Oxford University Press, New York.

Hansen, J., and Lebedeff, S. (1987). Global trends of measured surface air temperature. *J. Geophys. Res.* **92**, 13345–72.

Harmon, M. E., Ferrell, W. K., and Franklin, J. F. (1990). Effects on carbon storage of conversion of old-growth forests to young forests. *Science* **247**, 699–702.

Houghton, R. A. (1987). Biotic changes consistent with the increased seasonal amplitude of atmospheric CO_2 concentrations. *J. Geophys. Res.* **92**, 4223–4230.

Houghton, R. A. (1991a). Tropical deforestation and atmospheric carbon dioxide. *Climate Change* **19**, 99–118.

Houghton, R. A. (1991b). Terrestrial ecosystems in the global carbon balance: Changes over the last 135 years. *In* "The Global Carbon Cycle" (M. Heimann, ed.). NATO ASI, Il Ciocco, Italy, September 8–20, 1991. In press.

Houghton, R. A., Schlesinger, W. H., Brown, S., and Richards, J. F. (1985). Carbon dioxide exchange between the atmosphere and terrestrial ecosystems. *In* "Atmospheric Carbon Dioxide and the Global Carbon Cycle" (J. R. Trabalka, ed.), pp. 113–140. U.S. Department of Energy, Washington, D.C.

Houghton, R. A., Boone, R. D., Fruci, J. R., Hobbie, J. E., Melillo, J. M., Palm, C. A., Peterson, B. J., Shaver, G. R., Woodwell, G. M., Moore, B., Skole, D. L., and Myers, N. (1987). The flux of carbon from terrestrial ecosystems to the atmosphere in 1980 due to changes in land use: Geographical distribution of the global flux. *Tellus* **39B**, 122–139.

Houghton, R. A., and Skole, D. L. (1990). Carbon. *In* "The Earth as Transformed by Human Action" (B. L. Turner, W. C. Clark, R. W. Kates, J. F. Richards, J. T. Mathews, and W. B. Meyer, eds.), pp. 393–408. Cambridge University Press, UK.

Houghton, J. T., Jenkins, G. J., and Ephraums, J. J. (eds.) (1990). "Climate Change: The IPCC Scientific Assessment." Cambridge University Press, Cambridge.

Jarvis, P. G. (1992). Forests and atmospheric carbon dioxide. *In* "The Greenhouse Effect: Causes, Consequences and Correctives." Environmental & Scientific Services, London.

Keeling, C. D., Whorf, T. P., Heimann, M., and Mook, W. M. (1990). A three-dimensional model of atmospheric CO_2 transport based on observed winds. 1. Observational data and preliminary analysis. *AGU Monograph* **55**, 165–236.

King, A. W. (1991). Translating models across scales in the landscape. *In* "Quantitative Methods in Landscape Ecology" (M. G. Turner and R. H. Gardner, eds.), Springer-Verlag, New York.

King, A. W., DeAngelis, D. L., and Post, W. M. (1987). "The Seasonal Exchange of Carbon Dioxide between the Atmosphere and the Terrestrial Biosphere: Extrapolation from Site-Specific Models to Regional Models." Report ORNL/TM-10570, Oak Ridge National Laboratory, Oak Ridge, Tennessee.

King, A. W., O'Neill, R. V., and DeAngelis, D. L. (1989). Using ecosystem models to predict regional CO_2 exchange between the atmosphere and the terrestrial biosphere. *Global Biogeochem. Cycles* **3**, 337–361.

Kurz, W. A., Apps, M. J., Webb, T. M., McNamee, P. J., and Lekstrum, T. (1991). "The Carbon Budget of the Canadian Forest Sector: Phase I." Information Report for Canadian Govt. NOR-X-326 (Forestry Canada Inform. Report Series). Northwest Region, Northern Forestry Center.

Lanly, J.-P., Singh, K. D., and Janz, K. (1991). FAO's 1990 reassessment of tropical forest cover. *Nat. Resour.* **27**, 21–26.

Luxmoore, R. J., King, A. W., and Tharp, M. L. (1991). Approaches to scaling up physiologically-based soil-plant models in space and time. *Tree Physiol.* **9**, 281–292.

Maier-Reimer, E., and Hasselmann, K. (1987). Transport and storage of CO_2 in the ocean: An inorganic ocean-circulation carbon cycle model. *Climate Dynamics* **2**, 63–90.

Marland, G., Rotty, R. M., and Treat, N. L. (1985). CO_2 from fossil fuel burning: Global distribution of emissions. *Tellus* **37B**, 243–258.

Mascart, P., Pinty, J.-P., Taconet, O., and Ben Mehrez, M. (1991). Canopy resistance formulation and its effect in mesoscale models: A HAPEX perspective. *Agric. For. Meteorol.* **54**, 319–351.

Matthews, E., and Fung, I. Y. (1987). Methane emission from natural wetlands: Global distribution, area, and environmental characteristics of sources. *Global Biogeochem. Cycles* **1**, 61–86.

McNaughton, K. G. (1989). Regional interactions between canopies and the atmosphere. *In* "Plant Canopies: Their Growth, Form and Function" (G. Russell, B. Marshall, and P. G. Jarvis, eds.), pp. 63–81. Cambridge University Press, Cambridge.

McNaughton, K. G., and Jarvis, P. G. (1991). Effects of spatial scale on stomatal control of transpiration. *Agric. For. Meteorol.* **54**, 279–301.

McNaughton, K. G., and Spriggs, T. W. (1986). An evaluation of the Priestley and Taylor equation and the complementary relationship using results from a mixed-layer model of the convective boundary layer. *In* "Estimation of Areal Evapotranspiration" (T. A. Black, D. L. Spittelhouse, M. D. Novak, and D. T. Price, eds.), pp. 89–104. LAHS Press, Wallingford, Connecticut.

Myers, N. (1989). "Deforestation Rates in Tropical Forests and Their Climatic Implications." Friends of the Earth, London.

Parton, W., Tiessen, H., and Walker, B. (1991). A toy terrestrial carbon flow model. *In* "1990 Global Change Institute on Earth System Modelling." Office for Interdisciplinary Earth Sciences, Boulder, Colorado.

Prentice, K. C., and Fung, I. Y. (1990). The sensitivity of terrestrial carbon storage to climate change. *Nature* **346**, 48–51.

Raich, J. W., Rastetter, E. B., Melillo, J. M., Kicklighter, D. W., Steudler, P. A., and Peterson, B. J. (1991). Potential net primary production in South America: Application of a global model. *Ecol. Appl.* **1**, 399–429.

Rastetter, E. B., Ryan, M. G., Shaver, G. R., Melillo, J. M., Nadelhoffer, K. J., Hobbie, J. E., and Aber, J. D. (1991a). A general biochemical model describing the responses of the C and N cycles in terrestrial ecosystems to changes in CO_2 climate and N deposition. *Tree Physiol.* **9**, 101–126.

Rastetter, E. B., King, A. W., Cosby, B. J., Hornberger, G. M., O'Neill, R. V., and Hobbie, J. E. (1991b). Aggregating fine-scale ecological knowledge to model coarser-scale attributes of ecosystems. *Ecol. Appl.* **2**, 55–70.

Roeloffzen, J. C., Mook, W. G., and Keeling, C. D. (1990). "Trends and Variations in Stable Carbon Isotopes of Atmospheric Carbon Dioxide." International Symposium on the Use of Stable Isotopes in Plant Nutrition, Soil Fertility and Environmental Studies, Vienna, Austria, October 1–5, 1990.

Rotty, R. M. (1987a). A look at 1983 CO_2 emissions from fossil fuels (with preliminary data for 1984). *Tellus* **39B**, 203–208.

Rotty, R. M. (1987b). Estimates of seasonal variation in fossil fuel CO_2 emissions. *Tellus* **39B**, 184–202.

Sarmiento, J. L., Orr, J. C., and Siegenthaler, U. (1992). A perturbation simulation of CO_2 uptake in an ocean general circulation model. *J. Geophys. Res.*, in press.

Schlesinger, W. H. (1990). Evidence from chronosequence studies for a low carbon-storage potential of soils. *Nature* **348**, 232–234.

Setzer, A. W., and Pereira, M. C. (1988). Amazon basin burnings in 1987 and their tropospheric emissions. Unpublished manuscript.

Siegenthaler, U. (1983). Uptake of excess CO_2 by an outcrop-diffusion model of the ocean. *J. Geophys. Res.* **88**, 3599–3608.

Siegenthaler, U., and Oeschger, H. (1987). Biospheric CO_2 emissions during the past 200 years reconstructed by deconvolution of ice core data. *Tellus* **39B**, 140–154.

Takahashi, T. (1989). The carbon dioxide puzzle. *Oceanus* **32(2)**, 22–29.

Tans, P. P., Conway, T. J., and Nakazawa, T. (1989). Latitudinal distribution of the sources and sinks of atmospheric carbon dioxide derived from surface observations and an atmospheric transport model. *J. Geophys. Res.* **94(D4)**, 5151–5172.

Tans, P. P., Fung, I. Y., and Takahashi, T. (1990). Observational constraints on the global atmospheric CO_2 budget. *Science* **247**, 1431–1438.

Taylor, J. A. (1989). A stochastic Lagrangian atmospheric transport model to determine global CO_2 sources and sinks: A preliminary discussion. *Tellus* **41B**, 272–285.

Thornley, J. H. M., and Verberne, E. L. J. (1989). A model of nitrogen flows in grassland. *Plant Cell Environ.* **12**, 863–886.

Thornley, J.H. M., Fowler, D., and Cannell, M. G. R. (1991). Terrestrial carbon storage resulting from CO_2 and nitrogen fertilization in temperate grasslands. *Plant Cell Environ.* **14**, 1007–1011.

Watson, R., Rodhe, H., Oeschger, H., and Siegenthaler, U. (1990). Greenhouse gases and aerosols. *In* "Climate Change, the Scientific Assessment" (J. T. Houghton, G. J. Jenkins, and J. J. Ephraums, eds.), pp. 1–40. Cambridge University Press, Cambridge.

Wigley, T. M. L. (1991). A simple inverse carbon cycle model. *Global Biogeochem. Cycles* **5**, 373–382.

Wilson, S. R., and Ayers, G. P. (eds.) (1990). "Baseline Atmospheric Program (Australia) 1988." CSIRO, Newcastle.

13

Prospects for Scaling

Martyn M. Caldwell, Pamela A. Matson,
Carol Wessman, and John Gamon

I. Introduction

We have a natural fascination with detail and diversity. The tendency to probe increasingly finer scales of resolution and more basic principles is seemingly inherent in scientists. Although ecologists presumably always have recognized the need to extend their understanding of principles and specifics to more expansive scales of space and time, questions of current interest relating to regional perturbations and global change bring an urgency to studies of scale.

Discussions on scaling in this volume have dealt primarily with processes, for example, evapotranspiration or nutrient cycling, rather than with cartographic considerations or pattern analysis. The emphasis also has been on spatial rather than temporal scaling, although these concepts are often necessarily coupled. Scaling is not simply integration or aggregation of values at one level to achieve estimates at a more encompassing level of consideration. Rather, scaling represents the transcending concepts that link processes at different levels of space and time. It entails a change upward in scale that identifies major factors operational on a given scale of observation, their congruency with those on the lower scale, and the constraints and feedbacks on those factors. Scaling also involves not being distracted with those factors that are less important in the transitions among scales.

Although statistical tests have been adopted in ecological studies to determine the scale of particular phenomena, for example, the use of semivariogram analysis (see Chapter 2, Chapter 3), clear rules for scaling

223

are more elusive, as are generalizations about the scaling process. Approaches are necessarily somewhat specific to the question at hand. However, simply deferring to the refrain "It depends on the question being asked" is not so useful in contemplating the state of our art. Can some guidelines be formulated? How can the greatest progress be realized in the future? How are new tools and technologies contributing? Who should be involved and how should our science be structured for the greatest gains?

II. Approaches and Guidelines

A. Bottom-Up and Top-Down Models

Advocates and practitioners of bottom-up and top-down approaches recognize that both play important roles in scaling.

Bottom-up modeling, reckoning from lower to higher scales, involves extending calculations from an easily measured and reasonably well understood unit to processes at a more encompassing scale. As indicated in the title of this volume, the intact leaf is often the starting point. This beginning is, however, arbitrary. To the physiologist or biochemist, the leaf is a very complex entity with structural heterogeneity and gradients of light, CO_2, and sun- and shade-adapted chloroplasts. However, it is easily measured and its behavior is well characterized. By contrast, the individual root is less effectively used as a starting point of scaling. Bottom-up modeling to canopies has been remarkably successful considering the complexities of the canopy light environment and gradients of momentum, water vapor, and CO_2. Refinements are still needed (see Chapter 4, Chapter 5), but progress has been gratifying. Reckoning from canopy to region is more daunting, but some strides are being made (e.g., Chapter 6, Chapter 18).

Top-down approaches can be a matter of making the coarser scale measurements or estimates that set the boundary conditions, lead to problem identification and stimulate the testing of general relationships for specific cases. Some powerful generalizations have emerged, for example, the relationship between net production and absorbed photosynthetically active radiation (APAR) by plant canopies (Monteith, 1977) or the link between leaf area and stem sapwood of forest stands (Shinozaki et al., 1964; Waring et al., 1981). These generalizations are not without a need for adjustments (for example, for the ε coefficient that defines the system-specific ratio of carbon fixed per radiation intercepted in the first relationship; Russell et al., 1989), yet their applicability enjoys general appreciation. Such relationships have a mechanistic basis, but the amount of detail and diversity that is subsumed in these relationships is impressive. These examples take their starting point at the canopy scale and work down. Top-down approaches that begin at global or regional scales can be useful also. Global source strength calculations derived from inverse modeling (e.g., Chapter 10) and atmospheric isotopic information (Chapter 11) set constraints on processes and, at the same time, suggest the geographic regions and physiological processes in most critical need of study.

B. Guidelines for Scaling

Some guidelines for scaling might include (1) assessing the scale of the phenomenon in question, (2) identifying the boundary conditions and constraints, (3) searching for consistencies at different scales, (4) streamlining bottom-up models to incorporate only the salient features, (5) incorporating feedbacks (both positive and negative) that may operate on some scales but not necessarily on other scales, and (6) testing the results on different scales with independent estimates. Finally, in many cases, the usefulness and possibility of aggregating species into functional groups should be explored.

These guidelines are not meant to be an exclusive set of rules analogous to Koch's postulates that are employed rigorously to identify pathogenic agents (Koch, 1876), but as processes that have been employed successfully in different scaling efforts. As further progress is made in scaling, more definitive rules may emerge. Some elaboration of these guidelines follows:

1. Assessing the Scale Geostatistical analyses bring new insights into the interpretation of ecological spatial dependence (Rossi *et al.*, 1992). Spatial continuity tools, such as variograms and correlograms, provide useful means of testing the spatial scales on which phenomena manifest themselves (Chapter 2, Chapter 3). This is, thus, a necessary first step in the scaling process.

2. Boundary Conditions and Constraints Boundary conditions derived from top-down approaches include those crucial estimates of parameters that provide the working limits or targets of the phenomenon under study (e.g., the $\delta^{13}C$ of the atmosphere at particular latitudes; see Chapter 11) or estimates of region-scale evapotranspiration based on ^{36}Cl of large water bodies (Magaritz *et al.*, 1990). These parameters are useful for question identification as well (Chapter 10). Constraints are, in part, defined by the boundary conditions, but constraints also involve physical and biological principles that can streamline the scaling operation conveniently. In general circulation models (GCMs), conservation of mass and momentum are important constraints; they may apply in aspects of region- or canopy-scale models, but often to a lesser degree.

Although physical constraints, as mentioned, are well appreciated, a recognition of the biological constraints in scaling is not so well established. The processes of natural selection and evolution can offer guides for identifying constraints. Understanding the manner in which natural selection operates and how plant populations are molded can allow one to characterize salient properties at different scales. Natural selection is for fitness, not for production or resource acquisition. Although these properties may be correlated in many circumstances, when they are not it is important to recognize the priority for fitness. Also, the selective forces that cause plants to be more competitive may not necessarily be those that lead to optimal use of resources. For example, the height of trees may often contribute less to optimal light harvesting than to competitive advantage for the individual tree (Mäkelä, 1985; King, 1990). Phenology and life history characteristics also need to be recognized since they constrain plant performance. In the evolutionary context, game theory may be a useful tool in conceiving constraints.

3. Consistencies among Scales   Recognizing consistencies on different scales presents perhaps the greatest difficulties in the scaling process. Hierarchy theory (Simon, 1973; O'Neill *et al.*, 1986; Urban *et al.*, 1987) and recognition of the appropriate relaxation times of processes on different scales (de Wit, 1970) can be employed, as can spatial continuity tools, as mentioned earlier. Judgment based on a knowledge of ecophysiological principles is needed in identifying the salient consistencies.

4. Streamlining Models   An identification of consistencies on different scales provides a basis for streamlining bottom-up models. Such models often may be the core of the scaling process. Judicious omission of the less pertinent features of these models for spanning scales must be undertaken with great care. Sensitivity analyses with the model on one scale may be a guide for streamlining the model for use on a higher scale, but the choice and range of parameters used in the sensitivity analyses are also critical. The selection of feedback loops for different scales is important in the streamlining process.

5. Feedbacks   Feedback loops that operate on some scales and are comparatively unimportant on other scales present a particular challenge. If feedbacks on higher scales tend to be negative in nature, much detail can be subsumed and potential complications often are ameliorated (e.g., Chapter 6). On the other hand, positive feedback loops on higher scales suggest some sizable ramifications of control on a lower scale (e.g., Chapter 18). Identifying and understanding the nature of feedback loops is an integral component of the scaling process.

6. Testing   Testing the products of scaling efforts by corroborative measures on a larger scale is very similar to setting the boundary conditions, but may involve parameters that do not necessarily define boundaries or constraints. Validations and corroborations with independently measured parameters are highly desirable. Technical and conceptual advances in integrative measures are giving practitioners many new options, such as approaches using the natural abundance of stable isotopes (particularly useful for corroboration across temporal scales) and remote sensing based on optical principles and repeated coverage (for corroboration across both spatial and temporal scales).

7. Functional Groupings   Population biology vividly depicts the diversity of nature, the individualistic behavior of population members, and the complications that can arise from ignoring diversity (Chapter 14). However, at least for questions of nutrient and energy movement in ecosystems, a functional commonality of diverse species can be seen. Chapin and Fung (Chapter 16) suggest even simple categories such as plant size and relative growth rate may serve well to characterize sufficiently the manner in which species are contributing to trophic transfer, nutrient cycling, productivity, evapotranspiration, and their sensitivity to disturbance. Certainly there are major functional considerations that complicate such simple categorization, for example, nitrogen fixers vs. nonfixers or C_3 vs. C_4 species, and there are vivid demonstrations of species-specific alterations of production and nutrient cycling (e.g., Brown and Heske, 1990; Wedin and Tilman, 1990). Despite such examples, these species groupings are useful in questions of productivity and nutrient cycling. Issues of species interactions and the individualistic migration of species with perturbations usually do not lend themselves to the use of species groupings (Chapter 15). Although the use of functional groups may enhance the ease of scaling greatly for some questions, it clearly may hamper scaling efforts for others. Improper grouping easily could lead to erroneous conclusions. Clearly, assessment and application of functional groups must be done with judicious consideration of physical and biological principles.

C. New Tools

New technology and approaches for making integrative measurements are contributing and will continue to contribute substantially to progress in scaling. Whole-canopy flux measurements have provided a useful means of assessing canopy-scale energy and gas movements for several decades. Progress and refinement in this field is allowing such approaches to be extended to greater scales and to include sloping and heterogeneous terrain not previously considered suitable (Chapter 5). Advances in remote sensing have provided new means of scaling. Improved technology, studies focused on vegetation optical properties and radiative transfer, and other innovative uses of remote sensing are taking an important role in quantitative landscape analysis (Chapter 19), but the use of this technology is predicated on the assumption that we recognize the appropriate features to be scaled. The new technology and analytical approaches are making it possible to measure variables more closely associated with the physiological process of interest, and to track those variables from local to global scales.

Other experimental approaches have been driven more by conceptual progress than by technical advances. The application of stable isotope analysis to understanding ecophysiological processes, for example, the link between $\delta^{13}C$ and leaf internal CO_2 concentrations, represents approaches that benefitted much more from advances in theory than from the development of more refined ratio mass spectrometers (Farquhar *et al.*, 1982). In general, stable isotope assessments appear to be ripe for many new applications.

Some approaches, new to the ecologists at least, are largely analytical, for example, statistical innovations for grappling with large-scale perturbations (Matson and Carpenter, 1990), object-oriented programming (Chapter 7), geographic information systems, and geostatistics (Rossi *et al.*, 1992; Chapter 3).

Many of the methodological advances new to ecology have been borrowed in part from other fields, which only emphasizes the need for continued and expanded interdisciplinary collaborations.

D. Structuring Our Science

Many aspects of ecology, and certainly those activities involving scaling, will profit most from collaborations both among ecologists of different ilks and among members of widely separate disciplines, including unconventional combinations. The experiment of the First International Satellite Land Surface Climatology Project Field Experiment (FIFE) was as much an experiment in collaborations as in new science (Chapter 3). Teamwork of ecophysiologists, ecosystem ecologists, atmospheric researchers, and remote sensing specialists has proven successful in FIFE as well as in a number of other multidisciplinary projects.

Although scientists and even funding agencies are often willing to undertake and support collaborations, recognition and support of such teamwork is necessary from universities and federal agencies that employ research personnel. For example, bestowing tenure and other rewards in the university system must accommodate and even promote team research. Such a change necessarily includes innovative evaluation procedures and enlightened administrators. The traditional administrative approach of counting primarily senior-authored journal papers and principal-investigator National Science Foundation grants must end.

Ecological researchers also must be willing to take certain risks to venture into new collaborations and undertake new approaches that may not always have the blessing of the more conservative scientific community. At the same time, researchers must preserve the high standards of the mainstream scientific disciplines and avoid the dangers of capriciousness and dilettantism.

Ecologists also need to involve themselves in national and international committees and studies apart from research per se, through the International Geosphere–Biosphere Programme, the Scientific Committee on Problems of the Environment, or the United Nations Environmental Programme, to name a few. The coming decade is one of much opportunity for ecology, but one of considerable responsibility as well. Ecological approaches will be required to define not only the emerging scientific issues (Lubchenco *et al.*, 1991) but also the approaches directed at their solutions.

References

Brown, J. H., and Heske, E. J. (1990). Control of a desert-grassland transition by a keystone rodent guild. *Science* **250**, 1705–1707.

de Wit, C. T. (1970). Dynamic concepts in biology. *In* "Prediction and Measurement of Photosynthetic Productivity," pp. 17–23. Centre for Agricultural Publishing and Documentation. Wageningen, The Netherlands.

Farquhar, G. D., O'Leary, M. H., and Berry, J. A. (1982). On the relationship between carbon isotope discrimination and the intercellular carbon dioxide concentration in leaves. *Aust. J. Plant Physiol.* **9**, 121–137.

King, D. A. (1990). The adaptive significance of tree height. *Am. Nat.* **135**, 809–828.

Koch, R. (1876). Die Aetiologie der Milzbrand-Krankheit, begründet auf die Entwicklungsgeschichte des *Bacillus anthracis*. *In* "Gessammelte Werke von Robert Koch," Vol. 1. Leipzig.

Lubchenco, J., Olson, A. M., Brubaker, L. B., Carpenter, S. R., Holland, M. M., Hubbell, S. P., Levin, S. A., MacMahon, J. A., Matson, P. A., Melillo, J. M., Mooney, H. A., Peterson, C. H., Pulliam, H. R., Real, L. A., Regal, P. J. and Risser, P. G. (1991). The sustainable biosphere initiative: an ecological research agenda. *Ecology* **72**, 371–412.

Magaritz, M., Kaufman, A., Paul, M., Boaretto, E., and Hollos, G. (1990). A new method to determine regional evapotranspiration. *Water Resources Res.* **26**, 1759–1762.

Mäkelä, A. (1985). Differential games in evolutionary theory: Height growth strategies of trees. *Theoret. Popul. Biol.* **27**, 239–267.

Matson, P. A., and Carpenter, S. R. (1990). Statistical analysis of ecological response to large-scale perturbations. *Ecology* **71**, 2037.

Monteith, J. L. (1977). Climate and the efficiency of crop production in Britain. *Phil. Trans. R. Soc. London B Biol. Sci.* **281**, 277–294.

O'Neill, R. V., DeAngelis, D. L., Waide, J. B., and Allen, T. F. H. (1986). "A Hierarchical Concept of Ecosystems" (R. M. May, ed.), Vol. 23. Princeton University Press, Princeton, New Jersey.

Rossi, R. E., Mulla, D. J., Journel, A. G., and Franz, E. H. (1992). Geostatistical tools for the modeling and interpretation of ecological spatial dependence. *Ecol. Monogr.* **62**, 277–314.

Russell, G., Jarvis, P. G., and Monteith, J. L. (1989). Absorption of radiation by canopies and stand growth. *In* "Plant Canopies: Their Growth Form and Function" (G. Russell, B. Marshall, and P. G. Jarvis, eds.), pp. 21–39. Cambridge University Press, Cambridge.

Shinozaki, K., Yoda, K., Hozumi, K., and Kira, T. (1964). A quantitative analysis of plant form: The pipe model theory. II. Further evidence of the theory and its application in forest ecology. *Jpn. J. Ecol.* **14,** 133–139.

Simon, H. A. (1973). The organization of complex systems. *In* "Hierarchy theory" (H. H. Pattee, ed.), pp. 3–27. George Braziller, New York.

Urban, D. L., O'Neill, R. V., and Shugart, H. H., Jr. (1987). Landscape ecology: A hierarchical perspective can help scientists understand spatial patterns. *BioScience* **37,** 119–127.

Waring, R. H., Schroeder, P. E., and Oren, R. (1981). Application of the pipe model theory to predict canopy leaf area. *Can. J. For. Res.* **12,** 556–560.

Wedin, D. A., and Tilman, D. (1990). Species effects on nitrogen cycling: A test with perennial grasses. *Oecologia* **84,** 433–441.

IV

Functional Units
in Ecology

Clearly when addressing questions of scale, ecologists are forced to cross the boundary where organisms must be lumped and can no longer be considered individually. This is perhaps one of the greatest challenges to ecologists as they shift from considering the role of natural selection in affecting organisms (scale of the individual) to addressing changes in flux through ecosystems in response environmental perturbations. This section addresses the interface between plant diversity and the need for large-scale syntheses. In successive chapters, both Bazzaz and Clark develop population and community level perspectives on scaling, discussing the interface between analyses based on physiological processes and population dynamics. The following chapter by Chapin and the discussion by Dawson and Chapin assess the prospects for analyses that manage some of the complications arising from biodiversity by collecting species into groups based on morphological and functional characters.

14

Scaling in Biological Systems: Population and Community Perspectives

Fakhri A. Bazzaz

I. Introduction

Population and physiological ecologists react with some alarm to the development and use of global circulation models that ignore the enormous diversity of life and the intricacy of interactions within ecosystems. In these models, plants are aggregated and factors of the physical environment summarized as state variables expressing carbon and heat transfer. In the context of the global carbon flux, for example, an assumption is made, explicitly or implicitly, that all units of photosynthetic tissue in an ecosystem are equivalent, regardless of the genotype, species, developmental state, and local biotic and abiotic environment of the source plant. This "big leaf model" approach may be particularly successful in ecosystems with identifiable "dominant species" (e.g., *Spartina* in salt marshes). In such situations, an understanding of the intricacies of the population biology and physiology of the dominant species can lead to broad-scale predictions. However, such situations are not common. Spatial heterogeneity in both resource availability and species composition results in the "big leaf" actually being a mosaic of dissimilar patches, each patch greatly influencing the responses of the aggregate. Individuals and species can differ greatly in their responses to and effects on the environment. An understanding of specific component species or guild responses, and their effects on the aggregate ecosystem (i.e., the "big leaf"), would enable modelers to better predict carbon flux across the natural variation found throughout the ranges of ecosystems. More importantly, and perhaps less appreciated by modelers, the mosaic of patches within most ecosystems is temporally dynamic, that is, the "big leaf" changes through time. As components (e.g., species or guilds) within ecosystems interact with one another and respond to their changing local and global environments, their relative abundances and spatial distributions will change also. Unless an understanding of how each of the components of the ecosystem responds, and how these responses scale up to produce the net aggregate effect, is reached, modelers will be limited in making long-term predictions of natural ecosystem responses to climate change.

In this chapter, I explore ways in which population ecology can compromise the simplicity of aggregate modeling. I hope to identify some concepts and approaches from population and community ecology relevant to global change that should be considered in these modeling approaches. I consider the problem of scaling up from the individual level to the population, community, and ecosystem levels. The motivation to write the chapter was to highlight relevant approaches from population and community ecology, and to develop a framework for answering some of the pressing ecological questions raised by global change initiatives, notably in the intergovernmental Panel on Climate Change (IPCC) (Houghton *et al.*, 1990) and the Ecological Society of America's Sustainable Biosphere Initiative (SBI) (Lubchenco *et al.*, 1991).

II. Individual Plants as Members of Populations, Communities, and Ecosystems

Two obvious but critical points need to be stated at the outset. First, physiological processes are influenced by neighbors because plants most commonly exist as members of a population, both in the population biological sense and in the physical sense. Second, the responses of individuals and of populations to broad-scale environmental changes may be fundamentally different from each other in magnitude and direction. Despite the fact that individuals are almost always present as members of interacting populations, rarely has the coupling between the physiological response of the individual and the population or community been made. Demographic consequences of plant physiological traits have not been investigated adequately (see Bazzaz, 1984; Mooney, 1991). In addition to the aspects treated at the individual level, population biology emphasizes the number (density) of neighboring individuals, size and reproductive hierarchies, demography, and population genetic structure. All are directly relevant to the response of ecosystems to a changing environment.

Research in physiological ecology has focused particularly on carbon gain in response to various levels of resources, such as light, water, nutrients, and carbon dioxide concentrations. Growth and allocation, energy and water balance, reproduction, and chemical defense against herbivores also have been emphasized. Much of this research considered the average response of isolated or experimentally grown individuals (or a few individuals), but extrapolated these conclusions to represent the response of the "species" or the "ecotype." Customarily, more in-depth understanding of the behavior at the individual level has been sought by examining lower levels of biological organization, for example, at the leaf or cell level (Osmond *et al.,* 1980). In these approaches, the distinction between ecology, physiology, and biochemistry becomes blurred. The physiological approach has been extremely successful in elucidating how plants interact with their physical environment and how they might have evolved to cope with it (see Mooney, 1991). Less commonly, interactions among neighboring individuals have been considered in terms of how neighbors modify each others' environment, particularly by reducing resource availability for neighbors or by limiting the neighbor's ability to capture these resources. However, neighboring individuals can modify greatly each other's responses in terms of photosynthetic response, growth, and allocation (Caldwell, 1987) (Fig. 14.1). Behaviors and mechanisms by which plants increase their ability to acquire resources from a wide range of habitats, and therefore diminish habitat selection or avoid their neighbors (Bazzaz, 1991), also have been considered less frequently by physiological ecologists.

In structurally complex ecosystems, such as forests, which contain a mixture of plant sizes ranging from understory seedlings and saplings to mature trees, profiles of environmental factors such as light, carbon dioxide, air humidity, wind speed, and temperature exist. As a result, different patches of individuals may experience great variation in their exposure to these factors. Even within a given layer of vegetation, individuals have different sizes and are therefore found in microenvironments that can differ greatly in light, air humidity, wind, temperature, and CO_2 levels. For example, the annual herb *Impatiens capensis* can form much of the understory plant cover in some temperate forests in the northeastern United States. The stands usually have a well developed size hierarchy and a complex environment. Dominant and suppressed individuals that vary in size by only a few centimeters may receive very different light and CO_2 regimes, and therefore differ in physiological characteristics such as photosynthetic light response and water use efficiency (Fig. 14.2).

Scaling up from the responses of leaves or individual plants to populations and communities is a necessary prerequisite for making long-term predictions about the response of natural ecosystems to climate change. Thus, a major challenge to addressing how ecosystems respond to climate change is that, in nature, there is a large number of plant species (and genotypes), each with its own particular responses to the changing environment. It is essential, therefore, to group species into a smaller subset of "guilds" or "functional types" so we can address their response by groups. These groups can neither be fixed in the identity of their members nor can they be of predetermined size. Rather, ecological criteria should determine the type and extent of aggregation. Clearly, these groupings will vary in composition, depending on the questions asked and the level of refinement necessary and sufficient to reach useful answers. For example, at some level, recognition of large groups such as C_3, C_4, and CAM plants or tillering vs. nontillering grasses may be all that is needed whereas, at another level, CO_2-responsive and CO_2-nonresponsive genotypes within a single population may need to be identified. Irrespective of the level of detail required, aggregation is essential, because without some aggregation the task of assessing the response of ecosystems to climate change within the required time frame will be impossible.

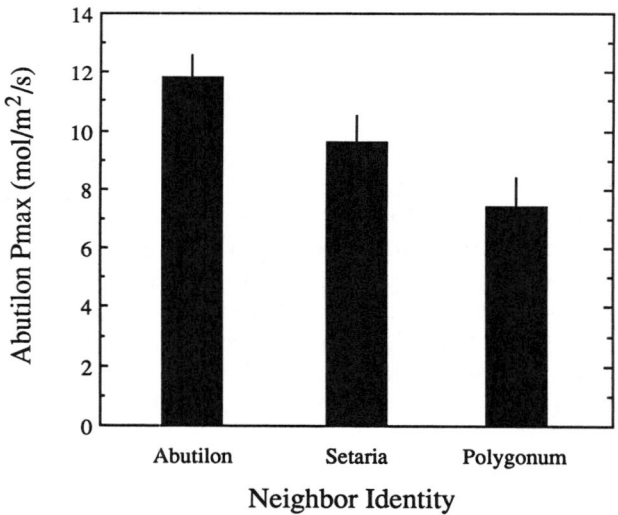

Figure 14.1 Maximum leaf level photosynthetic rate of *Abutilon* plants grown in the same environment but with different neighbors.

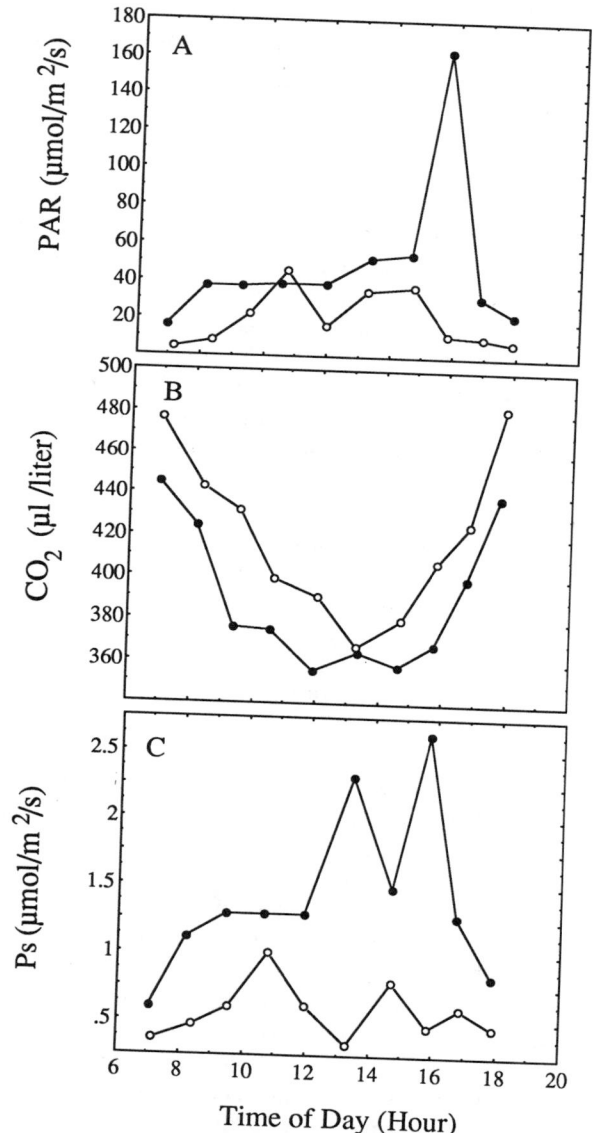

Figure 14.2 Daily patterns of (A) photosynthetically active radiation (PAR), (B) CO_2 concentration, and (C) photosynthetic rate of canopy (○) and subcanopy (●) individuals of a population of *Impatiens capensis* in the understory of a deciduous forest in New England. Unpublished data of Sean Thomas.

III. Global Change, Resource Augmentation, and the Response of Individuals and Populations: Are There General Patterns?

Theoretical and experimental evidence shows that individual responses to changes in environmental resources may be fundamentally different. When alone, an individual has sole access to available resources within its domain. More resources generally means that the individual gets larger. However, when in dense stands, each individual must compete with its neighbors and will therefore have access to only part of the resources in the neighborhood. Resource-phenomenological models in which resources may be incorporated as rate variables (e.g., Ford and Diggle, 1981; Firbank and Watkinson, 1985; Hara, 1986) predict a monotonic decrease in the number of surviving individuals with time and with increasing density. Further, resource-mechanistic models (e.g., Clark, 1990; Hilbert *et al.*, 1991) also predict a decrease in plant numbers with increasing soil fertility. In natural environments, many more seedlings often are recruited than can be supported to maturity and reproduction. Many individuals will die because of resource limitations caused by crowding, even in the absence of other causes of mortality. Thinning in natural experimental stands has been documented widely in both woody and herbaceous communities (e.g., Weller, 1987). The inverse relationship between mean size of individuals and density also has been well established (Harper, 1977).

Many experiments in controlled environments and in the field have shown the "Sukatschew effect": that the addition of nutrients to the soil medium enhances thinning in single species populations and increases the mean weight of individuals (e.g., Yoda *et al.*, 1963; Harper, 1977). Similar results have been found in two species mixtures (Bazzaz and Harper, 1976). The addition of water to naturally growing *Erodium* populations reduced their effective population sizes (Rice, 1990). Similarly, in a simulated global change experiment, elevated CO_2 levels and increased temperatures resulted in the reduction in the number of surviving individuals (S. Morse and F. Bazzaz, unpublished).

At the community level and in the field, several investigators working with communities distinctly different in composition and structure have found an inverse relationship between site fertility and species diversity (in a short grassland, Grime *et al.*, 1979; in tropical forests, Ashton, 1977; Huston, 1980). Raynal and Bazzaz (1975) added nutrients and water to field plots of annuals. They found that, despite only small changes in the mean weight of individuals, population sizes of the dominant species *Ambrosia artemisiifolia* declined with time in both plots amended with either water or nutrients. Further, the surviving individuals in the water addition treatment failed to reproduce, thereby decoupling reproduction from growth (Fig. 14.3). When different amounts of nitrogen were added to field plots, there was a general reduction in plant species diversity and a change in community composition with increasing fertility (Tilman, 1987). Different species in this community differed in their response to nutrient addition—some decreased and others increased. These patterns could be extended to consider genotypes within populations. With resource augmentation, less responsive genotypes or species probably will be replaced in these populations by more responsive ones. For example, in experimental populations of the annual plant *Phlox drummondii*, the addition of nutrients resulted in the elimination of certain genotypes from the population and a reduction in the contribution of others (Bazzaz *et al.*, 1982). In experimental communities of annual plants, changing nutrient levels, water availability, and temperature during the early part of the growth season led to small changes in total community biomass, but large changes in the number of individuals and especially in relative contributions of species to community biomass (Fig. 14.4). Thus, there can be changes in population size, effective population size, community composition, and diversity with climate change. Change in community structure may be the most critical result of climate change. How this change might, directly and indirectly, influence ecosystem function remains an area of active speculation.

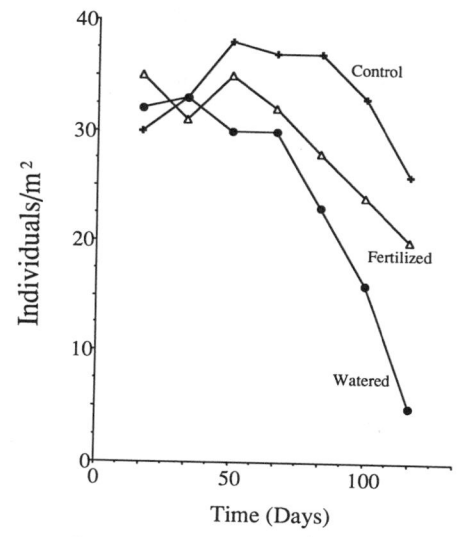

Figure 14.3 Response of *Ambrosia* population (+) to the addition of nutrient (△) and water (●) to field plots.

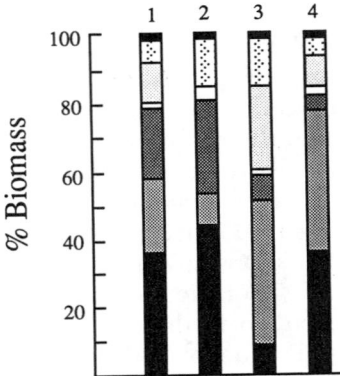

Figure 14.4 The influence of modification of environmental factors simulating control (1), a cold spring (2), a wet spring (3), and nitrogen limitation (4) on the contribution of co-occuring annual plants to community biomass. The species from the top down are *Datura, Chenopodium, Polygonum, Amaranthus, Abutilon, Ambrosia,* and *Setaria.*

A general model is that resource augmentation leads to increased growth and, therefore, increased competition among neighboring individuals (Harper, 1977; Grime, 1987; Tilman, 1988). On a population level that, in turn, results in the death of the disadvantaged individuals and, in some cases, the failure of some of the survivors to reproduce—a reduction both in population size (N) and in effective population size (Ne/N) (Fig. 14.5).

Future climates are predicted to entail large-scale global changes in the levels of plant resources and response controllers. Increased mean annual global temperature, particularly in higher latitudes, changes in

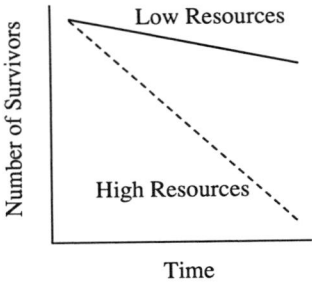

Figure 14.5 General model of the relationship between resource level and survivorship of plants in a populations.

precipitation patterns, and an overall increase in CO_2 concentrations likely will result largely from human activities. These changes can influence the responses of individuals and also the behavior of populations and communities differently. It is therefore critical to understand how these various levels of aggregation might be influenced by components of global change, particularly those that involve resource augmentation such as CO_2 increase, nitrogen deposition, and changed water availability through enhanced water use efficiency in a high CO_2 environment. This understanding is directly relevant to modeling and other scaling approaches.

"Global change" also very likely will involve simultaneous alterations in several environmental factors to which individuals, populations, and species also may respond, independently and differentially. For example, in experiments in which both CO_2 and air temperature were manipulated, increased temperature and CO_2 resulted in an increase in the rate and percentage of germination, but a decrease in the number of survivors of *Amaranthus*. Size inequality was not affected. In contrast, in *Abutilon*, a species that occurs in communities with *Amaranthus*, elevated temperatures in conjunction with high CO_2 levels reduced germination, did not influence mortality, but reduced seedling survivorship. Size inequality was lowered greatly under these conditions (S. Morse and F. A. Bazzaz, unpublished). Elevated temperatures at ambient CO_2 resulted in a reduction of above-ground biomass of plants grown as individuals and in populations. However, at elevated CO_2 levels, higher temperatures increased the biomass of populations but reduced the biomass of plants grown individually. With these complex and varied responses, a search for patterns clearly is difficult but necessary, lest we become paralyzed from making predictions in this enormous natural complexity.

IV. Models as Tools for Scaling: Single Individual and Single Species Models without Competition

Biological activities occur over a wide range of time, size, and spatial scales (Osmond *et al.*, 1980). In order to connect these levels and produce prediction at high levels of organization, modeling continues to be the most powerful tool for scaling biological data. Because of the immediate need for solutions to many environmental problems, emphasis in models necessarily will be placed on their predictive power and simplicity rather than on their traditional utility to highlight areas needing further investigation. Over the last decade, many physiologically based plant models have been developed that are used extensively in agricultural sciences. Several growth models with varying degrees of complexity are now available (see Chapter 7). They usually are based on measurements of certain responses at the leaf or whole-plant level and on the nature of the physical environment, especially light. Stand-level models also have been developed using this approach (Rauscher *et al.*, 1990; Wang and Jarvis, 1990a,b) These models range in complexity from those that use only a few parameters to predict yield to those that develop complex algorithms to describe the detailed behavior of the stand and to make predictions about many other characters in addition to plant yield. Remarkably, some models have shown that only a knowledge of the amount of absorbed photosynthetically active radiation (APAR) by the plant canopy and of the conversion efficiency of the species considered is necessary for good predictions of plant productivity in many single- and in some several-species stands (Russell *et al.*, 1989; Norman, 1991).

These models are based largely on notions about the behavior of forest plantations or annual crops, where interest is focused mainly on plant productivity or yield on a ground area basis, or on the premise that the stand will be started again from seed in the following growing season, usually from uniform genetic material. These models do not consider the enormous variation among individuals of natural population nor are they concerned with the differential fitness of these individuals and their potential contribution to future generations, that is, they do not have a demographic component. Therefore, they may be of limited utility in predicting how complex natural ecosystems may respond to climate change over several generations.

V. Models with Competition among Neighbors: A Step Closer to Natural Ecosystems

Although many of the available population models were intended for all organisms, they have been less useful for plants than for mobile animals. Many of these models are based on the Lotka–Volterra population growth equations. These models have many shortcomings associated with the logistic equation itself. There are, however, a few competition models based on the Lotka–Volterra formulation that were specifically developed for plants.

The simplest and most common of these plant-based models are those developed for annuals. These models treat plants as monocultures and emphasize yield–mean plant weight relationships or yield–plant density relationships (e.g., Yoda *et al.*, 1963; Hara, 1986; see Weller, 1987 for a review). Like these models, many models of multispecific competition also assume that light is the resource limiting plant growth and therefore use the pattern of light distribution within the canopy as the basis of the model (e.g., Barns *et al.,* 1990).

Several other competition models that concern the behavior of plant populations for more than one generation have been developed. Some consider potential seed production, density, and a measure of resource use efficiency to predict the number of seeds produced by a plant (e.g., Firbank and Watkinson, 1987). Neighborhood models (e.g., Weiner, 1982) are more complex. They consider distance between neighbors, the influence of an individual of one species on another individual of another species, the number of individuals of each species in a given neighborhood, the number of neighborhoods, the number of species present, the reproductive output of individuals, and the maximum seed production of an individual in the absence of competition. These and several other plant competition models (e.g., Connolly, 1987) do emphasize population level parameters. Tilman (1988) used a more mechanistic approach to model plant response. In his models, plants interact through the use of limiting resources, particularly light and nutrient availability and their influence on plant growth and allocation. Long-term predictions from the Tilman models correspond well with the observed changes in succession in a low nutrient grassland ecosystem. Despite the attractiveness and simplicity of these models, their generality to other ecosystems has been questioned (see Grace, 1990, 1991 Shipley and Peters, 1990). In particular, the models have no spatial component and therefore are less applicable in patchy environments (Pacala and Silander, 1990). Recent models such as the ecological field theory (Wu *et al.,* 1985; Clark, 1990) and those of Shugart (1987), Pastor and Post (1988), and others using FORET-type models promise to enhance our ability to predict the response of plants to their complex environment. Many of these models emphasize either the physiology or the population biology of plants, but less commonly both. Happily, some recent attempts in modeling (e.g., Luxmoore *et al.,* 1990; Hilbert *et al.,* 1991; Chapter 7) are trying to bridge the gap between these two approaches, particularly with respect to how plants respond to elevated CO_2 levels.

A physiological parameter that has direct influence on population dynamics and stand structure is seed germination. The pattern of seedling emergence can vary in different locations under different environmental combinations. This pattern sets the stage for the determination of stand structure (density, dispersion, genetic structure) and the hierarchies of height, weight, and reproductive output that result from the interaction of individuals with each other and with their environment (e.g., Harper, 1977). Therefore, seed germination patterns form a major link between the ecophysiology and population biology of plants. We have developed a simple simulation model that illustrates this approach while exploring the effects of elevated CO_2 levels on the population dynamics of an annual plant (Bazzaz et al., 1992). We grew populations of *Abutilon theophrasti* at four densities and two CO_2 levels and examined population level seed output and seed germinability. As in many other studies, individuals at low density produced more seeds under elevated CO_2 levels than those under ambient CO_2 levels. However, at high density, seed production fell to zero. A simple difference equation simulated population density in successive generations:

$$N_{t+1} = N_t f(CO_2, \text{density})\ g, \tag{1}$$

where N_t is the population density at time t, f is fecundity as a function of density and CO_2 environment, and g is the germination percentage. Because plants grown at elevated CO_2 levels produced more seeds at low density, population size rapidly rose. However, at high density the populations crashed due to a complete failure of seed set. Given the observed germinability of seeds from the various parental environments, the population at ambient CO_2 levels will grow to a stable initial population size but the population at elevated CO_2 levels will rise quickly to high density and then become extinct. If germination levels are varied, both populations exhibit a wide range of behavior, from stability to chaotic oscillations and extinction. However, the range of stability is narrower for the population at elevated CO_2 levels (Fig. 14.6).

The effect of a persistent seed pool (which is common in this and many other colonizers) was considered also using the paired equations

$$N_t = S_e g$$
$$S_{t+1} = N_t f(CO_2, \text{density}) + p(S_t - N_t), \tag{2}$$

where S_t is the size of the seed pool at the beginning of year t and p is the fraction of seeds remaining after germination that survive to the next year. In this model, a presistent seed pool stabilizes both populations under a wide range of conditions, although the elevated CO_2-grown population tends to be less stable under many combinations of germination and persistence parameters. Our model shows that an increase in reproductive output does not simply mean that the population size will be larger. Rather, population-level processes must be considered directly to understand the consequences of an essentially physiological response to the abiotic environment.

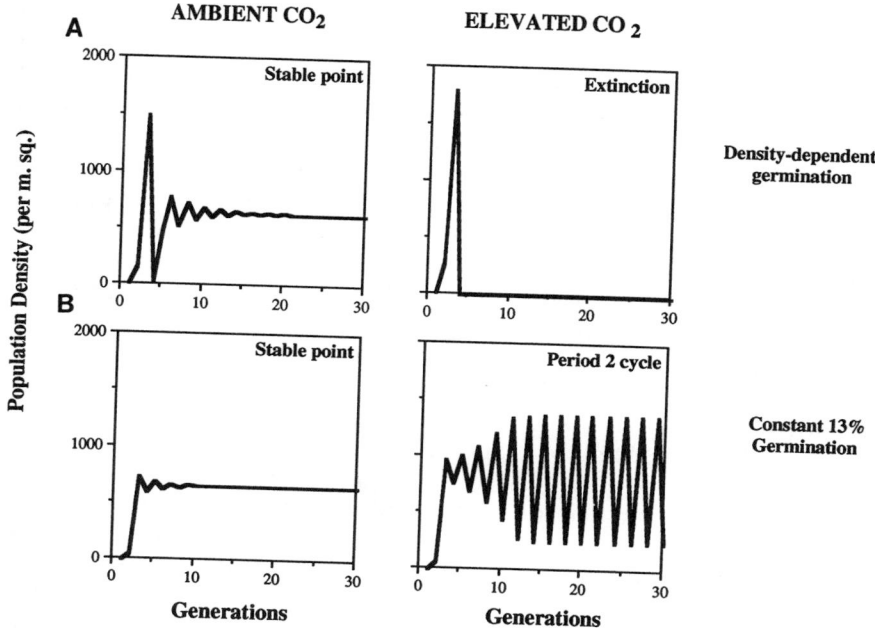

Figure 14.6 Simulated dynamics of *Abutilon theophrasti* populations based on density dependence of fecundity at ambient (left) and elevated (right) CO_2 concentrations. (A) Simulation I: Observed germination behavior. (B) Simulation II: Density-independent germination, 13%. See text for further explanation.

VI. Factors that Can Compromise the Simplicity of Models

To achieve maximum efficiency, simplicity, and power, modelers have attempted to reduce the number of environmental parameters considered and the potential responses of plants to that environment. In natural environments, however, several factors should be considered in the development and use of these models. There are, of course, situations in which simplicity is sufficient and others in which complexity is necessary. Thus, the challenge is to develop the most appropriate approach for the different circumstances. In addition to structural complexity, factors that can frustrate simple modeling efforts include environmental heterogeneity, incongruent resource availability, phenotypic variation, and complex stand structure. In natural habitats, these parameters are related intimately to each other, and in practice, they are difficult to separate. Each simultaneously affects and is influenced by the others.

A. Environmental Heterogeneity

Evolutionists and ecologists appreciate that natural environments are heterogeneous both in space and time (e.g., Levins, 1968; Kosola and Pickett, 1991). This heterogeneity occurs on many scales (Allen and Starr, 1982), elicits different responses from different species and genotypes (Bazzaz, 1987), and has significant short-term and long-term evolutionary consequences (Levins, 1968; Lewontin, 1974). However, scaling up from the responses of individuals to those at higher levels of biological organization has not considered this variation adequately (but see Pacala and Silander, 1990). Although environmental heterogeneity for many parameters now can be measured quite accurately and over fine spatial and temporal scales, plant responses to that variation have not been considered fully. Further, what is important is not the "measured" level of a particular parameter of the environment, but the level of variation "perceived" by the plant for a particular plant response. Collectively, this variation may be considered as an "ecologically relevant variation." The classic, but unfortunately underused, phytometer technique can be a powerful tool to assess the response of plants to the variation in their natural environment. Norms of reaction of individuals and populations on resource gradients, such as light, water, nutrient, and CO_2 (see Austin and Austin, 1980; Bazzaz, 1987; Tilman, 1988; Keddy, 1991), and niche breadth along several gradients can facilitate the identification of this variation. Aggregation and modeling based on these kinds of responses will lead to more realistic, long range predictions of ecosystem response to global change than will models based on average responses.

Most studies only consider plant response to one factor, or to a few, each determined separately and used to make models to predict plant behavior. There is a tendency to search for a single factor that controls the response of plants to their environment. Although the multifactorial nature of the environment of the plant has been recognized fully by ecologists for decades (e.g., Billings, 1952), much work still depends heavily on knowledge of plant responses to single factors (but see Chapin *et al.,* 1987; Chapin, 1991; Mooney 1991). Although in some cases this has proven sufficient, for example, the use of intercepted light quantity to predict stand productivity, some responses of plants to multiple factors may not be predictable from responses to individual factors, even in simple communities of annual plants (Parrish and Bazzaz, 1979; Zangerl and Bazzaz, 1983).

In ecosystems in which gap creation and filling is an active process, the resulting patchiness can introduce much uncertainty in commonly measured parameters such as carbon and water budgets, and can compromise predictions of future ecosystem responses to a changing environment greatly. In these situations, parts of the landscape can be a major sink for carbon (i.e., gaps with fast growing plants), whereas others may be at equilibrium with the atmosphere (i.e., mature phase patches), with a net flux near zero. In this case, ecosystem-level techniques for measuring carbon flux (e.g., eddy correlations, Chapter 5) and remote sensing spectrometric techniques (see Chapter 1) must contain all these patches in the appropriate mix of areal extend and biological activity if we hope to assess carbon flux at the ecosystem, landscape, or regional levels accurately.

B. Incongruent Availability of Resource

Even a good knowledge of how plants respond to relevant single environmental factors may be insufficient for scaling purposes and for the development of predictive models in some ecosystems. Over short distances, not only do individual resource availabilities vary, but so do the relative abundances and proportions of multiple resources. At all levels of organization, it is becoming clear that responses of ecologically relevant processes (e.g., leaf photosynthesis, whole tree growth, litter decomposition) to one resource or environmental variable often are modified by the status of other variables, sometimes in complex ways. For example, the response of temperate tree seedlings to elevated CO_2 concentrations is proportionally much greater under higher than under lower nutrient levels, but less under high than under low light. (Bazzaz *et al.*, 1990).

The relative proportions of resources not only vary spatially, but also temporally across the day, across the season, and over time scales of years. Within a day, for example, seedlings regenerating in gaps generally receive their highest levels of light at times when other critical factors important to photosynthesis, such as water status, humidity, and air temperature, are least favorable. This temporal incongruency of resources has been shown to be responsible for mid- and late-day depressions is seedling gas exchange (Tenhunen *et al.*, 1987). On the longer time scale, anthropogenically induced changes such as the relative proportions of SO_2 and CO_2 are also continuously changing. Research has demonstrated strong interactive effects of these two gases on plant photosynthesis and growth (Carlson and Bazzaz, 1982). Thus, global change modeling efforts must consider multiple resources, both in space and time, since the response to one factor can be largely contingent on the status of another factor.

C. Phenotypic Variation

Genetic variation, interacting with environmental heterogeneity, has the potential to generate much phenotypic variation in the physiological responses of individuals within a population. Thus, individuals of a population can differ greatly in their carbon gain capacity and their allocation of resources to various activities and functions. Under these situations, the response of individuals may have little correspondence with the responses of the population or its future in nature. In such cases, scaling from the individual to the population potentially leads to imprecise conclusions. This situation is best illustrated by examining the presumed photosynthetic response of sugar maple (*Acer saccharum*) to light intensity (see Fig. 14.7). Seedling responses have been studied under controlled conditions in a glasshouse, in the field where measurements were made on a few individuals for a short duration, and under field conditions in several microsites and over an entire growing season . Clearly, population level light-response curves are the nearest to what might happen in nature and give an accurate assessment of the responses of *Acer saccharum* populations to the light environment. Scaling efforts based on this type of response will have a better chance of more accurately predicting the behavior of a population in the spatially and temporally changing environments of the field.

To what degree, then, do we have to consider variation among individuals when we wish to make predictions about the response of ecosystems to global change? Are there rules that can help us arrive at such a decision? Theory and experiments over the last two decades have produced some relevant results. Response breadths of populations to environmental gradients are wider for pioneer and other invading species than for species in mature, more stable vegetation (Bazzaz, 1986,1987). Pioneer species are generally more plastic in their responses and are able to respond rather quickly to changing levels of resources (Bazzaz, 1979; Grime, 1979; Grime *et al.,* 1988). The degree of plasticity for various traits and the pattern of heterogeneity in the environment interact to determine the nature of the ecological responses of individuals within a population. If this response could be known with certainty, then the responses of individuals could be scaled correctly to that of the population. Although obtaining this knowledge for many species is time consuming, there exists enough knowledge to make a few generalizations. For example, species with individuals that tend to have broad response breadths (i.e., generalized niche) probably can be modeled more faithfully because the individuals are more similar to each other in terms of their response to their environment (Fig. 14.8). In contrast, narrow-niched individuals of a population can respond quite differently from each other; these differences among individuals may have to be considered in modeling. Another generalization at an even higher level of aggregation is the tendency of early successional plants to have lower variability in responses than do late successional plants (see Bazzaz, 1987).

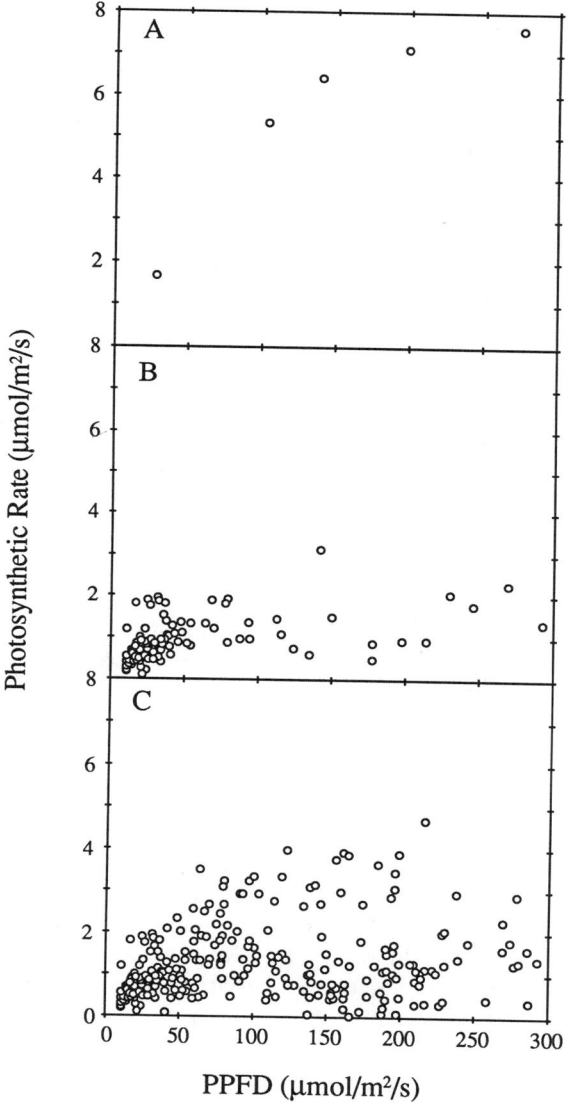

Figure 14.7 Photosynthetic rate of sugar maple (*Acer saccharum*) in relation to light. (A) "Ecophysiological response" of plants grown under glasshouse conditions (Loach, 1967). (B) Field data taken on a few individuals in a uniform parcel of a mixed forest (Weber *et al.*, 1985; "ecological response"). (C) Field data from many individuals taken over a growing season in various locations in the forest, including gaps of various sizes and under close canopy (T. W. Sipe and F. A. Bazzaz, unpublished data; "population-level response").

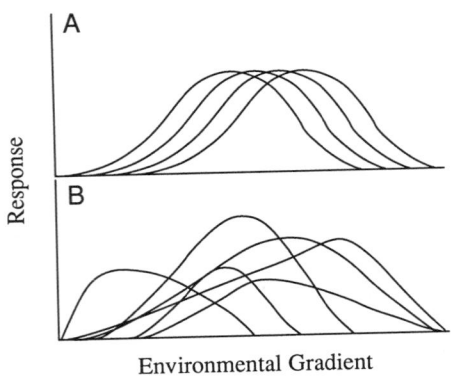

Figure 14.8   Two contrasting types of response to environmental gradients. In Population A, individuals differ slightly in their response to the gradient, whereas in Population B, individuals greatly differ in their response. We predict that modeling the response to global change of Population A would be easier, and perhaps more accurate, then modeling that of Population B.

Uncovering these sorts of general response patterns is absolutely necessary for ecologits to contribute significantly to the challenges of global environmental change. In a global change context, relevant generalizations are likely to emerge faster from the interplay of physiological, population, and community ecology than from various disciplines acting independently.

In conclusion, to understand the relationships between the physiological ecology of individuals and population and community response, particularly with respect to global change, the following points should be emphasized.

1. The "Big Leaf" concept, commonly used in modeling ecosystem responses to environmental change, has important biological limits. Because individuals in complex ecosystems are likely to differ greatly in response to changes in their environment, we need to consider these differences, especially in predicting long-term ecosystem response to global change.

2. Plants usually have neighbors that influence their environment and therefore influence their growth and function. These neighbors could be competitors, supporters, or both, but are always resource modifiers. Models that fail to incorporate neighborhood effects are therefore less useful in understanding responses of ecosystems to climate change.

3. Environmental resource modifications influence, and are influenced by, physiological, morphological, and architectural attributes of the

interacting individuals; a dialectic interaction that is changing with time.

4. These physiological, morphological, and architectural attributes have different potentials for change (plasticity and acclimation) when the environment changes, including acclimation or evolution through natural selection.

5. If information about Items 2 and 3 is available and the size structure of the population (or the ecosystem) is known, predictive models with a much higher degree of certainty can be developed.

6. Plant populations and communities are made up of individuals of different sizes, ages, and identities. Considering only the response of an "average individual" during modeling efforts may mask important variations caused by genotypic differences, environmental circumstances, and neighbor influences.

7. Position of individuals relative to each other may determine physiological performance, allocation patterns, survivorship probability, and, ultimately, effective population size.

8. Resource augmentation (e.g., CO_2, N, H_2O) may have different and often opposing consequences for individuals and populations. In general, resource addition leads to increased growth of the fewer surviving individuals. Excess resources may also decouple reproduction from enhanced growth and greatly reduce effective population size (Ne/N) and, therefore, the long-term future of the ecosystem.

9. Population-level considerations may not be necessary for short-term predictions about the effects of climate change on natural ecosystem function (e.g., stand productivity, nutrient cycling, water and carbon budgets), but are essential for long-term predictions. Population dynamics and community succession are likely to be modified by elements of the global change, and alterations in these components of biological systems may have a pronounced impact on ecosystem structure and function. Without their inclusion, ecosystem models will have less predictive power than they might otherwise have.

Acknowledgments

I thank Sean Thomas and David Ackerly for their contributions and Susan Bassow, Glenn Berntson, Eric Fajer, and Peter Wayne for reviewing this chapter and making many useful suggestions.

References

Allen, T. F. H., and Starr, T. B. (1982). "Hierarchy: Perspectives for Ecology Complexity." University Chicago Press, Chicago.

Ashton, P. S. (1977). A contribution of rainforest research to evolutionary theory. *Ann. Missouri Bot. Gard.* **64,** 694–705.

Austin, M. P., and Austin, B. O. (1980). Behaviour of experimental plant communities along a nutrient gradient. *J. Ecol.* **68,** 891–918.

Barns, P. W., Beyschlag, W., Ryel, R., Flint, S. D., and Caldwell, M. M. (1990). Plant competition for light analyzed with a multispecies canopy model. III. Influence of canopy structure in mixtures and monocultures of wheat and wild oat. *Oecologia* **82,** 560–566.

Bazzaz, F. A. (1979). Physiological ecology of plant succession. *Annu. Rev. Ecol. Systematics* **10,** 351–371.

Bazzaz, F. A. (1984). Demographic consequences of plant physiological traits: Some case studies. *In* "Perspectives in Plant Population Ecology" (R. Dirzo and J. Sarukhan, eds.) pp. 324–346. Sinauer Publishers, Sunderland, Massachusetts.

Bazzaz, F. A. (1986). Life history of colonizing plants: Some demographic, genetic and physiological features. *In* "Ecology of Biological Invasions" (H. A. Mooney and J. A. Drake, eds.), pp. 96–110. Springer-Verlag, New York.

Bazzaz, F. A. (1987). Experimental studies on the evolution of niche in successional plant populations: A synthesis. *In* "Colonization, Succession and Stability" (A. J. Gray, M. J. Crawley, and P. J. Edwards, eds.) pp. 245–272. Blackwell, Oxford.

Bazzaz, F. A. (1991). Habitat selection in plants. *Am. Nat.* **137,** S116–S130.

Bazzaz, F. A., Ackerly, D. D., Woodward, I., and Rochefort, L. (1992). Effects of CO_2 enrichment on the density-dependence of reproduction in an annual plant and a simulation of population dynamics. *J. Ecol.* (in press).

Bazzaz, F. A., Coleman, J. S., and Morse, S. R. (1990). Growth response of seven major cooccurring tree species of the northeastern United States to elevated CO_2. *Can. J. For. Res.* **20,** 1479–1484.

Bazzaz, F. A., and Harper, J. L. (1976). Relationship between plant weight and numbers in mixed populations of *Sinapsis alba* (L.) Rabenh. and *Lepidium sativum* L. *J. Appl. Ecol.* **13,** 211–216.

Bazzaz, F. A., Levin, D. A., and Schmierbach, M. R. (1982). Differential survival of genetic variants in crowded populations of *Phlox*. *J. Appl. Ecol.* **19,** 891–900.

Billings, W. D. (1952). The environmental complex in relation to plant growth and distribution. *Q. Rev. Biol.* **27,** 251–265.

Caldwell, M. M. (1987). Plant architecture and response competition. *In* Potential and Limitations of Ecosystem Analysis, Ecological Studies" (E. D. Schulze and H. Zwolfer, eds.) Vol. 61, pp. 164–179. Springer-Verlag, Berlin.

Carlson, R. W., and Bazzaz, F. A. (1982). Photosynthetic acclimation to variability in the light environment of early and late successional plants. *Oecologia* **54,** 76–79.

Chapin, F. S. (1991). Integrated responses of plants to stress. *BioScience* **41,** 29–36.

Chapin, F. S., Bloom, A. J., Field, C. B., and Waring, R. H. (1987). Plant responses to multiple environmental factors. *BioScience* **37,** 49–57.

Clark, J. S. (1990). Integration of ecological levels: individual plant growth, population mortality and ecosystem processes. *J. Ecol.* **78,** 275–299.

Connolly, J. (1987). On the use of response models in mixture experiments. *Oecologia* **72,** 95–103.

Firbank, L. G., and Watkinson, A. R. (1985). a model of interference within plant monocultures. *J. Theor. Biol.* **115,** 291–311.

Firbank, L. G., and Watkinson, A. R. (1987). On the analysis of competition at the level of the individual plant. *Oecologia* **71,** 308–317.

Ford, E. D., and Diggle, P. J. (1981). Competition for light in a plant monoculture modelled as a spatial stochastic process. *Ann. Bot.* **48,** 481–500.

Grace, J. B. (1990). On the relationship between plant traits and competitive ability. *In* "Perspectives on Plant Competition" (J. B. Grace and D. Tilman, eds.), pp. 51–65. Academic Press, San Diego.

Grace, J. B. (1991). A clarification of the debate between Grime and Tilman. *Funct. Ecol.* **5,** 583–587.

Grime, P. J. (1979). "Plant Strategies and Vegetational Processes. Wiley Chichester, UK.

Grime, P. J. (1987) Dominant and subordinate components of plant communities: Implications for succession, stability and diversity. *In* "Colonization, Succession and Stability" (A. J. Gray, M. J. Crawley, and P. J. Edwards, eds.). Blackwell, Oxford.

Grime, J. P., Hodgson, J. G., and Hunt, R. (1988). "Comparative Plant Ecology: A Functional Approach to Common British Species," p. 742. Unwin Hyman, London.

Hara, T. (1986). Effects of density and extinction coefficient on size variability in plant populations. *Ann. Bot.* **57,** 885–892.

Harper, J. L. (1977). "Population Biology of Plants." Academic Press, London.

Hilbert, D. W., Reynolds, J. F., and Bazzaz, F. A., (1991). Effects of carbon dioxide enrichment on plant communities. I. Models of competition in species mixtures based on scaling single plant responses. *Bull. Ecol. Soc.* **72,** 142.

Houghton, J. T., Jenkins, G. J., and Ephraums, J. J. (eds.) (1990). "Intergovernmental Panel on Climate Change. Climate Change: The IPCC Scientific Assessment." Press Syndicate of the University of Cambridge, Cambridge.

Huston, M. (1980). Soil nutrients and tree species richness in Costa Rican forests. *J. Biogeo.* **7,** 147–157.

Keddy, P. A. (1991). Working with heterogeneity: An operator's guide to environmental gradients. *In* "Ecological Heterogeneity." (J. Kolasa and S. T. A. Pickett, eds.), pp. 181–201. Springer-Verlag, New York.

Kosola, J., and Pickett, S. T. A. (eds.) (1991). "Ecological Heterogeneity." Springer-Verlag, New York.

Levins, R. (1968). "Evolution in Changing Environments." Princeton University Press, Princeton, New Jersey.

Lewontin, R. C. 1974. "The Genetic Basis of Evolutionary Change." Columbia Univ Press, New York.

Loach, K. (1967). Shade tolerance in tree seedlings. I. Leaf photosynthesis and respiration in plants raised under artificial light. *New Phytologist* **69,** 273–286.

Lubchenco, J., *et al.* (1991). The sustainable biosphere initiative: An ecological research agenda. *Ecology* **72,** 371–412.

Luxmoore, R. J., King, A. W., and Tharp, M. L. (1990). Approaches to scaling up physiologically based soil-plant models in space and time. *Tree Physiol.* **9,** 281–292.

Mooney, H. A. (1991). Plant physiological ecology: Determinants of progress. *Func. Ecol.* **5,** 127–135.

Norman, J. M. (1989). Synthesis of canopy processes. *In* "Plant Canopies: Their Growth, Form and Function" (G. Russell, B. Marshall, and P. G. Jarvis, eds.), pp. 161–175. Cambridge University Press, Cambridge.

Osmond, C. B., Bjorkman, O., and Anderson, D. J. (1980). "Physiological Preccesses in Plant Ecology toward a Synthesis with Atriplex." Springer-Verlag, Berlin.

Pacala, S. W., and Silander, J. A., Jr. (1985). Neighborhood models of plant population dynamics. I. Single-species models of annuals. *Am. Nat.* **125,** 385–411.

Pacala, S. W., and Silander, J. A., Jr. (1987). Neighborhood interference among velvet leaf *Abutilon theophrasti*, and pigweed, *Amaranthus retroflexus. Oikos* **48,** 217–224.

Pacala, S. W., and Silander, J. A. Jr. (1990). The application of plant population dynamic models to understanding plant competition. *In* "Perspectives on Plant Competition" (J. B. Grace and D. Tilman, eds.), pp. 67–91. Academic Press, San Diego.

Parrish, J. A. D., and Bazzaz, F. A. (1979). Difference in pollination niche, relationships in early and late successional plant communities. *Ecology* **60**, 597–610.

Pastor, J. and Post, W. M. (1988). Response of northern forests to CO_2-induced climate change. *Nature* **334**, 55–58.

Rauscher, H. M., Isebrands, J. G., Host, G. E., Dickson, R. E., Dickmann, D. I., Crow, T. R., and Michael, D. A. (1990). ECOPHYS: An ecophysiological growth process model for juvenile poplar. *Tree Physiol.* **7**, 255–281.

Raynal, D. J., and Bazzaz, F. A. (1975). Interference of winter annuals with *Ambrosia artemisiifolia* in early succession fields. *Ecology* **56**, 35–49.

Rice, K. (1990). Reproductive hierarchies in Erodium: Effects of variation in plant density and rainfall distribution. *Ecology* **71**, 1316–1322.

Russell, G., Jarvis, P. G., and Monteith, J. L. (1989). Absorption of radiation by canopies and sand growth. *In* "Plant Canopies: Their Growth, Form and Function" (G. Russell, B. Marshall, and P. G. Jarvis, eds.), pp. 21–39. Cambridge University Press, Cambridge.

Shipley, B., and Peters, R. H. (1990). A test of the Tilman model of plant strategies: Relative growth rate and biomass partitioning. *Am. Nat.* **136**, 139–153.

Shugart, H. H. (1987). Dynamic ecosystem consequences of tree birth and death patterns. *BioScience* **37**, 596–602.

Silvertown, J. (1987). "Introduction to Plant Population Biology," 2nd Ed. Longman, London, England.

Tenhunen, J. D., Pearcy, R. W. and Lange, O. L. (1987). Diurnal variation in leaf conductance and gas exchange in natural environments. *In* "Stomatal Function" (E. Ziegler, G. D. Farquar, and I. R. Crowan, eds.), pp. 323–351. Stanford Univ. Press, Stanford, California.

Tilman, D. (1987). Secondary succession and the pattern of plant dominance along experimental nitrogen gradients. *Ecol. Monogr.* **57**, 189–214.

Tilman, D. (1988). "Plant Strategies and the Structure and Dynamics of Plant Communities." Princeton Univ. Press, Princeton, New Jersey.

Wang, Y. P., and Jarvis, P. G. (1990a). Description and validation of an array model-MAESTRO. *Agric. For. Meteorl.* **51**, 257–280.

Wang, Y. P., and Jarvis, P. G. (1990b). Influence of crown structural properties on PAR absurphon, photosynthesis, a transpiration in Sitka spruce: Application of a model (MAESTRO). *Tree Physiol.* **7**, 297–316.

Watkinson, A. (1980). Density dependence in a single species population of plants. *J. Theor. Biol.* **83**, 345–357.

Watkinson, A. (1981). Interference in pure and mixed populations of *Agrostemona githago. J. Appl. Ecol.* **18**, 967–976.

Weber, J. A., Jurik, T. W., Tenhunen, J. D., and Gates, D. M. (1985). Analysis of gas exchange in seedlings of *Acer saccharum:* Integration of field and laboratory studies. *Oecologia* **65**, 338–347.

Weiner, J. (1982). A neighborhood model of annual plant interference. *Ecology* **63**, 1237–1241.

Weller, D. E. (1987). A re-evaluation of the -3/2 power rule of plant self-thinning. *Ecol. Monogr.* **57**, 23–43.

Wu, H., Sharpe, P. J. H., Walker, J., and Penridge, L. K. (1985). Ecological field theory: A spatial analysis of resource interference among plants. *Ecol. Model.* **29**, 215–243.

Yoda, K., Kira, T., Ogawa, H., and Hozumi, K. (1963). Self-thinning in over-crowed pure stands under cultivated and natural conditions. *J. Biol.* (Osaka City Univ.) **14**, 107.

Zangerl, A. R., and Bazzaz, F. A. (1983). Responses of an early and a late successional species of *Polygonum* to variations in resource availability. *Oecologia* **56**, 397–404.

15

Scaling the Population Level: Effects of Species Composition and Population Structure

James S. Clark

I. Introduction

Scaling involves problems related to transferring information among spatial domains, temporal domains, and levels of complexity. For example, leaf gas exchange is affected by dynamics of individual plants, populations, ecosystems, and landscapes. Scaling the exchange of gases through stomates to whole canopies produces bias as a result of interactions among leaves that may become obvious only on the new scale. These interactions involve a feedback of transpiration rate on itself through its contribution to boundary-layer humidity (Jarvis and McNaughton 1986, Jarvis, Chapter 6). The bias contained in a predicted whole-canopy transpiration rate might be attributed to a failure to incorporate adequately the contributions and responses of leaves to the changing environment they help produce. Identifying the cause for biases that result from such interactions becomes increasingly difficult as the numbers of indirect effects and feedbacks increase.

Populations are of potential importance for processes like gas exchange, nutrient cycling, and hydrology, because these processes may involve microbially mediated transformations and plant organs on the one hand and ecosystems on the other (Fig. 15.1). Consequences of microbial or leaf activities on the landscape level may depend on how those activities are organized at the intermediate scales by population-level phenomena. Yet it may not always be desirable to consider population-level phenomena when seeking to understand processes that depend on local and regional scales. Population processes may involve consider-

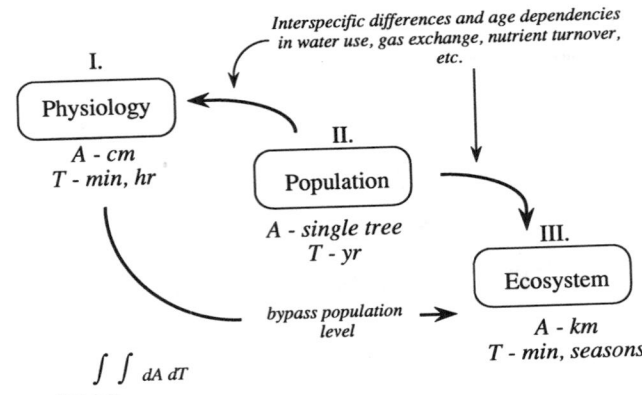

Figure 15.1 Population-level processes influence processes at other spatial (*A*) and temporal (*T*) scales and levels of complexity through species differences and population structure (e.g., age, density). These effects may be sufficiently large that they must be treated explicitly in ecophysiology models. If effects are small or dependencies weak, however, it may be possible to scale up to ecosystems or biomes without explicit consideration of population-level processes.

ation of a new set of scales and, thus, a great deal of added complexity. Tree population modelers have adopted their own favorite scales, the size of a mature individual tree and a time step of 1 year. These scales nicely accommodate processes most frequently considered by population biologists, including demography, life history traits, and disturbance. They are not always the best scales for describing many physiological responses and ecosystem processes. Assessment of the need to detail population-level contributions to biophysical problems should be among the first questions addressed by studies of landscape processes.

Responses of populations to changing environments also involve questions of scale. Preservation of biotic diversity, habitat quality, forest products, and fire management concern how populations react to their changing physical and biotic settings. Understanding these responses requires an understanding of how to transfer environmental effects on physiological and ecosystem processes to population-level responses involving resource use, life history, and demography. The spatial and temporal domains that best describe environmental influences on forests at leaf or landscape scales often contrast with those used to describe demography.

In this chapter, I focus on simplification and analysis to understand relationships as we move among scales. I first consider the question of whether to include population-level phenomena at all in studies that target other levels of complexity. For the reaminder of the chapter, I assume an established need to consider population processes, and I mention aspects of population models that have relevance to scaling. I focus on population structure and species composition as the most likely candidates for population-level variables that influence processes at other levels of complexity. I use examples of changes in population-level processes that probably influence how one might parameterize biophysical models. Because the error that arises in the process of scaling is often difficult to explain, the bulk of this chapter deals with the relationships among processes that allow for the simplification required if one is to identify the source of scaling error. I suggest that detailed species-specific parameterizations may often be less important in models than is the inherent heterogeneity contained in all tree populations. Although landscape-level responses to changing species composition sometimes may be modest, ignoring the higher moments of distributions of population structure may introduce large error at higher levels of organization.

II. When to Consider the Population Level in the Context of Scaling

The increasing need to understand landscape and subcontinental responses to environmental change presents an imperative for models that consider large areas (Bretherton, 1987). What factors must such models include? Although it is tempting to incorporate the effects of many processes, the level of detail contained in such models is limited because of the decreased understanding that attends complexity. With the increasing model size required to accommodate added realism, comprehensiveness, or complexity, the potential for analysis decreases, and those analyses that are conducted have ever-diminishing generality. The loss of generality results from the dependence of a result on a large number of functions and parameters that may be difficult to obtain.

In this section, I consider when it is necessary for such efforts to deal with the population level explicitly. Ecosystem or biome studies might choose to ignore population-level phenomena, on the grounds that increased complexity does not add understanding. Clearly, populations play a role at the ecosystem level as a result of gas exchange, hydrology, and nutrient cycling effects on growth, reproduction, and mortality, and vice versa. However, scaling from leaves to canopies, for example, might not benefit from a full demographic analysis for several reasons (Fig. 15.1). First, the population level requires knowledge of many life-history and demographic parameters. Translation of leaf-level measurements of gas exchange to canopies might best ignore all the complexity associated with the demography of populations bearing those leaves. Parameter estimation may be impossible and sensitivity analysis impractical.

Second, the time scales describing photosynthesis and respiration are similar at leaf and canopy levels, whereas population-level phenomena, such as recruitment and mortality rates, are best described in terms of years. One must consider the error introduced simply by shifting back and forth between time scales. Third, detail may prove superfluous if dependencies are weak or effects are averaged at some higher level of complexity. Because leaf production and display have a potential to respond to localized light availabilities, population density can vary over orders of magnitude while leaf area remains rather constant. Thus, it may be useful to pass directly from leaves to canopy (e.g., by treating the canopy as a large leaf; Running and Coughlan, 1988), bypassing the intermediate population level. Ideally, such decisions about the amount of detail to include are guided by information concerning sensitivity of results at one level to those processes that are best described at some other level. If the effects of demography on ecosystem- or biome-level variables under study are averaged out in space and time, it is often best to omit them.

Species composition is another population-level variable that often is ignored at higher levels of complexity. Weak dependencies, or an averaging effect, allow ecophysiology models to arrive at reasonable results using a single parameter set that ignores species differences (e.g., Running and Coughlan, 1988; Schimel *et al.,* 1991). If differences among species are small or parameter dependencies are weak, there is no need to complicate models with added detail.

There are instances in which population-level processes require explicit treatment in ecosystem studies; it is important to recognize when errors that arise in the process of scaling might derive from failure to consider species composition and population structure. Parameterizations of many relationships in ecophysiological models can depend importantly on species. If climate changes are sufficient to force changes in species composition, as has occurred frequently in the past, then water use, gas exchange, and nutrient cycling all might be altered in ways that could have important feedback effects on model parameterizations through effects on species composition.

There are many examples of substantial changes in species composition with recent climate changes that have potentially important feedback effects on landscape processes. The abrupt and dramatic increases of humic accumulation in lakes with the expansion of *Picea* in Labrador 6000 years ago (Engstrom and Hansen, 1985) is a transition of sufficient magnitude to warrant reparameterization of canopy interception and transpiration, litter quality, and leaf area. The increased humic levels suggest important changes in carbon cycling and water balance that likely are influenced by the low quality of *Picea* litter (Van Cleve *et al.*, 1983; Prescott *et al.*, 1989a,b) and its effects on hydrology (Nordén, 1990). The widespread decline of *Tsuga* throughout North America 4800 years ago occurred within a period of decades (Allison *et al.*, 1986). Litter quality and water use differences between *Tsuga* and the hardwood-dominated forests (Pastor *et al.*, 1984; McClaugherty *et al.*, 1985; Hill and Shackleton, 1989) that followed in many regions suggest a need for parameterizations responsive to changes in species composition. A rapid expansion of *Alnus* in many parts of eastern North America in the early Holocene Period likewise may have had important consequences for many ecosystem processes, including nitrogen cycling (Whitehead and Jackson, 1990). Minnesota is the site of a host of paleo studies suggesting rapid, dramatic changes in species composition that influence hydrology (Grimm, 1983), fire behavior (Grimm, 1984; Clark, 1989), landscape heterogeneity (McAndrews, 1966; Jacobson, 1979; Grimm, 1984; Clark, 1990b), and nutrient cycling (Dean *et al.*, 1984; Clark, 1990b; Almendinger, 1990). Population models for this region suggest that feedback effects of species composition on nutrient cycling could become extremely important if doubled CO_2 climate scenarios for the region (essentially warmer and drier) are realized (Pastor and Post, 1988). These model results support the notion that climate effects on species composition and nutrient cycling at the population scale would have dramatic repercussions for landscape heterogeneity. Responses of changing species composition to environmental change as climate and land use shift also appear to have important impacts in desert regions, where feedbacks among vegetation cover, nutrient cycling, and atmospheric processes can produce complex ecosystem-level responses (Schlesinger *et al.*, 1990). Jacobson *et al.* (1987) summarized rapidly changing species composition throughout the last millennium in eastern North America, including transitions that may result from changing temperature and precipitation (Gajewski, 1988).

Continual change in species composition at the biome level means that climate–vegetation relationships derived from modern landscapes may represent transient conditions. Models of environmental effects on ecosystems that do not accommodate changing species composition may retain parameterizations that become outdated as species composition responds to the environment. The transitions just mentioned occur on time scales comparable to those forecast for a doubling of atmospheric CO_2 concentration, for example, and each transition is of sufficient magnitude to suggest different parameterizations of ecophysiological models.

Models that fail to acknowledge the changing parameter values that arise as species composition responds to climate might produce misleading results in some cases.

There is no general rule about when ecosystem or biome models must consider population structure and species composition explicitly. Model sensitivity is an obvious criterion, but full sensitivity analyses of demographic models containing hundreds of parameters and fraught with nonlinearities are impractical and have not been done. In the following sections, I focus on simplification of the problem and the model structure, to explore how demography and species composition influence results at higher levels of complexity.

III. Patchiness and the Gap Paradigm

Scaling at any level must accommodate discontinuities in spatial and temporal domains. Population structure can be patchy, consisting of a mosaic of cohorts characterized by different histories (Watt, 1947; Bormann and Likens, 1979; Oliver and Larson, 1990). Processes that produce patches are generally stochastic and have probabilities that can change with time since the last event (Clark, 1989). Many landscapes may experience several such processes, some of which may be interdependent. An example of such interdependent processes on different scales that can be common in forests are tree-fall gaps that occur on relatively small areas and depend on the time elapsed since the last fire. Fire represents a second process that affects larger areas, thereby tending to synchronize patches created by tree-fall over larger spatial areas.

Patchiness can result from environmental variance or from aspects of the biota itself. Edaphic factors, climate, and land use produce spatial variability in vegetation. Some of these factors (e.g., soil texture) are not time dependent on the scales of interest in population modeling, although their *effects* still may vary with stand age and structure. Other environmental factors, such as disturbance, operate in an episodic fashion. If the occurrences of these events are independent of the biota, then the probability of an event may not change from year to year in a systematic fashion.

The probability of disturbance can change over time when that probability generally depends on characteristics of the biota itself. Fire depends on accumulation of fuels (Heinselman, 1973). Windthrow risk increases with tree size (King, 1986; Foster, 1988; Webb, 1988). Even cultural practices, such as forest harvest and slash-and-burn agriculture, become more likely with increased time elapsed since the last such event as a consequence of the changes that occur between events. The way in which this probability changes with time is important, because it defines landscape heterogeneity through its control over patch age structure (Clark 1991a). Because so many important processes depend on patch age (e.g., consider that population-model output generally is presented as a time series), it is important to understand something about this changing probability and whether the ability to describe it accurately is of consequence as we scale from a patch to a landscape.

Canopy gaps produced by adult mortality are among the more common types of disturbance with age-dependent probability. Across a forested landscape, much of the spatial variance in composition and structure is concentrated on scales roughly equivalent to the size of a large individual tree. The basis for this pattern lies in the sessile habit and the directional supply of one of the key resources that limits plant growth, that is, light. Mortality of one or more large individuals is often necessary for recruitment to occur (Watt, 1947; Grubb, 1977; Harper, 1977; Denslow, 1980; Runkle, 1982; Canham and Loucks, 1985). This recruitment phase is of limited duration, because resources on the forest floor are drawn down rapidly as gaps are closed by regrowth. Recruitment is low throughout a subsequent period of intense thinning (Oliver, 1981; Peet, 1981; Sprugel, 1985; Oliver and Larson, 1990), when density-dependent mortality may be greater than 5% annually, depending on plant growth rates. Only when growth rates slow does the probability of a new canopy gap increase substantially (Oliver and Larsen, 1990; Clark, 1991b). Even on an edaphically homogeneous landscape, a high degree of spatial heterogeneity develops as individual growth rates of trees decrease with age and canopy gaps form (Watt, 1947; Peet, 1981; Shugart, 1984). The landscape has an age structure, which in turn is responsible for variance in a host of response variables of potential interest at a landscape level.

Heterogeneity produced by canopy gaps plays an important role in simulation models of forest tree population ("gap") models. The variance concentrated on the scale of a mature tree makes it reasonable to assume this spatial scale in models that hope to capture some of the important dynamics of population-level processes. Gap models simulate forest tree dynamics as a collection of such patches (Botkin *et al.*, 1972; Shugart, 1984). On an individual patch, demography is episodic; recruitment of most species is concentrated in rather short episodes separated by periods of thinning (Fig. 15.2).

One of the more common ways to scale observations or model results from local scales to landscapes is to extrapolate from local averages. These averages do not capture population structure because that structure depends on the disturbance probability and the time development of the response variable between disturbances, both of which can be importantly nonlinear. This structure may be responsible for some of

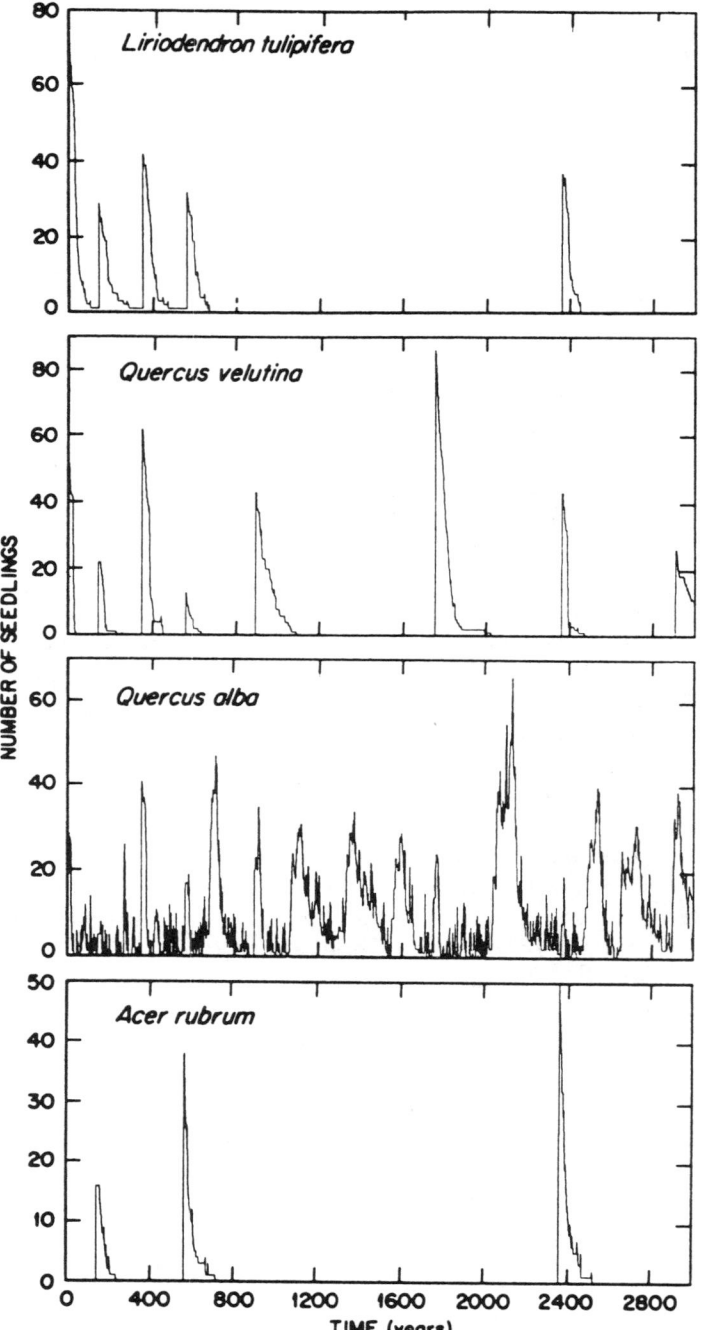

Figure 15.2 Gap model results for density on a single plot at the scale of an individual mature tree. The episodic nature of recruitment determines landscape structure, much of which is not well described by the mean. Figure reprinted, with permission, from Shugart (1984).

the problems that arise in the process of scaling. There have been few efforts to analyze this structure. Thus, it is poorly understood. Analysis of the role of this structure at different scales and levels of complexity might be facilitated by (1) retention of the higher moments of variable responsible for this structure and (2) simplification, to allow for analysis of their effects.

IV. Why Simplify?

Simplification is necessary when models produce results we do not understand. Unexpected developments that arise in the process of shifting among scales and levels of complexity are examples of such results. Large numerical models potentially allow for much realism and for experimentation with processes that develop too slowly, too rapidly, or on too small or too large an area to be observed directly. Models containing much realism include many indirect effects and nonlinearities that frustrate analysis. Gap models provide the potential realism necessary for exploring how a combination of variables with many indirect effects may influence populations and ecosystems. Like any model containing dozens of functions and hundreds of parameters, sensitivity analyses can be conducted only piecemeal. Answers to many questions regarding cause will require a combination of complementary models at several levels of complexity. The complementary analytical models must capture the effects of these nonlinearities on structure but retain the simplicity that permits analysis.

V. How to Simplify

I began this chapter by saying that at least two aspects of plant population dynamics play an important role in the scaling process: species composition and population structure. The potential for simplification exists for both of these aspects for different reasons.

A. Species Composition

We would like to limit the number of species included in models at the population level because each species implies a whole new set of parameters. If a large degree of overlap exists in parameter values among or within species (Noble and Slatyer, 1980; Shugart, 1984), a reduced species set or fewer parameters per species would reduce the complexity of the problem and increase the potential for analysis. Redundant information in parameter sets is a consequence of interdependencies (i.e., parameter correlations). Such correlations result from the fact that many species are faced with similar challenges (Noble and Slatyer, 1980; Tilman, 1982). Trade-offs posed by the need to allocate finite amounts of energy to several activities means that investment in one activity diminishes the energy that can be devoted to another activity. Indeed, a degree of correlation among life-history traits and competitive abilities in plants has been the subject of much attention. Negative correlations between competitive abilities for resources above and below ground have been observed for many species (reviewed by Smith and Huston, 1989); these correlations are primary components of the conceptual models of Tilman (1982, 1985, 1988) and Huston and Smith (1987; Smith and Huston, 1989). The fact that high maximum growth rate may be achieved at the cost of reduced competitive ability (Werner and Platt, 1976; Tilman, 1988) or longevity (Loehle, 1988) provides a link between life history and competition. Correlations between maturation time and longevity are postulated (Schaffer and Gadgil, 1975; Kozlowski and Weigert, 1986) and observed (Harper and White, 1974; Loehle, 1988) for plants, reflecting competing carbon sinks for finite energy over the life-span of a plant. More energy devoted to growth at a particular age necessarily means less energy available for reproduction. It is also possible that additional correlations could be found among other relevant variables, such as those that govern nutrient cycling (Chapin, 1980; Vitousek, 1982) and stress responses (Chapin, 1991).

These correlated traits provide a potentially useful tool for analysis. Simplified models could operate from a reduced parameter set, in which redundancies that result from such correlations are limited. Charnov and Berrigan (1990) argue that life-history theory of animals might focus on dimensionless parameters that incorporate these correlations among traits. A similar approach might provide a valuable way to simplify population models of plants for purposes of focusing on ecosystem-level responses to environmental change. Again, the degree to which such simplification can be accomplished depends on the particular application of the model. The usefulness of the approach rests with the potential to analyze and assess the importance of species differences on various scales. If ecosystem- or biome-level responses are weakly dependent on observed variability in parameters describing life history and nutrient use, then the complexity of such models might be reduced.

B. Population Structure

The patchiness conferred by population structure is described by functions that are nonlinear in patch age, producing effects that are difficult to assess in existing numerical models. An important aspect of plant population structure that aids the process of abstraction and analysis is the fact that many processes are linked through their relationships to the growth of individual plants. Indeed, these plant-growth dependencies often can be so strong that processes such as density-dependent mortality (Ford, 1975; Aikman and Watkinson, 1980; Ford and Diggle, 1981; Tait, 1988; Valentine, 1988), net ecosystem production (Clark, 1990a, 1991b), gap formation, the shifting mosaic structure (Clark, 1991a), and the self-thinning rule (Yoda *et al.*, 1963; White, 1981; Weller, 1987; Norberg, 1988) can be represented in terms of the parameters that describe individual plant growth. Spatial structure likewise can be approximated from knowledge of dispersal characteristics and distance-dependent effects of individual plants on limiting resources (Pacala and Silander, 1985; Clark, 1991c). These relationships provide important tools, because a common parameter set that applies to each of these processes allows analysis of dependencies across scales.

This approach has led to the development of analytical models of population-level phenomena in herbaceous monocultures (Ford, 1975; Aikman and Watkinson, 1980) and forest stands (Khil'mi, 1962; Tait, 1988; Valentine, 1988; Clark, 1990a). Clark (1990a, 1991b) has extended these efforts to include a number of ecosystem-level processes. These models express variables such as population density, net primary production, net ecosystem production, gap area, and the self-thinning rule in terms of stand age and parameters that describe individual plant growth (see Box 15.1). Because these relationships are defined by parameters describing individual plant growth, it is possible to analyze how growth at the level of individual plants influences processes on other spatial and temporal scales and at different levels of complexity. Box 15.1 contains an example in which population thinning (dx/da), net primary production (NPP), net ecosystem production (NEP), and the "self-thinning rule" all are expressed in terms of plant weight w. Clark (1990a, 1991b) used an example in which plant size is described by a two-parameter nonlinear (logistic) function. All these processes can be expressed in terms of the same two parameters and an additional parameter α, describing plant shape. Such simple functions accommodate the realtionships among these variables observed in the real world. (Fig. 15.3) and allow for analysis of interdependencies (Clark, 1991b).

Scaling these local (i.e., intracohort) dynamics of population and ecosystem processes to landscapes requires a model that accommodates the mosaic structure of populations. Gap models make the assumption that patches resulting from the stochastic nature of gap formation are independent. This assumption allows landscape-level dynamics to be treated as a collection of patches developing out of phase. Analytical models that

Box 15.1 Simplifying population- and ecosystem-level processes by exploiting their mutual relationships to individual plant growth.

Many of the processes that occur in a forest stand depend in some way on plant growth. In a crowded even-aged stand, for example, the exclusive crown area $A(a)$ of an average tree at age a can be related to density x:

$$\frac{1}{x}\frac{dx}{da} = -\frac{1}{A}\frac{dA}{da}.$$ (1.1)

(For reference see Clark, 1990a, and Fig. 15.3D.) Thus, we can learn much about density simply by understanding individual plant growth $A(a)$, because density can be expressed in terms of the parameters that describe growth. Since plant weight w can be related to crown area A as

$$w \propto A^\alpha,$$ (1.2)

where α generally assumes values in the range $1 < \alpha < 2$ (Norberg, 1988), we can extend the analysis to include such ecosystem-level processes as NEP and NPP. Standing crop y is the product of individual plant weight and density, $y = wx$, and NPP and NEP can be written as

$$\text{NPP} = x\frac{dw}{da}$$ (1.3)

and

$$\text{NEP} = \frac{dy}{da} = x\frac{dw}{da} + w\frac{dx}{da}.$$ (1.4)

We have seen already that we can solve for density in terms of plant size, leading to some rather simple expressions for these quantities that depend on plant growth:

$$\text{NPP} \propto w^{-1/\alpha}\frac{dw}{da}$$ (1.5)

and

$$\text{NEP} \propto w^{-1/\alpha}\frac{dw}{da}\left(1 - \frac{1}{\alpha w}\right).$$ (1.6)

The approach can be extended to analysis of the self-thinning rule, which relates stand density x to individual plant weight w and to standing crop per area y as

$$w \propto x^{b_w} \tag{1.7}$$

$$y \propto x^{b_y}. \tag{1.8}$$

Here again, the relationship of density to individual plant growth allows us to solve for the unknown coefficients b_w and b_y solely in terms of individual plant growth using the relationships

$$b_w = \frac{\partial w}{w} \Big/ \frac{\partial x}{x} \tag{1.9}$$

$$b_y = \frac{\partial y}{y} \Big/ \frac{\partial x}{x}. \tag{1.10}$$

(For reference, see Clark, 1991b.)

---

assume a similar population structure can be used for analysis of how local population dynamics (which themselves depend on individual plant growth) depend on regional disturbance regimes, and vice versa (Clark, 1991a). The numerical gap models and simpler analytical models represent complementary approaches to understanding forest population dynamics.

Examples of how these simpler models facilitate analysis of assumptions at one level for dynamics at another is illustrated in Box 15.2. The landscape distribution of the response variable, $f(x)$, is related to dynamics of a local response variable $x(a)$ for two different disturbance regimes, represented by patch-age distribution $\omega(a)$. $x(a)$ is the local response variable that develops with patch age a (the elapsed time since the last disturbance on the patch); $f(x)$ and $\omega(a)$ are probability density functions in x and a, respectively. The structure (distribution) is the product of two quantities that depend on the changing probability of disturbance [summarized by $\omega(a)$] and on the dynamics of the response variable between disturbances [summarized by $x(a)$]. Box 15.2 uses an example in which the response variable $x(a)$ is the density of a small cohort in a newly disturbed patch. The resultant analytical solutions for density structure $f(x)$ can be analyzed for effects of individual plant growth rate ($\approx \rho$) and the average disturbance interval λ on structure of the metapopulation on a landscape scale. We also can change the assumptions about how disturbance probability changes over time, as is done in Section VI to

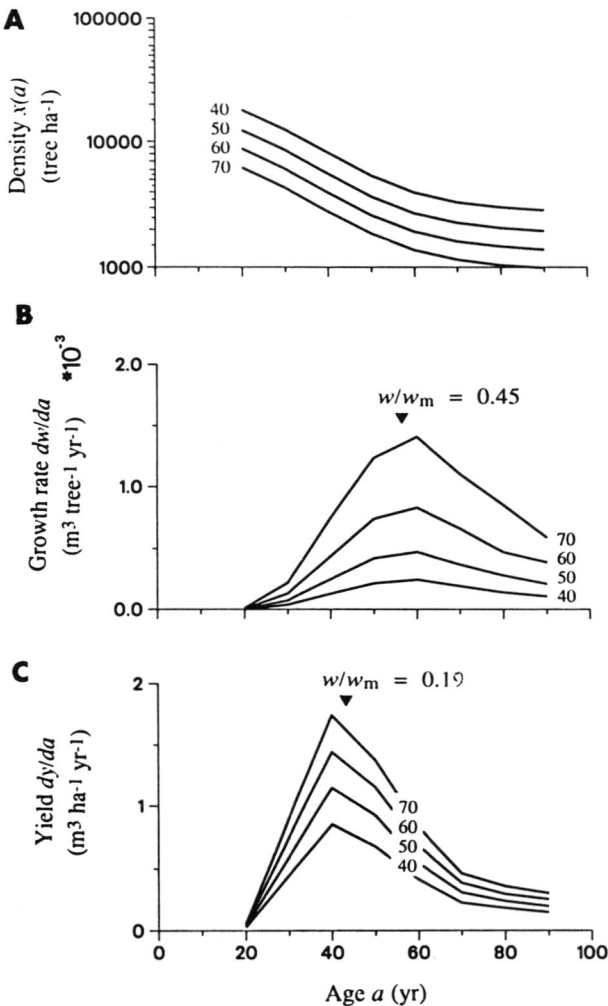

Figure 15.3 Relationships of population and ecosystem-level processes to individual plant growth using an example of *Abies balsamea* on four different site indices, that is, the age at which the average tree attains a height of 50 ft (data from Meyer, 1929). Observed constant per-capita mortality rates (A) are predicted until individual plant growth rate begins to decline (see D). The relationships between growth and mortality also can be used to predict relationships between maximum growth increment (B) and maximum NEP (C), which closely match predicted values, w/w_m (Clark, 1990a). These maxima are predicted to be independent of growth rates, as observed here, despite strong dependencies of growth on fertility (site index); maxima are higher on more fertile sites, but they are obtained at the same time during stand development. If per-capita mortality rate is equal to per-area rate of crown-area increase in crowded stands, as predicted in Box 15.1, then the slope of this plot should be -1, as observed once canopy closure is obtained (D). This relationship holds despite large differences in growth rates on different site indices (B).

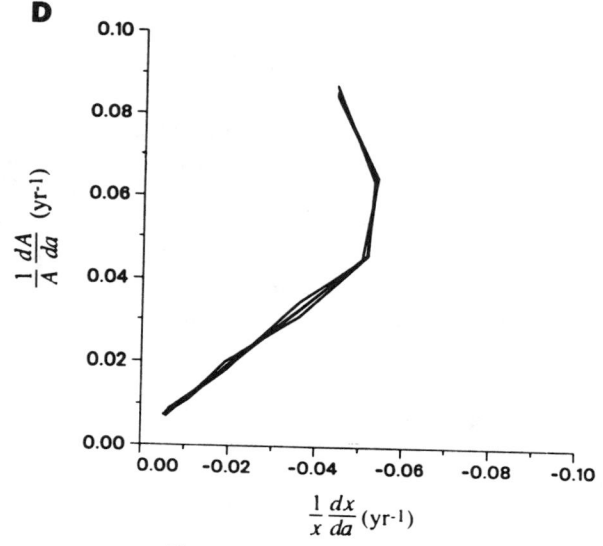

Figure 15.3—*Continued*

determine how this aspect of a disturbance regime influences density structure (e.g., the two different distributions and expectations for density in Box 15.2). The same approach could be used for other response variables such as plant growth, NEP, NPP, and leaf area. Table 15.1 contains examples of three types of responses under two disturbance types. The table lists some processes that might be summarized by each of the response types, examples of which are presented in Section VI, along with effects of the different disturbance types.

The usefulness of these approaches derives from the simplicity that is obtained without loss of the effects imposed by nonlinear responses and changing disturbance probability. The analysis of these models helps determine when population structure has sufficiently important effects on a response variable that it should be treated explicitly (see Section VI). These analytical and numerical approaches are complementary, because the analytical models could be parameterized from the numerical models or from real data. Clark (1991a) presents examples using more flexible disturbance distributions that can accommodate diverse types of disturbance, such as tree-fall and fire.

VI. Spatial and Temporal Dependencies

Scaling involves model results or measurements that are made on units that may be connected in space and time. These results are transferred

Box 15.2 Landscape distribution of response variables on a shifting-mosaic landscape.

General model The structure of a response variable can be described by a probability density function that follows from two components: (1) the changes that occur in that variable over time on an individual patch and (2) the distribution of events that initiate new patches. Suppose that the response variable (e.g., cohort density, tree height, NEP) changes with cohort age a according to some continuous function $R(a)$, and the distribution of patch age classes is given by

$$\omega(a) = S(a)/\mu, \qquad (2.1)$$

where $S(a)$ is the probability that a new patch does not occur before age a, and μ is the average interval between formation of new patches. (Of course, knowing $S(a)$, we can solve for μ.) Then the landscape distribution of a monotonic response variable $R(a)$ is given by

$$f(R) = \omega(R)|dR/da|^{-1}, \qquad (2.2)$$

where $\omega(R)$ is solved from $\omega(a(R))$ and $a(R)$ from $R(a)$.

Example Suppose that the response variable of interest is cohort density, that is, let $R(a) \equiv x(a)$. In an even-aged stand that becomes established following disturbance, this response often is well described by an exponential function

$$x(a) = x_0\, e^{-\rho a}, \qquad (2.3)$$

where ρ is the mortality rate and x_0 is initial density. The corresponding rate equation is $dx/da = -\rho x$. This expression is related to individual plant growth as $\rho = (1/x)dx/da \approx -(1/A)dA/da$ (see Box 15.1). We can use this relationship to analyze how a disturbance regime $S(a)$ affects population density x. For example, "How is the distribution $f(x)$ influenced by the assumptions of constant probability or periodic disturbance?"

For constant disturbance probability, the appropriate function is $S(a) = e^{-\lambda a}$, where $\lambda = 1/\mu$ is a rate constant (the frequency of disturbance on any given piece of ground). Inverting Eq. 2.3, we find $a(x) = \ln(x_0/x)/\rho$. Using this result, with Eq. 2.1, $\omega(x) = \lambda(x/x_0)^{\lambda/\rho}$, and Eq. 2.2 gives

$$f(x) = \frac{\lambda}{\rho x}\left[\frac{x}{x_0}\right]^{\lambda/\rho} \text{ for } 0 < x \le x_0. \qquad (2.4)$$

If we were to model density under these conditions, we would estimate an average value of x for this landscape of nearly

$$E[x] = \frac{\lambda\, x_0}{\lambda + \rho}$$

$$(2.5)$$

the expected mean density value across this landscape (Table 15.1).

How does this result compare with the result we would obtain if patches were formed periodically? To compare these two assumptions, we repeat the process using the same average interval between disturbances. For the periodic case, this interval always occurs at patch age $a = \mu = 1/\lambda$. Note that this "average" interval is the same as that used in the preceding case. Then the survivor function is given by

$$S(a) = \begin{cases} 1, & a < 1/\lambda, \\ 0, & a \geq 1/\lambda. \end{cases}$$

$$(2.6)$$

Equation 2.1 is used to show that the distribution of patch ages is uniform:

$$\omega(a) = \begin{cases} \lambda, & a < 1/\lambda, \\ 0, & a \geq 1/\lambda. \end{cases}$$

$$(2.7)$$

Given this distribution of patch ages and the same exponential response $x(a)$ used earlier, the distribution of the response variable is given by

$$f(x) = \frac{\lambda}{\rho x} \text{ for } x_0 e^{-\rho/\lambda} \leq x < x_0.$$

$$(2.8)$$

Solving for the moments of this distribution, we find that a simulation model of this process would converge to a time- and space-averaged density of

$$E[x] = \frac{\lambda\, x_0}{\rho} (1 - e^{-\rho/\lambda}).$$

$$(2.9)$$

to another scale or level of complexity by summation (if a large number of measurements has been made) or by extrapolation of a smaller number of measurements to all elements of the landscape to which those measured values are believed to apply. Errors can arise in the process of scaling as a result of temporal correlations within and spatial correlations among these units. Failure to incorporate the interactions among spatially discrete or continuous elements that produce spatial correlation or the inherent memory in the system that produces temporal correlation amounts to an implicit assumption of independence. The feedback effect of transpiration rate on itself (Jarvis and McNaughton, 1986) results

Table 15.1 Distributions of Response Variables under Assumptions of Constant Probability vs Periodic Disturbance

Response variable[a]	dR/da	Constant probability			
		$\omega(R)$	$f(R)$	Limits[b]	$E[R]$[c]
Exponential $R(a) = R_0 e^{-\rho a}$	$-\rho R$	$\lambda(R/R_0)^{\lambda/\rho}$	$\dfrac{\lambda(R/R_0)^{\lambda/\rho}}{\rho R}$	$(0, R_0)$	$\dfrac{R_0 \lambda}{\lambda + \rho}$
Examples: thinning (density)					
Saturating $R(a) = R_m(1 - e^{-\rho a})$	$\rho(R_m - R)$	$\lambda(1 - R/R_m)^{\lambda/\rho}$	$\dfrac{\lambda}{\rho R_m}(1 - R/R_m)^{\lambda/\rho - 1}$	(O, R_m)	$R_m\left(1 - \dfrac{\lambda}{\lambda + \rho}\right)$
Examples: leaf area, organic matter accumulation, standing crop					
Logistic $R(a) = \dfrac{R_m}{1 + be^{-\rho a}}$	$\rho R\left(1 - \dfrac{R}{R_m}\right)$	$\lambda\left(\dfrac{R_m/R - 1}{b}\right)^{\lambda/\rho}$	$\dfrac{\lambda R_m\left(\dfrac{R_m}{R} - 1\right)^{\lambda/\rho - 1}}{\rho R^2 b^{\lambda/\rho}}$	(R_0, R_m)	

Note. Listed here are functions used in Box 15.2 for different types of responses. $f(R)$ are probability density functions of response variable R across a landscape that experiences processes $\omega(a)$ (see Box 15.2).

[a] Variables and parameters are a, time, in years, since the last disturbance; b, integration constant for logistic equation; λ, average disturbance frequency (a rate constant); ρ, rate constant for response variable; $R(a)$, value of response variable in year a; R_0, boundary condition for response variable; R_m, asymptotic value for response variable.

[b] Integration limits of $f(R)$.

[c] Expectation of $f(R)$. An analytical solution could not be found for logistic $R(a)$ having constant disturbance probability.

from the fact that stomatal behavior is "correlated" among leaves, since they mutually influence humidity. Leaves "interact" through their effects on a shared atmospheric environment. The same principle applies to populations, that is, higher-level patterns in population density and NEP can be forecast on the basis of individual tree growth only if the effects of individuals on one another (e.g., growth, mortality) are taken into account. The scaling rules might represent these spatial and temporal correlations. These interactions among units can be much more difficult to assess than the composite behavior of an individual unit because of the many ways in which units can influence one another. Recognition of the role of these interdependencies can guide efforts to assess when they can be ignored and when the interactions must play a key role in the scaling process.

A. Spatial Dependence

Spatial correlations in population models are important in neighborhood interactions among individuals, in the ways in which dynamics in one patch affect resource availability in neighboring patches, and in population processes that link patches over larger landscapes. Neighborhood interactions involve competition among plants (Harper, 1977; Weiner,

Table 15.1—*Continued*

	Periodic disturbance		
$\omega(R)$	$f(R)$	Limits[b]	$E[R]$[c]
λ	$\dfrac{\lambda}{\rho R}$	$(R(1/\lambda), R_0)$	$\dfrac{R_0\lambda}{\rho}[1 - e^{-\rho/\lambda}]$
λ	$\dfrac{\lambda}{\rho(R_m - R)}$	$(0, R(1/\lambda))$	$R_m\left[1 - \dfrac{\lambda}{\rho}(1 - e^{-\rho/\lambda})\right]$
λ	$\dfrac{\lambda}{\rho R(1 - R/R_m)}$	$(R_0, R(1/\lambda))$	$R_m\left\{1 + \dfrac{\lambda}{\rho}\ln\left[\dfrac{R_0(1 + be^{-\rho/\lambda})}{R_m}\right]\right\}$

1985; Goldberg, 1987). Gap models simplify this process by assuming that resource availability is averaged over a small area within which plants intercept light and take up water and nitrogen (Pastor and Post, 1986). Analytical models of this process use some simpler rules, for example, that thinning is a necessary consequence of plant growth (Tait, 1988; Valentine, 1988; Clark, 1990a) without explicitly considering resource supply and uptake. Both approaches miss the finer scale variability in resource availability (Zinke, 1962; Robertson *et al.*, 1988; Boerner and Koslowsky, 1989). Because numerical (Botkin *et al.*, 1972; Shugart, 1984) and analytical (Clark, 1991a) models treat patches as independent, they miss the effects of gap formation and resource availability in one patch on that in neighboring patches (e.g., Canham, 1988). Both types of models also have assumed a global seed availability as a means of simplifying models. Landscape-level processes, such as seed dispersal and fire, will require models that contain spatial dependence among patches to assess the role of these interactions when results are scaled from patches to landscapes. Clark (1991a) relaxed the assumption of independent patches to a degree, and numerical models of stand dynamics that contain patch interdependencies now are being developed in several labs.

Pacala (1988; Pacala and Silander, 1985) has taken a complementary approach to simplifying population dynamics, one that targets spatial dynamics of individuals and of resources. These models circumvent the complexity required by spatially explicit models of each individual (e.g., Schaffer and Leigh, 1976) by working, instead, from probability distributions of distances to neighbors. Further development of these models should aid understanding of patch-type models that simplify spatial dynamics by making them discrete.

Spatial dependencies in population models are of potential importance for models of biosphere–atmosphere interactions. Patches of different aged plants influence roughness length and concentration gradients of gases, which, in turn, influence heat flux and gas exchange. Moreover, patches "interact" as a result of vegetation effects on humidity and wind resistance, and are linked by subsurface flow and groundwater. Models that average the output of patches simulated independently miss these landscape interactions that may influence results importantly (Hicks, 1989). Other chapters deal with this problem in more detail. It is important to mention here that population-level processes are a source of this heterogeneity and therefore might be an important consideration when attempting to link landscape models of ecosystem processes.

B. Temporal Dependence

This chapter has emphasized the importance of temporal correlations in population models, because response variables change with time elapsed since the most recent disturbance and because the probability of disturbance changes over time. A number of ecosystem processes vary with patch age, for example, leaf area, net primary production, and nutrient turnover and pool sizes. This strong dependence on patch age is part of the motivation for gap models (Botkin *et al.*, 1972; Shugart, 1984; Pastor and Post, 1986; Huston and Smith, 1987) that simulate landscapes as a mosaic of patches the size of mature individual trees. The spatial structure permits retention of this temporal dependence of population- and ecosystem-level processes, a dependence that is established on local scales. A population process, such as thinning, is expressed as changing density since the time of establishment. Ecosystem-level variables, such as leaf area and NPP, similarly are expressed in terms of elapsed time since the last disturbance (Sprugel, 1984,1985). A landscape containing a large number of such patches is characterized by an age distribution of patches, which, in turn, determines the mean, variance, and higher moments of the response variable (e.g., leaf area, NPP) across the landscape. Here I consider (1) how assumptions concerning the patch-formation process influence the landscape structure predicted by the model, (2) how these assumptions affect the predicted response at the landscape level, and (3) what is lost by ignoring the higher moments of the response-variable distribution.

Population models of landscape processes either (1) assume the distribution of patch ages rather directly, through the use of some "disturbance regime", or (2) produce one in a more indirect way, through the model structure and parameterizations. Common examples of the first case include patch models of fire (Shugart and Noble, 1981; Keane *et al.*, 1990) and "disturbance" (Huston, 1979; Comins and Noble, 1985; Armstrong, 1989; Hastings and Wolin, 1989). The second case includes tree-fall in forest gap models, where death of a large individual results in the initiation of a new cohort.

One of two common assumptions generally is used regarding the nature of the patch-formation process: that disturbances occur with *constant probability* or that they are *periodic*. A *constant* disturbance probability is used in stochastic models that flip a coin at each time increment to determine whether or not a new patch replaces an old patch. The probabilities of heads and tails differ, but their values do not change over time. The long-term result of this process is a distribution of waiting times between formations of new patches that is exponential; short waiting times are more common than long waiting times. Long waiting times are rare simply because they occur only when a run of coin tosses of "no new patch" occurs.

A periodic process does not allow for uncertainty—a new patch is formed at a given time interval. Rather than flip coins, the model simply keeps track of how long ago each patch was formed. A new patch always replaces an old patch at some predesignated time interval. These two types of processes can be viewed as lying along a continuum; a constant probability process is stochastic with a coefficient of variation (CV) of unity and a periodic process has zero variance. It is important to note that both processes could be modeled with the same average waiting time, differing only in the higher moments of the distribution.

Since neither assumption is typically qualified, and both may be unrealistic, it is important to ask how these assumptions influence results. Specifically, how does the disturbance process influence the value of the response variable? The answer can be demonstrated using the two different assumptions about the distribution of disturbance events, given the same average interval between events used in Box 15.2. In addition to our assumption about the nature of the disturbance process on this landscape, we also need to know how the response variable of interest changes over time on any given patch (Table 15.1). The time development of this response might be measured, some examples of which are given here, or it might be modeled.

Population density serves as a simple example. Box 15.2 shows that the average population densities $E(x)$ across the mosaic landscape characterized by constant probability and periodic disturbance are clearly different. Is this difference important? In many cases, it is a rather large difference. Consider the case in which the disturbance process and the response variable are described by similar time scales, that is, $\rho \sim \lambda$. This is the case for disturbances in forests such as blowdowns or, in some cases, fires (Clark, 1991a). Under these conditions, the periodic disturbance process results in an average for the response variable that is more than 25% higher than that resulting from a constant disturbance probability. Figure 15.4 compares distributions of population density under constant and periodic disturbance regimes, both having average disturbance intervals $(1/\lambda)$ of 20 years, for modest thinning rates $(\rho = 0.02 \text{ year}^{-1})$ and arbitrary x_0 of 100. In this case, the density mean for periodic disturbance (μ_p) exceeds that for constant probability (μ_c) by 15%, but differences in higher moments of these distributions are far more dramatic. Clearly, the higher moments of the disturbance process can have an important influence on the landscape structure of the response variable.

Biases introduced by the assumptions concerning higher moments of the disturbance distribution depend on the time scales describing the disturbance process and the response variable. These time scale effects can be demonstrated by contrasting a "fast" process with a "slow" process (Fig. 15.5A) on landscapes with different disturbance distributions but equal means (Fig. 15.5B). "Fast" and "slow" refer to the relationship between rates of change in these processes and the disturbance process (e.g., land use, fire). The example of a "fast" process is recovery of leaf area index (LAI) $(m^2 \cdot m^{-2})$ after disturbance. This recovery is much more rapid (large ρ) than is the recovery of above-ground standing crop, a relatively "slow" process (small ρ). For example, Boring and Swank (1986) report changes in leaf area after disturbance that could be described by

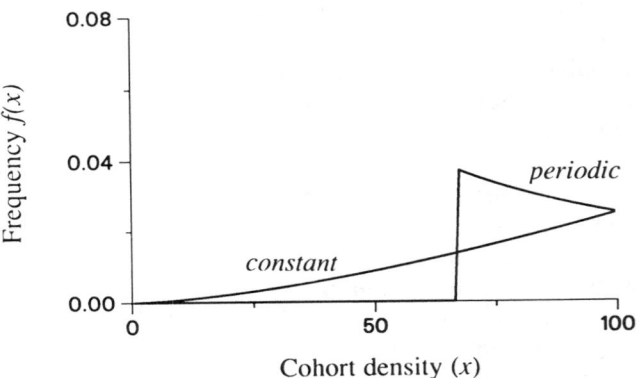

Figure 15.4 Distributions of population density under two disturbance regimes having the same average disturbance interval (see Box 15.2). $f(x)$ is relative importance of density x on the landscape. Mean value for the constant probability distribution is given by $\mu_c = 71$, and for the periodic disturbance distribution as $\mu_p = 82$. Parameter values are $x_0 = 100$, $\rho = 0.02$, and $\lambda = 0.05$.

a simple saturating function with R_0 ranging from 5 to 6 and ρ from 0.20 to 0.32 for dry and wet sites, respectively. Table 15.2 compares average values of LAI across landscapes in which constant and periodic disturbances occur on average every 20 and 100 years using the distribution for a saturating response (Table 15.1). Because recovery is rapid (Fig. 15.5A), assumptions regarding higher moments of the disturbance distribution (Fig. 15.5B) have little effect on LAI distributions (Fig. 15.5C).

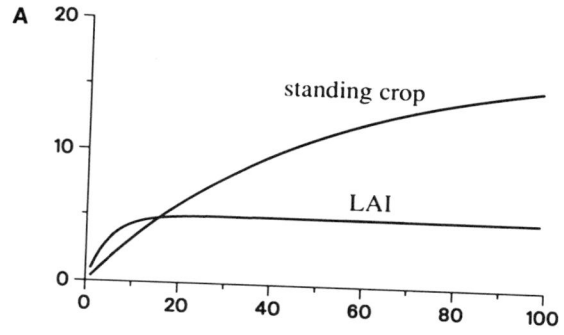

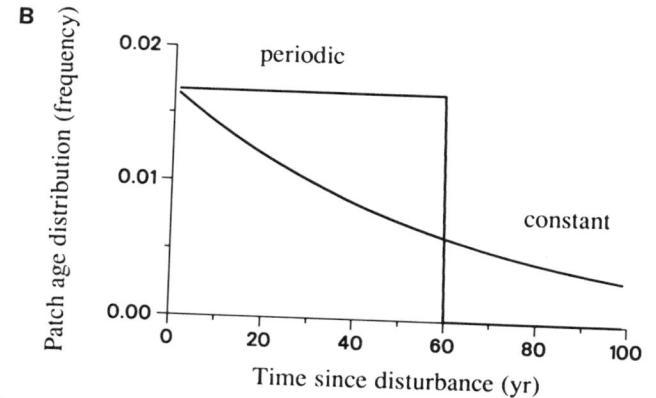

Figure 15.5 Effects of disturbance distribution on structure of "fast" and "slow" processes, as measured by their distributions. (A) LAI ($m^2 \cdot m^{-2}$; Boring and Swank, 1986) recovers rapidly following disturbance, whereas standing crop ($kg \cdot m^{-2}$; Sprugel, 1984) increases more slowly. (B) Constant probability compared with periodic disturbance regimes produce large differences in distributions of a variable that responds more slowly (D), but have little effect on one that recovers rapidly (C). The constant probability case is represented by the distributions that extend to higher values in (C) and (D).

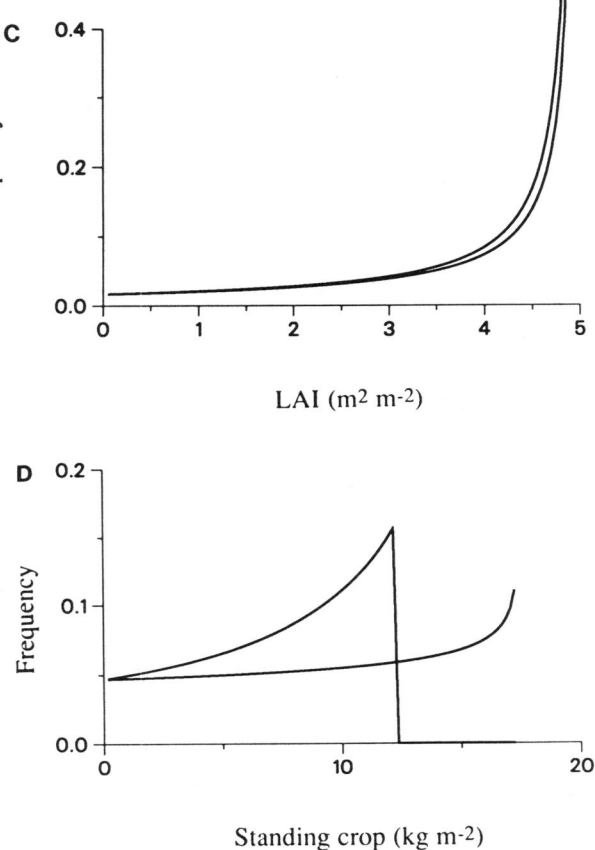

LAI (m² m⁻²)

Standing crop (kg m⁻²)

Figure 15.5—*Continued*

In contrast, the slower rate of accumulation of biomass in standing crop (Fig. 15.5A) means that the two assumptions concerning higher moments of the disturbance distribution have large effects on the average standing crop in this landscape (Fig. 15.5D). Thus, the effects of assumptions regarding the disturbance distribution depend on the rate at which processes develop after disturbance relative to the disturbance frequency. Rapid response and rare disturbance means that these effects are minimal.

These comparisons are instructive because both processes are important for understanding global climate change. LAI is used to drive landscape-level models of forest processes, hydrology, and gas exchange (Running and Coughlan, 1988). Standing crop and its rate of change

Table 15.2 Average Values of Leaf Area Index and Standing Crop across Landscapes Having the Same Average Disturbance Intervals, but Different Variances, Where Both Response Variables Are Described by Simple Saturating Functions[a]

$E[R]$	20-year intervals ($\lambda = 0.05$ yr^{-1})	100-year intervals ($\lambda = 0.01$ yr^{-1})
Leaf area index recovery following clearcutting[b]		
Constant	4.0, 5.2	4.8, 5.8
Periodic	3.8, 5.0	4.8, 5.8
Percentage difference[c]	5.2, 4.0	0.0, 0.0
Biomass accumulation in fir waves[d]		
Constant	5.1	12
Periodic	3.1	10
Percentage difference[c]	65	20

[a] See Table 15.1 for simple saturating functions.
[b] Dry sites (first LAI value given): $R_m = 5$ m$^2 \cdot$ m^{-2} and $\rho = 0.2$ year^{-1}. Wet sites (second LAI value given): $R_m = 6$ m$^2 \cdot$ m^{-2} and $\rho = 0.32$ year^{-1}.
[c] Calculated as

$$\left[\frac{|\text{Constant} - \text{Periodic}|}{\text{Periodic}} \times 100 \right].$$

[d] $R_m = 17.4$ kg $\cdot$ m^{-2} and $\rho = 0.205$ year^{-1}. Data are from Sprugel (1984).

(NEP) influences the terrestrial carbon pool, which in turn influences atmospheric CO_2. Comparison of the time scales of these processes with those describing the prevailing disturbance regime can facilitate decisions about when the structure imposed by disturbance and temporal response must be understood and incorporated into biophysical models for large regions.

The interaction between fire and tree life history also is affected importantly by these assumptions. Prediction of when the next fire may occur is a goal of fire management and forest ecology and may have important effects on carbon cycling (Crutzen and Andreae, 1990). Unless fire were periodic (i.e., waiting times with zero variance), the average interval may be misleading and provide insufficient information for realistic modeling of a landscape. Figure 15.6 compares the probability of an interval between fires of 20 years in northwest Minnesota using actual fire regimes with those implied by constant probability and periodic assumptions. The interval of 20 years is of interest, because it represents the amount of time required for the dominant tree species, *Pinus resinosa*, to become resistant to fire; saplings are killed by fire that occur before they reach that age. The errors introduced by assuming constant disturbance probability range from 11% too high to 33% too low. Periodic disturbances yield results from 100% too low to 54% too high. Thus, if

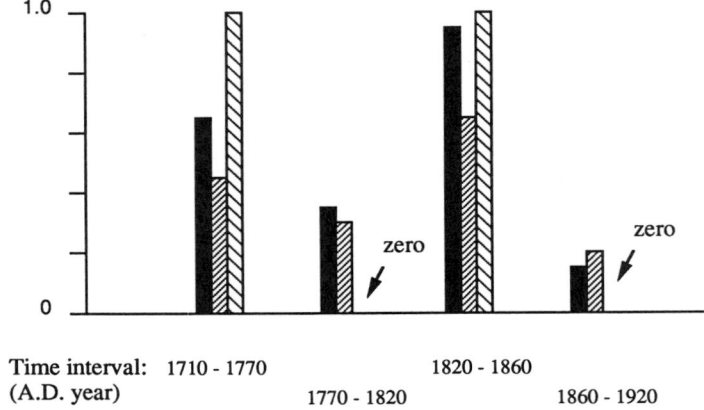

Time interval: 1710 - 1770   1820 - 1860
(A.D. year)   1770 - 1820   1860 - 1920

Figure 15.6 Comparison of probability of a 20-year interval without fire from different time periods with fire distributions having different higher moments. "Actual" probabilities (solid bars) are calculated using a Weibull distribution and parameter estimates in Clark (1989) from old-growth forests in northwestern Minnesota. "Constant" probability (first hatched bars) assumes that fires occur stochastically with constant probability equal to $1/\lambda$ and coefficient of variation equal to 1. "Periodic" (open bars) assumes that fires occur periodically at a time $a = 1/\lambda$ since the last fire, with zero variance. Probability under periodic conditions is zero in the intervals 1770–1820 and 1860–1920.

we wish to model population dynamics of *Pinus resinosa* using the right average interval between fires but the wrong distribution, our results may be wildly inaccurate. We easily could predict extinction using a periodic model when, in fact, fire regimes may have been suitable for continued establishment. These few examples demonstrate (1) the usefulness of simple models in assessing the potential importance of local dynamics in understanding system behavior at higher levels of complexity and (2) that these effects can be large.

VII. Future Directions

Clearly, more information will lead to increased understanding of relationships across scales. Models are the tools required for analysis of complex problems, and their development is a necessary step toward integrating processes that unfold over different spatial and temporal domains. It does not necessarily follow, however, that more complicated models will provide this increased understanding. Faster computers have made possible increased mechanistic detail and experimentation with scales we commonly do not observe. They have not solved the fundamental conceptualization problem of viewing relationships in more than several dimensions, nor can they solve problems that result from insufficient data. Each new parameter, dependency, and feedback can represent a new dimension with effects that may complicate system behavior.

A partial answer to the complexity problem lies in the potential for abstraction and analysis, with greater focus on spatial and temporal correlations. Gap and other numerical models and analytical theory traditionally have been used by separate investigators, although the situation is changing. The complementary nature of the models suggests greater interaction. A more balanced approach will foster understanding of how spatial and temporal correlations influence results as measurements or model outputs are scaled up to larger regions and time steps.

Acknowledgments

I thank Christopher Field and Chantal Reid for their helpful comments. Preparation of this manuscript was supported by NSF Grant BSR-8818355.

References

Aikman, D. P., and Watkinson, A. R. (1980). A model for growth and self-thinning in even-aged monocultures of plants. *Ann. Bot.* **45,** 419–427.

Allison, T. D., Moeller, R. E., and Davis, M. B. (1986). Pollen in laminated sediments provides evidence for a mid-Holocene forest pathogen outbreak. *Ecology* **67,** 1101–1105.

Almendinger, J. C. (1990). The decline of soil organic matter, total-N, and available water capacity following the late Holocene establishment of jack pine on sandy mollisols, north-central Minnesota. *Soil Sci.* **150,** 680–694.

Armstrong, R. A. (1989). Fugitive coexistence in sessile species: Models with continuous recruitment and determinant growth. *Ecology* **70,** 674–680.

Boerner, R. E. J., and Koslowsky, S. D. (1989). Microsite variations in soil chemistry and nitrogen mineralization in a beech-maple forest. *Soil Biol. Biogeochem.* **21,** 795–801.

Boring, L. R., and Swank, W. T. (1986). Hardwood biomass and net primary production following clearcutting in the Coweeta Basin. *In* "Proceedings of the 1986 Southern Forest Biomass Workshop" (R. T. Brooks, ed.), pp. 16–19. Tennessee Valley Authority, Knoxville, Tennessee.

Bormann, F. H., and Likens, G. E. (1979). Catastrophic disturbances and the steady state in northern hardwood forests. *Am. Sci.* **67,** 660–669.

Botkin, D. F., Janak, J. F., and Wallis, J. R. (1972). Some ecological consequences of a computer model of forest growth. *J. Ecol.* **60,** 849–872.

Bretherton, F. P. (1987). A closer view. *Earthquest* **1,** 1–2. (Office for Interdisciplinary Earth Studies, Boulder, Colorado)

Canham, C. D. (1988). An index for understory light levels in and around canopy gaps. *Ecology* **69,** 1634–1637.

Canham, C. D., and Loucks, O. L. (1984). Catastrophic windthrow in the presettlement forests of Wisconsin. *Ecology* **65,** 803–809.

Canham, C. D., and Marks, P. L. (1985). The response of woody plants to disturbance patterns of establishment and growth. *In* "The Ecology of Natural Disturbance and Patch Dynamics" (S. T. A. Pickett and P. S. White, eds.), pp. 197–217. Academic Press, Orlando, Florida.

Clark, J. S. (1989). Ecological disturbance as a renewal process: Theory and application to fire history. *Oikos* **56,** 17–30.

Clark, J. S. (1990a). Integration of ecological levels: Individual plant growth, population mortality, and ecosystem processes. *J. Ecol.* **78,** 275–299.

Clark, J. S. (1990b). Landscape interactions among nitrogen mineralization, species composition, and long-term fire frequency. *Biogeochemistry* **11,** 1–22.

Clark, J. S. (1991a). Disturbance and population structure on the shifting mosaic landscape. *Ecology* **72,** 1119–1137.

Clark, J. S. (1991b). Relationships between individual plant growth and the dynamics of populations and ecosystems. *In* "Populations, Communities, and Ecosystems: An Individual Perspective" (D. DeAngelis and L. Gross, eds.). Academic Press, New York.

Clark, J. S. (1991c). Shifting mosaic population dynamics. *In* "Patch Dynamics: Lecture Notes in Biomathematics" (J. Steele, S. Levin, and T. Powell, eds.). Springer-Verlag, New York.

Chapin, F. S., III. (1980). The mineral nutrition of wild plants. *Ann. Rev. Ecol. Syst.* **11,** 233–260.

Chapin, F. S. (1991). Integrated Responses of Plants to Stress. *BioScience* **41,** 29–36.

Charnov, E. L., and Berrigan, D. (1990). Dimensionless numbers and life history evolution: Age of maturity versus the adult lifespan. *Evol. Ecol.* **4,** 273–275.

Comins, H. N., and Noble, I. R. (1985). Dispersal, variability, and transient niches: Species coexistence in a uniformly variable environment. *Am. Nat.* **126,** 706–723.

Crutzen, P. J., and Andreae, M. O. (1990). Biomass burning in the tropics: Impact on atmospheric chemistry and biogeochemical cycles. *Science* (in press).

Dean, W. E., Bradbury, J. P., Anderson, R. Y., and Barnosky, C. W. (1984). The variability of Holocene climate change: Evidence from varved lake sediments. *Science* **226,** 1191–1194.

Denslow, J. S. (1980). Gap partitioning among tropical rain-forest trees. *Biotropica* **12,** 47–55.

Engstrom, D. R., and Hansen, B. C. S. (1985). Postglacial vegetational change and soil development in southeastern Labrador as inferred from pollen and chemical stratigraphy. *Can. J. Bot.* **63,** 543–561.

Ford, E. D. (1975). Competition and stand structure in some even-aged plant monocultures. *J. Ecol.* **63,** 311–333.

Ford, E. D., and Diggle, P. J. (1981). Competition for light in a plant monoculture modelled as a spatial stochastic process. *Ann. Bot.* **48,** 481–500.

Foster, D. R. (1988). Disturbance history, community organization and vegetation dynamics of the old-growth Pisgah forest, south-western New Hampshire, U.S.A. *J. Ecol.* **76,** 105–134.

Gajewski, K. (1988). Late Holocene climate changes in eastern North America estimated from pollen data. *Quat. Res.* **29,** 255–262.

Goldberg, D. E. (1987). Neighborhood competition in an old-field plant community. *Ecology* **68,** 1211–1223.

Grimm, E. C. (1983). Chronology and dynamics of vegetation change in the prairie-woodland region of southern Minnesota, U.S.A. *New Phytol.* **93,** 311–350.

Grimm, E. C. (1984). Fire and other factors controlling the Big Woods vegetation of Minnesota in the mid-nineteenth century. *Ecol. Monogr.* **54,** 291–311.

Grubb, P. J. (1977). The maintenance of species richness in plant communities. The importance of the regeneration niche. *Biol. Rev.* **52,** 107–145.

Harper, J. L. (1977). "Population Biology of Plants." Academic Press, New York.

Harper, J. L., and White, J. (1974). The demography of plants. *Annu. Rev. Ecol. Syst.* **5,** 419–463.

Hastings, A., and Wolin, C. L. (1989). Within-patch dynamics in a metapopulation. *Ecology* **70,** 1261–1266.

Heinselman, M. L. (1973). Fire in the virgin forest of the boundary waters canoe area, Minnesota. *Quat. Res.* **3,** 329–382.

Hicks, B. B. (1989). Regional extrapolation: Vegetation-atmosphere approach. *In* "Exchange of Trade Gases between Terrestrial Ecosystems and the Atmosphere" (M. O. Andreae and D. S. Schimel, eds.), pp. 109–118. Wiley, Chichester.

Hill, A. R., and Shackleton, M. (1989). Soil N mineralization and nitrification in relation to nitrogen solution chemistry in a small forested watershed. *Biogeochemistry* **8,** 167–184.

Huston, M. (1979). A general hypothesis of species diversity. *Am. Nat.* **113,** 81–101.

Huston, M., and Smith, T. (1987). Plant succession: Life history and competition. *Am. Nat.* **130,** 168–198.

Jacobson, G. L. (1979). The palaeoecology of white pine (*Pinus strobus*) in Minnesota. *J. Ecol.* **67,** 697–726.

Jacobson, G. L., Webb, T., and Grimm, E. C. (1987). Patterns and rates of vegetation change during the deglaciation of eastern North America. *In* "North America and Adjacent Oceans during the Last Deglaciation" (W. F. Ruddiman and H. E. Wright, eds.), pp. 277–288. The Geological Society of America, Boulder, Colorado.

Jarvis, P. G., and McNaughton, K. G. (1986). Stomatal control of transpiration: Scaling up from leaf to region. *Adv. Ecol. Res.* **15,** 1–49.

Keane, R. E., Arno, S. F., Brown, J. K., and Tomback, D. F. (1990). Modelling stand dynamics in whitebark pine (*Pinus albicaulis*) forests. *Ecol. Model.* **51,** 73–95.

Khil'mi, G. F. (1962). "Teoreticheskaya biogeofizika lesa (Theoretical Forest Biogeophysics)." Israel Program for Scientific Translations, Jerusalem.

King, D. (1986). Tree form, height growth, and susceptibility to wind damage in *Acer saccharum. Ecology* **67,** 980–990.

Kozlowski, J., and Weigert, R. G. (1986). Optimal allocation of energy to growth and reproduction. *Theoret. Pop. Biol.* **29,** 16–37.

Loehle, C. (1988). Tree life history strategies: The role of defenses. *Can. J. For. Res.* **18,** 209–222.

McAndrews, J. H. (1966). Postglacial history of prairie, savanna, and forest in northwestern Minnesota. *Memoirs Torrey Bot. Club* **22,** 1–72.

McClaugherty, C. A., Pastor, J., Aber, J. D., and Melillo, J. M. (1985). Forest litter decomposition in relation to soil nitrogen dynamics and litter quality. *Ecology* **66,** 266–275.

Meyer, W. H. (1929). "Yields of Second-Growth Spruce and Fir in the Northeast." U.S. Department of Agriculture Technical Bulletin Number 142. U.S. Govt. Printing Office, Washington, D.C.

Noble, I. R., and Slatyer, R. O. (1980). The use of vital attributes to predict successional changes in plant communities subject to recurrent disturbances. *Vegetatio* **43,** 5–21.

Norberg, Å. (1988). Theory of growth geometry of plants and self-thinning of plant populations: Geometric similarity, elastic similarity, and different growth modes of plant parts. *Am. Nat.* **131,** 220–256.

Nordén, L. G. (1990). Depletion and recharge of soil water in two stands of Norway spruce (*Picea abies* (L) Karst). *Hydrol. Proc.* **4,** 197–213.

Oliver, C. D. (1981). Forest development in North America following major disturbance. *For. Ecol. Mgmt.* **3,** 153–168.

Oliver, C. D., and Larson, B. C. (1990). "Forest Stand Dynamics." McGraw-Hill, New York.

Pacala, S. W. (1988). Competitive equivalence: The coevolutionary consequences of sedentary habit. *Am. Nat.* **132**, 576–593.

Pacala, S. W., and Silander, J. A. (1985). Neighborhood models of plant population dynamics. I. Single-species models of annuals. *Am. Nat.* **125**, 385–411.

Pastor, J. J., Aber, J. D., McClaughtery, C. A., and Melillo, J. M. (1984). Aboveground production and N and P cycling along a nitrogen mineralization gradient on Blackhawk Island, Wisconsin. *Ecology* **65**, 256–268.

Pastor, J., and Post, W. M. (1986). Influence of climate, soil moisture, and succession on forest carbon and nitrogen cycles. *Biogeochemistry* **2**, 3–27.

Pastor, J., and Post, W. M. (1988). Response of northern forests to CO_2-induced climate change. *Nature* **334**, 55–58.

Peet, R. K. (1981). Changes in biomass and production during secondary forest succession. *In* "Forest Succession: Concepts and Management" (D. C. West, H. H. Shugart, and D. B. Botkin, eds.), pp. 324–338. Springer-Verlag, New York.

Prescott, C. E., Corbin, J. P., and Parkinson, D. (1989a). Input, accumulation, and residence times of carbon, nitrogen, and phosphorus in four Rocky Mountain coniferous forests. *Can. J. For. Res.* **19**, 489–498.

Prescott, C. E., Corbin, J. P., and Parkinson, D. (1989b). Biomass, productivity, and nutrient-use efficiency of aboveground vegetation in four Rocky Mountain coniferous forests. *Can. J. For. Res.* **19**, 309–317.

Robertson, G. P., Huston, M. A., Evans, F. C., and Tiedje, J. M. (1988). Spatial variability in a successional plant community: Patterns of nitrogen availability. *Ecology* **69**, 1517–1524.

Runkle, J. R. (1982). Patterns of disturbance in some old-growth mesic forests of the eastern United States. *Ecology* **63**, 1533–1546.

Running, S. W., and Coughlan, J. C. (1988). A general model of forest ecosystem processes for regional applications. I. Hydrological balance, canopy gas exchange and primary production processes. *For. Ecol. Mgmt.* **42**, 125–154.

Schaffer, W. M., and Gadgil, M. D. (1975). Selection for optimal life histories in plants. *In* "The Ecology and Evolution of Communities" (M. Cody and J. Diamond, eds.). Harvard University Press, Cambridge.

Schaffer, W. M., and Leigh, E. G. (1976). The prospective role of mathematical theory in plant ecology. *Syst. Bot.* **1**, 209–232.

Schlesinger, W. H., Reynolds, J. F., Cunningham, G. L., Huenneke, L. F., Jarrell, W. M., Virginia, R. A., and Whitford, W. G. (1990). Biological feedbacks in global desertification. *Science* **247**, 1043–1048.

Shugart, H. H. (1984). "A Theory of Forest Dynamics: The Ecological Implications of Forest Succession Models." Springer-Verlag, New York.

Shugart, H. H., and Noble, I. R. (1981). A computer model of succession and fire response of the high altitude Eucalyptus forests of the Brindabella Range, Australian Capital Territory. *Aust. J. Ecol.* **6**, 149–164.

Smith, T., and Huston, M. (1989). A theory of the spatial and temporal dynamics of plant communities. *Vegetatio* **83**, 49–69.

Sprugel, D. G. (1984). Density, biomass, productivity, and nutrient-cycling changes during stand development in wave-regenerated balsam fir forests. *Ecol. Monogr.* **54**, 165–186.

Sprugel, D. G. (1985). Natural disturbance and ecosystem energetics. *In* "The Ecology of Natural Disturbance and Patch Dynamics" (S. T. A. Pickett and P. S. White, eds.), pp. 335–352. Academic Press, New York.

Tait, D. E. (1988). The dynamics of stand development: A general stand model applied to Douglas fir. *Can. J. For. Res.* **18**, 696–702.

Tilman, D. (1982). "Resource Competition and Community Structure." Princeton University Press, Princeton, New Jersey.

Tilman, D. (1985). The resource-ratio hypothesis of plant succession. *Am. Nat.* **125,** 827–852.

Tilman, D. (1988). "Plant Strategies and the Dynamics and Structure of Plant Communities." Princeton University Press, Princeton, New Jersey.

Valentine, H. T. (1988). A carbon-balance model of stand growth: A derivation employing pipe-model theory and the self-thinning rule. *Ann. Bot.* **62,** 389–396.

Van Cleve, K., Oliver, L, Schlentner, R., Viereck, L. A., and Dyrness, C. T. (1983). Productivity and nutrient cycling in taiga forest ecosystems. *Can. J. For. Res.* **13,** 747–766.

Vitousek, P. M. (1982). Nutrient cycling and nutrient use efficiency. *Am. Nat.* **119,** 553–572.

Watt, A. S. (1947). Pattern and process in the plant community. *J. Ecol.* **35,** 1–22.

Webb, S. L. (1988). Windstorm damage and microsite colonization in two Minnesota forests. *Can. J. For. Res.* **18,** 1186–1195.

Weiner, J. (1985). Size hierarchies in experimental populations of annual plants. *Ecology* **66,** 743–752.

Weller, D. E. (1987). Self-thinning exponent correlated with allometric measures of plant geometry. *Ecology* **68,** 813–821.

Werner, P. A., and Platt, W. J. (1976). Ecological relations of co-occurring goldenrods (*Solidago:* Compositae). *Am. Nat.* **110,** 959–971.

White, J. (1981). The allometric interpretation of the self-thinning rule. *J. Theoret. Biol.* **89,** 475–500.

Whitehead, D. R., and Jackson, S. T. (1990). The Regional Vegetational History of the High Peaks (Adirondacks Mountains), New York. New York State Museum Bulletin No. 478, Albany, New York.

Yoda, K., Kira, T., Ogawa, H., and Hozumi, K. (1963). Self-thinning in overcrowded pure stands under cultivated and natural conditions (intraspecific competition among higher plants). *J. Biol. (Osaka City Univ.)* **14,** 107–129.

Zinke, P. J. (1962). The pattern on influence of individual forest trees on soil properties. *Ecology* **43,** 130–133.

16

Functional Role of Growth Forms in Ecosystem and Global Processes

F. Stuart Chapin III

I. Introduction

Traditionally, plant physiological ecology has documented the physiological mechanisms with which plants cope with their physical environment (Schimper, 1898; Chabot and Mooney, 1985). The field has taken a new direction, one of understanding the controls that plants exert over community and ecosystem processes (Chapin *et al.*, 1992). However, the biological diversity that provided such a fertile laboratory for studying physiological adaptation to the environment now presents an overwhelming array of complexity. Physiological ecologists often compare adaptive strategies to simplify patterns of physiological adaptation to the environment. However, adaptive strategies are relative and may not always be based on traits that exert strongest control over ecosystem processes.

In this chapter, I suggest that growth forms provide a logical link between physiological strategies and ecosystem or global processes because (1) species of a given growth form often share a similar physiology, (2) traits by which growth forms are classified can have important ecosystem consequences, and (3) many growth forms can be recognized by remote sensing to provide an index of ecosystem function on regional and global scales. After discussing the physiological basis of adaptive strategies and the ecological factors that control their distribution, I ask which plant traits are key features of both physiological strategies and growth forms, allowing us to catagorize plants in groups that have both physiological and ecosystem relevance.

II. Physiological Basis of Adaptive Strategies

A. RGR and Suites of Physiological Traits

The task of developing a growth-form classification with both physiological and ecosystem relevance is simplified, because most important physiological traits are interdependent, so we can recognize suites of traits that consistently co-occur and constitute physiological strategies (Grime, 1977; Chapin, 1980). Plant relative growth rate (RGR, i.e., the growth rate per unit biomass) is a central feature of plant physiological strategies. To support a high RGR, a plant must have high rates of resource capture from the environment (Chapin, 1980; Fig. 16.1), which are achieved by high capacities for photosynthesis and nutrient absorption per gram of tissue, which in turn require high tissue-nitrogen concentrations (Chapin, 1980; Field and Mooney, 1986; Field, 1991). The relationship between photosynthetic potential and growth rate is clearly evident in comparisons among growth forms (Fig. 16.2), but occurs less consistently within growth forms or species (Schulze and Chapin, 1987; Field, 1991). Within a growth form, a high biomass allocation to leaves may be more important than high photosynthetic rate per unit leaf in supporting a high growth rate (Field and Mooney, 1986; Tilman, 1988). The photosynthetic potential of a plant and its stomatal conductance are linked tightly, because both are adjusted physiologically to simultaneously limit photosynthesis (Farquhar and Sharkey, 1982). Consequently, transpiration rate also is linked closely to photosynthetic rate within and among growth forms (Collatz *et al.*, 1991).

Rates of photosynthesis and nutrient approach absorption tend to decline as tissues age, so continued high rates of resource capture require high rates of root and leaf turnover (Fig. 16.1). For example, evergreen leaves tend to have lower photosynthetic rates than leaves of deciduous species. Although leaf turnover causes loss of about half the nutrients contained in the leaf (Chapin, 1980; Chapin and Kedrowski, 1983), and root turnover may lose all the nutrients in the root (Nambiar and Fife, 1987), these nutrients can be replaced at relatively low cost in a high-resource environment (Bloom *et al.*, 1985; Chapin *et al.*, 1987).

A high RGR also requires a large allocation of carbon and nutrients to production of resource-acquiring tissues. Under conditions of high nutrient and water availability or low light availability, a large leaf allocation maximizes RGR, whereas under low nutrient and water availability or high light availability a large root allocation maximizes RGR (Davidson, 1969). Species differences in allocation may be a major factor causing differences in growth rate (Tilman, 1988), which determines rates of photosynthesis and nutrient absorption through source–sink interactions (Chapin, 1991a).

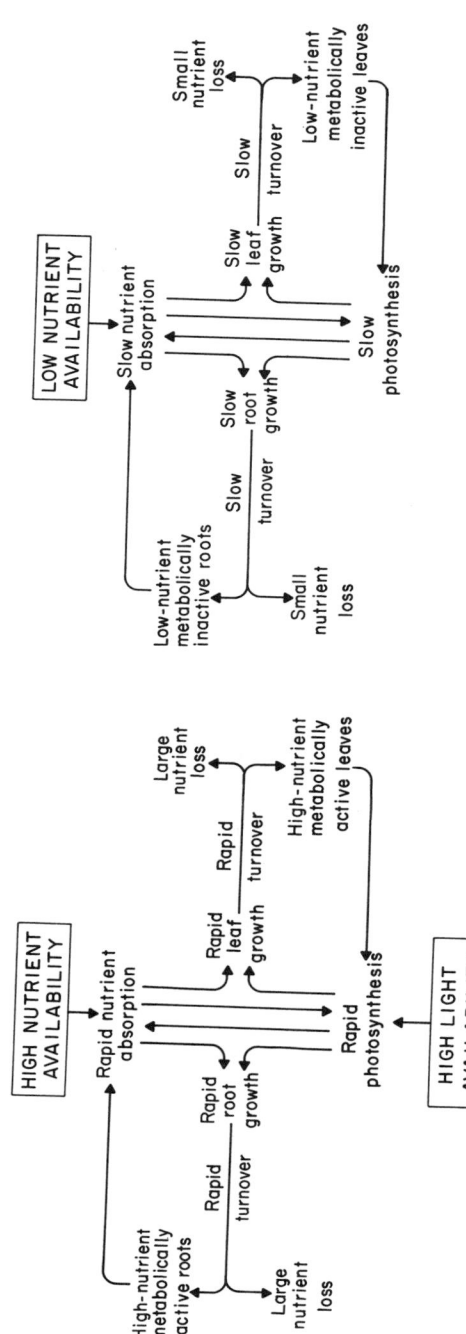

Figure 16.1 Relationship among physiological traits in plants from high- and low-resource environments (modified from Chapin, 1980).

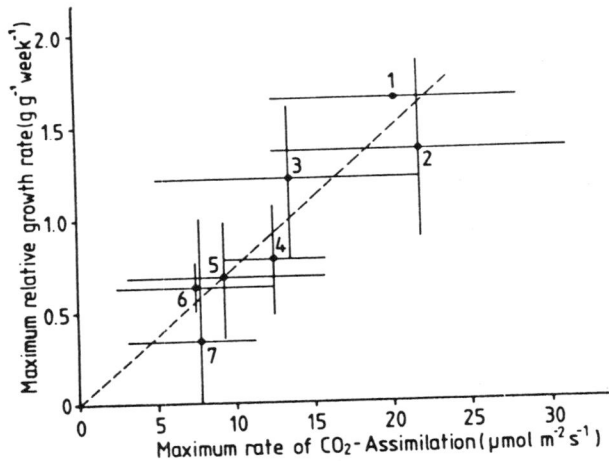

Figure 16.2 Relationship between photosynthetic rate and relative growth rate for major plant growth forms. 1, Agricultural crop species; 2, herbaceous sun species; 3, grasses and sedges; 4, summer deciduous trees; 5, evergreen and deciduous dwarf shrubs; 6, herbaceous shade species and geophytes; 7, evergreen conifers. Figure reprinted, with permission, from Schulze and Chapin (1987).

B. Trade-offs with Growth Rate

Any allocation to woody support structures, storage, plant defense, or reproduction diverts resources from production of resource-acquiring tissues and leads to a lower RGR (Fig. 16.3). The consistent negative correlation between size and RGR is seen in comparisons of growth forms (Grime and Hunt, 1975; Tilman, 1989), species within a growth form (Chapin *et al.*, 1989), and developmental stages within an individual (Cook and Evans, 1983). Similarly, large-seeded species grow more slowly than small-seeded species because large seeds produce large seedlings, which have a low RGR (Chapin *et al.*, 1989). In each case, larger plants presumably allocate more biomass to structure and less to growth.

It is important to distinguish between *relative* growth rate $(g \cdot g^{-1} \cdot day^{-1})$, a measure of the efficiency with which each unit of biomass produces additional biomass, and *absolute* growth rate $(g \cdot day^{-1})$, the absolute biomass increment per day. Large plants have a high *absolute* growth rate (and therefore high productivity) because they have large amounts of resource-acquiring tissues. However, large plants have a low *relative* growth rate because of the trade-offs just discussed.

Trade-offs also exist between storage and RGR. Annuals grow more rapidly than perennials because they allocate fewer resources to storage

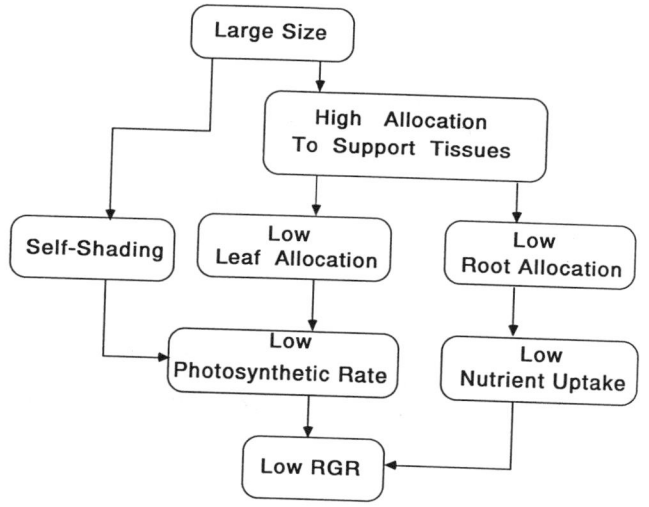

Figure 16.3 Causal links between large size and low RGR.

(Chapin *et al.*, 1990). Among annuals (Chiariello and Roughgarden, 1984) and perennials (Rappoport and Loomis, 1985), plants that allocate less to storage grow more quickly. However, in the long term, allocation to storage enables plants to continue growth when the environment becomes less favorable or to replace tissue loss after dormancy or herbivory.

A negative relationship between defensive allocation and RGR often is observed within and among species (Fig. 16.4; Bryant *et al.*, 1983; Coley *et al.*, 1985; Coley, 1988), because defensive allocation uses resources that might otherwise support growth.

There is often a negative relationship between growth and reproduction (Snow and Whigham, 1989; Chapin *et al.*, 1990), although in other cases there is no detectable cost of reproduction in terms of reduced growth (Reekie and Bazzaz, 1987; Horvitz and Schemske, 1988).

In summary, RGR is the central feature of plant adaptive strategies. It is correlated positively with the potential of the plant to acquire resources through photosynthesis and nutrient uptake and to lose water by transpiration. High rates of resource capture require high turnover rates. In contrast, there is a strong negative correlation between RGR and plant size and storage because of allocation trade-offs. Defense and reproduction often show a similar negative relationship with RGR.

The practical consequence of the strong correlations described here is that any of these correlates can be used as an index of other features.

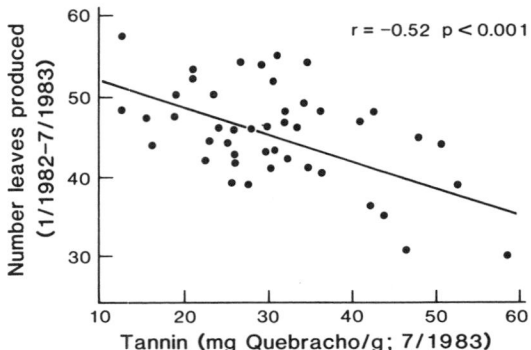

Figure 16.4   Relationship between rate of leaf production (growth rate) and leaf tannin concentration in a tropical tree, *Cecropia peltata*. Figure reprinted, with permission, from Coley (1986).

Because of the overwhelming importance of plant size in controlling ecosystem characteristics (Section IV), I suggest that plant size be used as the major yardstick for growth-form classification (Table 16.1). Depending on the purpose of the growth-form classification, other traits such as woodiness (a prerequisite for large size), leaf turnover rate (evergreen vs. deciduous), and growth rate, all of which are intercorrelated, can be used to subdivide a size-based classification (Table 16.1). Other traits useful in growth-form classification are more independent of RGR, for example, resource specialization (C_3 or C_4, sun or shade, Nitrogen–fixing or nonfixing) and life-history traits (annual/perennial, sprouter or nonsprouter).

III. Ecological Controls over Adaptive Strategies

Resource supply in the environment and time elapsed since disturbance determine the suite of physiological traits that is successful in an environment (Grime, 1977; Chapin, 1980; Tilman, 1988). High-resource environments can support a high productivity. If the time since disturbance is short, as in early succession, plant biomass will be small, so high productivity is achieved through a high RGR and its associated high rates of photosynthesis, transpiration, nutrient absorption, and tissue turnover (Grime, 1977; Chapin, 1980). As succession proceeds in a high-resource environment, shrubs replace herbs and trees replace shrubs (Fig. 16.5).

Table 16.1 Representative Growth-Form Classification and Its Relationship to Selected Parameters of Adaptive Strategy[a]

	Representative genus	Parameter[b]			
		RGR	Size	Resource acquisition	Chemical defense
Tree					
Deciduous					
Needle-leaf	*Larix*	− −	+ +	−	+ +
Broadleaf	*Populus*	+	+ + +	+ +	−
Evergreen	*Picea*	− − −	+ + + +	− − −	+ + + +
Shrub					
Deciduous					
Low RGR	*Vaccinium*	− −	−	−	+
High RGR					
N fixing	*Alnus*	+	+	+	+
Nonfixing	*Salix*	+ +	+	+ +	+
Evergreen	*Ledum*	− −	−	− −	+ +
Herb					
Annual	*Chenopodium*	+ + + +	−	+ + + +	− − − −
Perennial					
Graminoid	*Carex*	−	−	−	− −
Forb					
Sun	*Epilobium*	+ + +	−	+ + +	− − − −
Shade	*Pyrola*	−	− − −	+	+
Moss					
Feathermoss	*Hylocomium*	− −	− −	+ +	− −
Peatmoss	*Sphagnum*	− − −	−	+	+ + +
Lichen					
N fixing	*Peltigera*	− − −	− − − −	− − −	+ + + +
Nonfixing	*Cladonia*	− − − −	− − − −	− − − −	+ + +

[a] Data are from taiga forests of interior Alaska (Viereck *et al.*, 1983; Chapin, 1986).
[b] Parameter values range from high (+ + + +) to low (− − − −).

This replacement series involves progressively greater allocation to woody support structures, so RGR declines with time (Tilman, 1988). Within a given growth form at any stage of succession, RGR is greater in high-resource than in low-resource environments (Grime, 1977; Chapin, 1980; Tilman, 1990).

In contrast, in low-resource (dry or infertile) environments there are inadequate resources to support rapid growth, so plants are constrained to grow slowly and are likely to achieve a small size. Because of low tissue nutrient concentrations, plants in low-resource environments have low potentials to photosynthesize, transpire, and absorb nutrients. Tissue

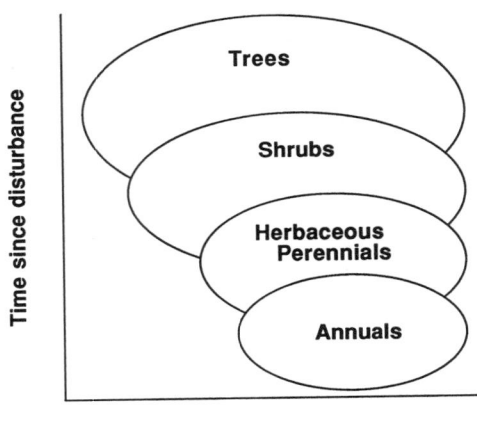

Figure 16.5 Growth-form dominance as a function of resource availability and time since disturbance.

turnover rates are low, causing tissue nutrients to be retained for a longer time (Chapin, 1980; Berendse *et al.,* 1988; Fig. 16.1). Leaves are chemically well defended in these environments because (1) tissue turnover is slow so the probability of herbivory during the life of the leaf increases, (2) nutrients lost to herbivores are costly to replace so there is a greater selective premium in minimizing tissue loss (Coley *et al.,* 1985), and (3) plants accumulate carbon in excess of immediate demands for growth in such environments, so carbon investments in defense do not detract greatly from growth (Bryant *et al.,* 1983; Tuomi *et al.,* 1984).

IV. Ecosystem Consequences of Growth Forms

A. Individuals versus Ecosystems

Most ecosystems and global processes depend on pool sizes and fluxes that are functions of either the mass of a component in the system (e.g., the biomass of vegetation or soil) or the surface area for exchange (e.g., the areal extent of a forest that exchanges gases with the atmosphere). In either case, these parameters are measured most usefully in units of ground area ($g \cdot m^{-2}$; $g \cdot m^{-2} \cdot year^{-1}$). The role of individual plants in ecosystem processes depends, therefore, on size and density; thus, we must know the relative importance of size and density in determining stand biomass. Communities that are in approximate equilibrium with their environment show an inverse relationship between ln(plant size) and ln(density), with a slope of about $-\frac{3}{2}$ (Fig. 16.6a; Harper, 1977). Increased resource supply in the environment generally increases the size and/or density of individuals that can be supported but does not change the $-\frac{3}{2}$ relationship between size of individuals and their density (Fig. 16.6a). Given this $-\frac{3}{2}$ law, increasing stand biomass (density × individual biomass) must be characterized by an increase in plant size and a decrease in plant density if the community is in equilibrium with its resource base (Fig. 16.6b,c):

$$\ln(b) = -\tfrac{3}{2}\ln(d).$$

Therefore:

$$b = (d)^{-3/2} \text{ or } d = (b)^{-2/3},$$
$$B = bd$$
$$= d^{-1/2}$$
$$= b^{1/3}$$

where b is individual biomass, d is density, and B is stand biomass.

Plants in equilibrium with their resource base move along the $-\frac{3}{2}$ thinning line by growing bigger at the expense of reduced density, as smaller individuals die (Harper, 1977; Peet and Christensen, 1987). Only by radically changing species composition (e.g., from forest to grassland with a change in disturbance regime) can this size–density relationship

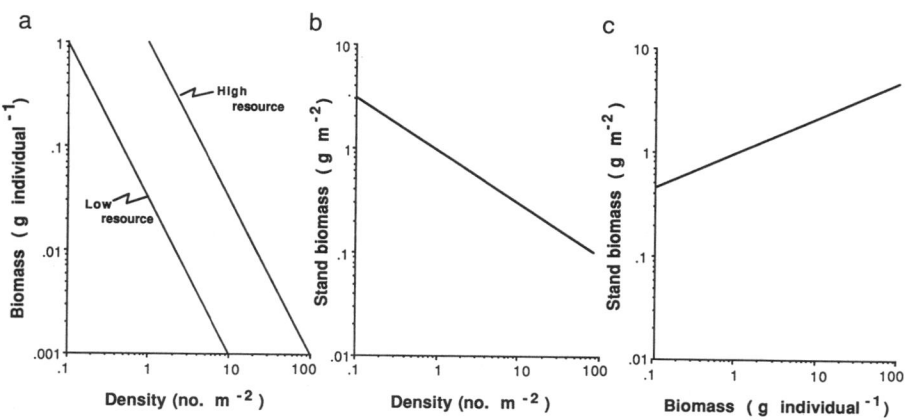

Figure 16.6 (a) Relationship between biomass of individual plants and density in communities in equilibrium with a low resource supply or a high resource supply. The relationships of stand biomass to (b) individual biomass and (c) stand density are shown also.

be altered. Although there is overlap, most large-statured communities such as forests have greater biomass than do small-statured communities such as grasslands, despite lower plant densities in forests (Fig. 16.7; Table 16.2).

B. Energy and Water Exchange

The major factors governing energy exchange between the land and the atmosphere are evapotranspiration, albedo, and surface roughness, each of which changes with plant size or biomass (Table 16.3). As discussed earlier (Section III), differences in resource availability are ultimately responsible for patterns of plant size and biomass, but here I emphasize the plant traits that link vegetation with ecosystem processes.

In all environments, evapotranspiration is a major factor influencing energy exchange between land and atmosphere and is the primary biotic control over terrestrial hydrologic budgets. Leaf area index (LAI, i.e., the amount of leaf area per unit ground area) is the strongest determinant of evapotranspiration because it determines (1) the amount of precipitation that is intercepted by the canopy and quickly evaporates after a rain and (2) the size of the transpiring surface (Fig. 16.8; Running and Coughlan, 1988). LAI is determined primarily by plant biomass, which is a function of plant size, because large plants have more leaf area than small plants. LAI is determined secondarily by phenology, because evergreen species have an active transpiring surface present for more of the year (Schulze et al., 1977). Evergreen species have a larger proportion of biomass in leaves than do deciduous species of the same size, but this attribute often is counterbalanced by a lower stomatal conductance, so evergreen and deciduous trees of a given size may have a similar rate of water loss under the same environmental conditions. Plant biomass indirectly influences evapotranspiration because of its correlation with the quantity of litter on the soil surface, which strongly influences the partitioning of water between surface runoff and infiltration into the soil. Surface runoff is negligible in forests and other communities with a well-developed litter layer (Running and Coughlan, 1988).

In dry environments, stomatal conductance and rooting depth exert additional influence over evapotranspiration. Drought-adapted species keep their stomata open at times of lower water availability and, thus, support greater evapotranspiration during dry periods than do plants from more mesic environments (Schulze and Hall, 1982). Tall plants such as trees generally transpire more water than herbs because of their more extensive root systems and greater leaf area. Consequently, forest clearing reduces evapotranspiration and increases runoff (Bormann and Likens, 1979; Shukla et al., 1990). The effect of forest clearing on runoff is most pronounced during the season of active plant growth (Bormann

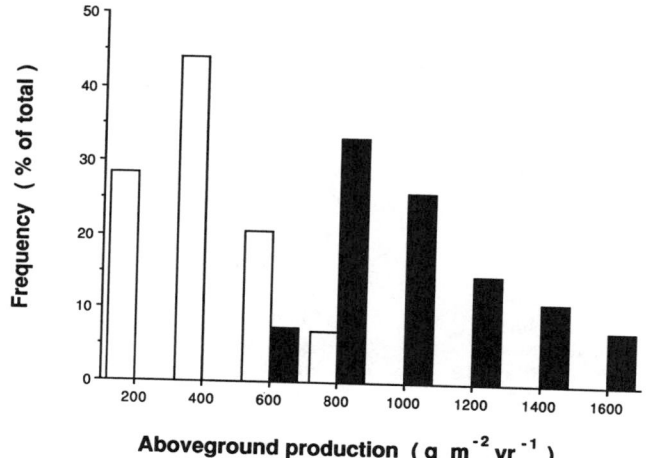

Figure 16.7 Frequency distribution of aboveground production of forests (solid bars) and grasslands (open bars). [Calculated from Cole and Rapp (1981) and Sala *et al.*, (1988)].

and Likens, 1979). In summary, plant size, which is a function of resource availability in the environment, is the major determinant of biomass and is the primary plant trait governing canopy water loss, although the response of stomatal conductance to plant water status becomes important under dry conditions. Any vegetation change that alters

Table 16.2 Aboveground Biomass, Production, and Nitrogen Flux[a] in Major Temperate Ecosystem Types, Maximum Height, and Relative Growth Rate of Species Typical of These Ecosystem Types

Parameter	Grassland	Shrubland	Deciduous forest	Evergreen forest
Aboveground biomass $(kg \cdot m^{-2})$[b]	0.3 ± 0.02	3.7 ± 0.5	15 ± 2	31 ± 8
Aboveground production $(kg \cdot m^{-2} \cdot year^{-1})$[b]	0.3 ± 0.02	0.4 ± 0.07	1.0 ± 0.08	0.8 ± 0.08
N flux $(g \cdot m^{-2} \cdot year^{-1})$[b]	2.6 ± 0.2	3.9 ± 1.6	7.5 ± 0.5	4.7 ± 0.5
Canopy height (cm)[c]	100	400	2200	2200
Field RGR $(year^{-1})$[d]	1.0	0.1	0.07	0.03
Laboratory RGR $(week^{-1})$[c]	1.3	0.8	0.7	0.4

[a] Mean ± SE.

[b] From Bokhari and Singh (1975), Gray and Schlesinger (1981), Cole and Rapp (1981), and Sala *et al.* (1988).

[c] From Grime and Hunt (1975) and Tilman (1988).

[d] Aboveground production/aboveground biomass.

Table 16.3 Major Ecosystem Processes and the Plant
Physiological Traits that Conrol Them[a]

Ecosystem process	Physiological basis
Energy exchange	
Albedo	Size, phenology
Evapotranspiration	Size, phenology, conductance
Roughness	Size
Carbon flux	
Productivity	Size, RGR
Decomposition	Size, chemistry, RGR
Net balance	Chemistry
Nutrient flux	
Nutrient uptake	Size, RGR, storage
Mineralization	Size, chemistry, RGR
Net balance	N nutrition, phenology
Trophic transfer	Size, chemistry, RGR
Sensitivity to disturbance	Size, storage, life history
Community change	Life history

[a] Physiological traits for each ecosystem process are listed in order of
decreasing importance. See text for explanation.

evapotranspiration can change the linkages with ecosystems that are
downwind (Schimel *et al.*, 1991) by altering surface temperature, wind,
and precipitation (Segal *et al.*, 1988; Sato *et al.*, 1989; Shukla *et al.*,
1990).

In temperate and boreal regions, albedo (reflectance) is even more
important than evapotranspiration in regulating annual energy exchange
between the land and the atmosphere. Snow and ice have a much higher
albedo (0.7–0.9) than vegetated or bare-soil surfaces (0.2–0.4). In winter,
plants that are taller than the snowpack reduce albedo and increase
energy transfer to the land. Plant size influences albedo even among
growth forms covered by winter snow because tall plants melt out of the
snow more quickly (thereby reducing albedo and speeding snowmelt).

Differences in albedo among snow-free vegetated surfaces are more
subtle and, therefore, have less impact on energy exchange between land
and atmosphere. Vegetation differences in albedo are governed largely
by leaf phenology. Leafing out in deciduous ecosystems increases albedo
if the soil surface is dark, because leaves are highly reflective in the
infrared zone (Gates, 1980). If branches and surface litter are light-
colored, leafing out reduces albedo. Evergreen communities show less
seasonal change in albedo.

The roughness of the canopy surface determines the degree of coupling between plants and the atmosphere, that is, the extent to which

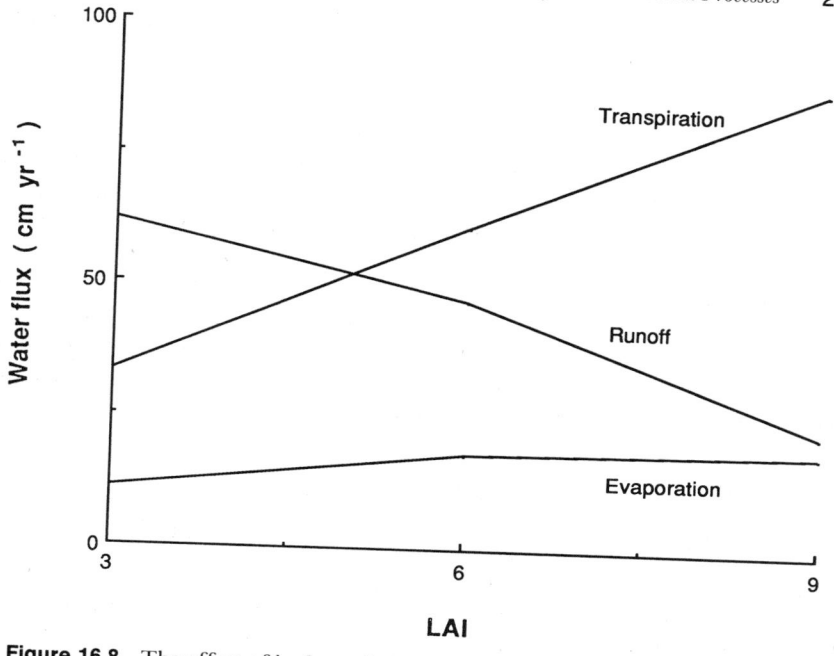

Figure 16.8 The effect of leaf area index (LAI) on simulated evaporation, transpiration, and runoff in a Florida forest. Data from Running and Coughlan (1988).

plants control energy exchange through evapotranspiration and convective heat exchange. In most areas, roughness is determined primarily by topography and patch structure of vegetation rather than by traits of individual plants. However, on broad flat landscapes such as prairies, low-statured vegetation has a well-developed boundary layer (a layer of relatively still air close to the vegetation surface) that is more important than stomatal conductance in governing water-vapor transport to the atmosphere when stomata are open (McNaughton and Jarvis, 1991). Under dry conditions, when stomatal conductance is low, the canopy heats up enough that convective exchange disrupts the boundary layer sufficiently for stomatal conductance to control water loss. In contrast, forest trees are large enough that turbulent eddies regularly penetrate their canopies, reducing the thickness of the boundary layer close to leaves and coupling plants more effectively with the atmosphere. Here, stomatal conductance consistently influences transpiration rate. In summary, plant size (when associated with a uniform canopy) is the major plant trait that governs canopy roughness and other components of water and energy exchange between vegetation and the atmosphere. Because energy exchange between vegetation and the atmosphere occurs at the top of the plant canopy, we need to characterize only the dominant canopy species of the community.

C. Carbon Flux

Carbon flux between the land and the atmosphere is determined largely by the balance between net primary production (NPP) and decomposition. Plant traits directly determine NPP and indirectly influence decomposition. NPP of an ecosystem is constrained ultimately by resource availability in the environment. However, the main plant traits that govern NPP in a given environment are size and growth rate (Tables 16.2, 16.3). Forests are generally more productive than shrublands, which in turn are often more productive than grasslands. If two plants have a similar RGR, the plant with greater biomass will, by definition, have greater NPP:

$$NPP = Biomass \times RGR$$

$$(g \cdot m^{-2} \cdot year^{-1}) = (g \cdot m^{-2}) \times (g \cdot g^{-1} \cdot year^{-1}).$$

Similarly, in ecosystems of similar biomass, communities dominated by rapidly growing species are more productive than those dominated by slowly growing species (Table 16.2). There are, however, trade-offs between size and RGR (Section II,B) because large plants allocate more biomass to structure and, therefore, grow more slowly than small plants (Table 16.2; Tilman, 1988; Chapin *et al.*, 1989). These trade-offs reduce the differences in potential NPP of ecosystems dominated by different growth forms.

Mortality of plants and plant parts is an inevitable consequence of NPP. After plant litter is transferred to the soil, decomposition is the major process of carbon loss from ecosystems and is influenced strongly by both plant traits and environmental factors such as temperature, moisture, and nutrient availability. NPP (a function of plant size) strongly influences carbon loss by decomposition because large-statured communities produce more litter (and therefore more substrate for decomposition) than do small-statured communities. The decomposition rate of a given quantity of litter is promoted by high nitrogen concentration and inhibited by plant secondary metabolites such as lignin (Aber and Melillo, 1982). Consequently, woody litter of trees and shrubs, which is high in lignin and low in nitrogen, decomposes more slowly than does litter of herbs. In summary, plant size, chemistry, and woodiness (which is a function of size and chemistry) strongly influence decomposition (Table 16.2).

It is logical to focus attention on the canopy species in predicting community production and decomposition, because these species generally have greatest production and generate most of the litter. However, understory species are more important in productivity and nutrient cycling than their biomass would suggest, because larger proportions of their biomass and nutrient pools turn over annually than of canopy species, which have greater allocation to long-lived structural support (Yarie, 1980; Chapin, 1983).

Carbon exchange between the land and the atmosphere is the balance between NPP and decomposition. In most communities, NPP is not balanced by decomposition, so there is a net change in carbon storage. Under these circumstances the growth-form effects on production and decomposition (see previous text) will govern net carbon exchange also. In general, I expect that woody plants would allow greater carbon storage than herbaceous plants because (1) the high C:N ratio of wood allows plants to store more carbon per unit of plant nitrogen and (2) the high C:N and lignin:N ratios of woody litter retard decomposition (Aber and Melillo, 1982). Forest soils do indeed have higher C:N ratios than grassland soils, but also store less N, so carbon storage is similar in cool temperate forest soils ($12-17$ kg C $\cdot$ m^{-3}) and temperate grassland soils (13 kg C $\cdot$ m^{-3}) (Post *et al.*, 1985).

D. Nutrient Cycling

In a fashion analogous to carbon exchange, rates of nutrient cycling through plants are governed by the rates of nutrient uptake and loss. Nutrient uptake is determined by productivity (nutrient requirement) and the extent to which this nutrient requirement is supported by new uptake rather than stored reserves. Therefore, the plant traits that directly influence nutrient cycling are plant size, RGR, and storage (Tables 16.2, 16.3). Nutrient loss from the plant is governed by litter fall (related to size and RGR) and the proportion of nutrients returned to storage after resorption from senescing leaves. After litter fall, the nutrients are released by mineralization, which is coupled chemically to decomposition and therefore related to tissue chemistry.

Nutrient uptake is linked to mineralization more tightly than production is linked to decomposition because the plant draws on nutrients largely from within the confines of the ecosystem rather than from a large atmospheric pool. Because there is only a limited pool of nutrients in the ecosystem, mineralization of nutrients such as nitrogen and phosphorus limits uptake and cycling. In highly fertile early successional environments, plant traits governing productivity may drive nutrient uptake, so plant size and RGR are the key traits. However, in most environments, mineralization, as governed by lignin, nitrogen, and physical environment, governs nutrient uptake, which drives production. Thus plant chemistry exerts a strong indirect effect on both nutrient uptake and productivity through its effect on nutrient availability.

Net nutrient exchange by the ecosystem with the atmosphere and hydrosphere is governed largely by successional status (Vitousek and Reiners, 1975). In early primary succession, low-nitrogen and high-light availability favor a nitrogen-fixing growth form (Vitousek and Walker, 1987), so presence or absence of this growth form strongly influences ecosystem nitrogen budgets (Crocker and Major, 1955; Vitousek and Walker, 1989). In early secondary succession after disturbances such as fire or plowing, the phenology of plant growth is the most important factor determining whether nutrients are retained by vegetation or lost to groundwater (Stark, 1977; Chapin and Van Cleve, 1981). Similarly, in ecosystems dominated by annual plants, for example, many Mediterranean grasslands, the first autumn rains often leach nutrients from the ecosystem because there is little plant biomass present to absorb the nutrients (Jackson et al., 1988), again pointing out the importance of phenology in governing net nutrient loss from ecosystems. In contrast, in most ecosystems with continuous plant cover (other than those in late succession), plants absorb most nutrients that become available through mineralization and weathering, so leaching loss of nutrients from the ecosystem is minimal, regardless of the nature of vegetation (Vitousek and Reiners, 1975).

E. Trophic Transfer

Trophic transfer, that is, the amount of primary production that moves through the herbivore-based food chain rather than transfer directly from plants to detrivores is determined by the first step in this process, that is, herbivory. The magnitude of this transfer may be determined by productivity (Oksanen, 1990; and thus by plant size and RGR) and by palatability of plants to herbivores, which is determined largely by plant defensive compounds and by nitrogen (Bryant et al., 1983; Coley et al., 1985). Rapidly growing plants often have low concentrations of defensive compounds (Bryant et al., 1983, Fig. 16.4), so RGR and chemistry are closely interrelated. The compounds that determine digestion of food by herbivores are often the same ones that deter decomposition, because both processes are mediated microbially, so well-defended or woody plants generally limit carbon loss through both decomposer- and herbivore-based food chains.

F. Sensitivity to Disturbance

The sensitivity of an ecosystem to disturbance is difficult to predict, because it depends on the type and magnitude of disturbance. Generally, a severe disturbance requires a longer recovery time in an ecosystem dominated by large plants than in a system dominated by small plants, because it takes longer to regenerate large plants. The capacity of species in a community to recover from disturbance depends on quantities of stored reserves to recover vegetatively or potential to produce new individuals through sexual reproduction. The propagules that give rise to new individuals might be stored in the soil, stored on the plant (e.g., serotinous cones), or produced following disturbance. Although the individuals that colonize a given disturbance may arise from outside the disturbed site, size, storage, and life-history traits related to reproduction and establishment are the major traits that determine the sensitivity of an area to disturbance on a regional scale.

G. Community Change

Life-history traits such as dispersal ability, reproductive potential, and degree of incorporation into the seed bank are probably the most important traits governing community change, both through succession (Walker and Chapin, 1987) and during long-term vegetation change (Davis, 1981). In early primary succession, presence or absence of nitrogen fixers also strongly influences the rate and pathway of succession (Vitousek and Walker, 1989).

In summary, plant size is the trait that most frequently influences ecosystem processes. RGR, tissue chemistry, phenology, and storage are other traits of major importance. Any growth-form classification that links ecophysiological and ecosystem characteristics should, therefore, incorporate these traits (e.g., Table 16.1). However, to predict community change we need to know much more about life-history traits that determine dispersal and establishment. Perhaps even more important, we need to know how the species respond to human activity (e.g., harvest, disturbance, and dispersal) because changes in human activity probably will be more important than climate in determining future vegetation composition over the coming decades.

V. Growth Form–Ecosystem Feedbacks

As discussed earlier, environment certainly influences the growth-form composition of a site. However, growth form or adaptive strategy also has a substantial impact on ecosystem function, leading to a system of feedbacks between vegetation and the environment (Vitousek, 1990). This concept will be developed with a few examples.

In the tundra, *Sphagnum* and other mosses are common in wet sites. Here, they insulate the soil so the soil thaws to a shallower depth and keeps the water table close to the soil surface. The high soil moisture resulting from the high water table promotes further moss growth. This positive feedback retards decomposition because the inherently low decomposition rate of *Sphagnum* is further reduced by the cold anaerobic soils beneath the moss. For this reason, *Sphagnum* is an important genus contributing to peat deposits in northern areas.

As discussed previously, fertile environments are dominated by herbaceous plants or deciduous trees that grow quickly. Leaves that are produced in this environment have high tissue nitogren and phosphorus concentrations and low concentrations of lignin and other plant secondary metabolites, traits that maximize mineralization rate (Aber and Melillo, 1982; White, 1988). This produces a positive feedback to support high productivity (Fig. 16.9). Even differences among species of a single growth form can cause substantial differences in nitrogen mineralization rate (Wedin and Tilman, 1990). In contrast, infertile sites are dominated by plants such as evergreen trees and shrubs that produce litter with lower tissue nitrogen and phosphorus concentrations and higher concentrations of lignin and secondary metabolites. These traits lead to initial nitrogen and phosphorus immobilization in the litter (reducing nutrient availability in the soil), causing a slow rate of release of these nutrients to the soil. This slow rate of nutrient release reinforces the low nutrient availability in the site.

Herbivores tend to magnify these initial differences in soil fertility, because they consume a higher proportion of plant production in high fertility sites. In these sites, they consume high quality plant material before plants resorb nutrients at senescence. Herbivores return nitrogen and phosphorus to the soil in highly available forms as feces and urine, short-circuiting decomposition (Fig. 16.9). Homeotherm herbivores are particularly effective in this nutrient recycling because the process of homeothermy consumes more energy, so less organic matter is returned to the soil. In contrast, low-resource sites are dominated by plants whose well-defended leaves deter herbivores and retard decomposition (Chapin and McNaughton, 1989; Pastor and Naiman, 1992). Thus, plant traits magnify inherent differences in soil fertility among sites, and the presence or absence of herbivores further reinforces these differences.

In wetlands, methane transport from soil to the atmosphere occurs largely through the aerenchyma of plants. Hence, in wetlands that are occupied by a mixture of growth forms, only aerenchymatous graminoids appear to transport methane. For example, in tundra, microsites occupied by these graminoids have high methane fluxes, whereas microsites occupied by mosses (which harbor methane oxidizers) have lower methane fluxes (Whalen and Reeburgh, 1988). The relative abundance of graminoids and mosses, therefore, strongly influences ecosystem methane flux.

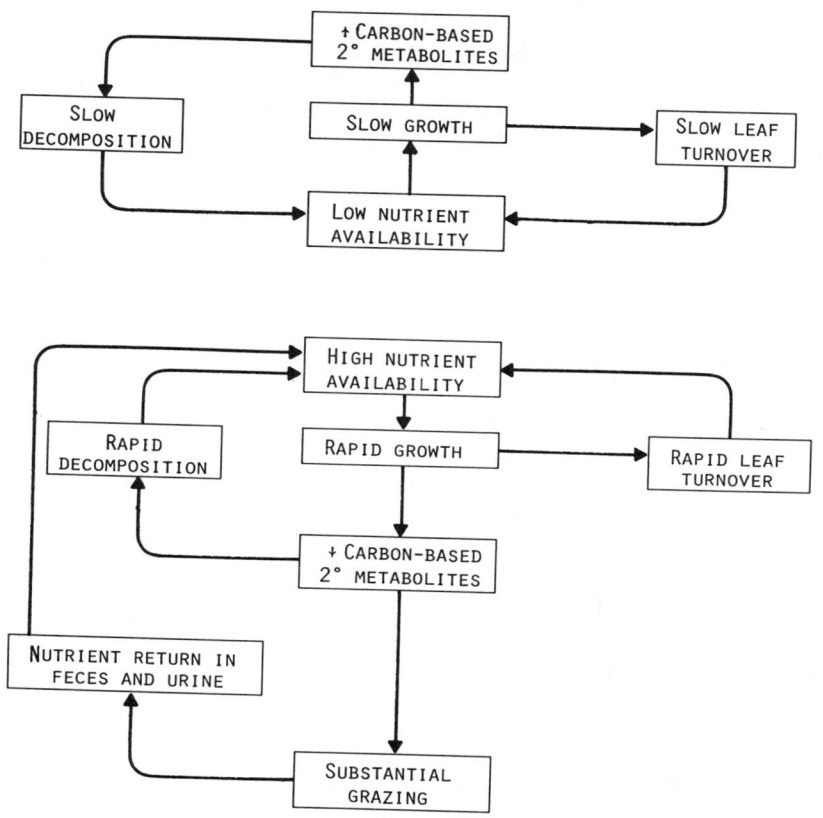

Figure 16.9 Feedbacks by which grazing and plant defense magnify differences in soil fertility between fertile and infertile sites. Figure reprinted, with permission, from Chapin (1991b).

VI. Remote Sensing of Growth Forms and Ecosystem Function

The dominant growth forms of an ecosystem often can be identified by remote sensing from aircraft or satellites. For example, evergreen communities have higher infrared albedo during the inactive season than do deciduous communities; graminoid vegetation can be distinguished from dicot-dominated vegetation. Synthetic-aperture radar (SAR) can distinguish more detailed structural characteristics and shows great promise for revealing size and relative proportions of boles, branches, and leaves, allowing differentiation of forests from shrublands (Ulaby *et al.*, 1990). Many of the environmental factors that govern patterns of growth-form distribution also directly control ecosystem processes such as nitrogen mineralization and methane flux. Moreover, the growth-form composition of vegetation exerts strong feedbacks on ecosystem function. Growth forms are, therefore, an effective measure of ecosystem function (Matson *et al.*, 1989; Matson and Vitousek, 1990), so remote sensing of growth forms allows mapping of ecosystem functions on a regional or global basis. For example, we can estimate nitrogen, water, or carbon fluxes across a landscape (Reiners *et al.*, 1989; Running *et al.*, 1989) or total methane fluxes for the wetlands of the globe (Matthews and Fung, 1987).

In other cases, remote sensing gives a direct measure of ecosystem processes. For example, normalized difference vegetation index (NDVI) is a measure of the absorbed photosynthetically active radiation (APAR). When measured at the level of leaves, canopies, or regions, APAR is correlated strongly with photosynthetic carbon gain, regardless of whether the variation in carbon gain is caused by differences in growth form, environmental stress, or phenological stage (Field, 1991). The seasonal integral of NDVI, therefore, provides a reasonable estimate of annual production, although NDVI may be a poor predictor of photosynthesis on any given date (Sellers *et al.*, 1986; Running, 1990; Field, 1991; Schimel *et al.*, 1991). For this reason, remote sensing of NDVI has permitted mapping of aboveground production for grasslands (Tucker *et al.*, 1984) and forests (Nemani and Running, 1989). Once production is estimated by remote sensing, it can be used to estimate rates of correlated processes such as nitrogen cycling (Fig. 16.10).

At a more subtle level, the nitrogen and lignin concentrations of the canopy can be sensed remotely (Wessman *et al.*, 1988; Peterson and Running, 1989). Because of the tight association of tissue quality with herbivory, decomposition, mineralization, RGR, photosynthesis, and transpiration, remote sensing of tissue quality can be used to predict many patterns of ecosystem processes, especially if the interrelationships among these processes are specified by simulation models (Matson and Vitousek, 1990). Because we can sense both nitrogen stress (Wessman *et al.*, 1988) and water stress (Pierce *et al.*, 1990) remotely, the stage is set to study the interaction between ecosystem processes and their environmental controls over broad areas of the globe.

VII. Conclusions

Plant growth forms provide an effective means of simplifying the vast array of physiological diversity in natural ecosystems, so plant-level physiology can be scaled to regional and global levels. Resource levels in

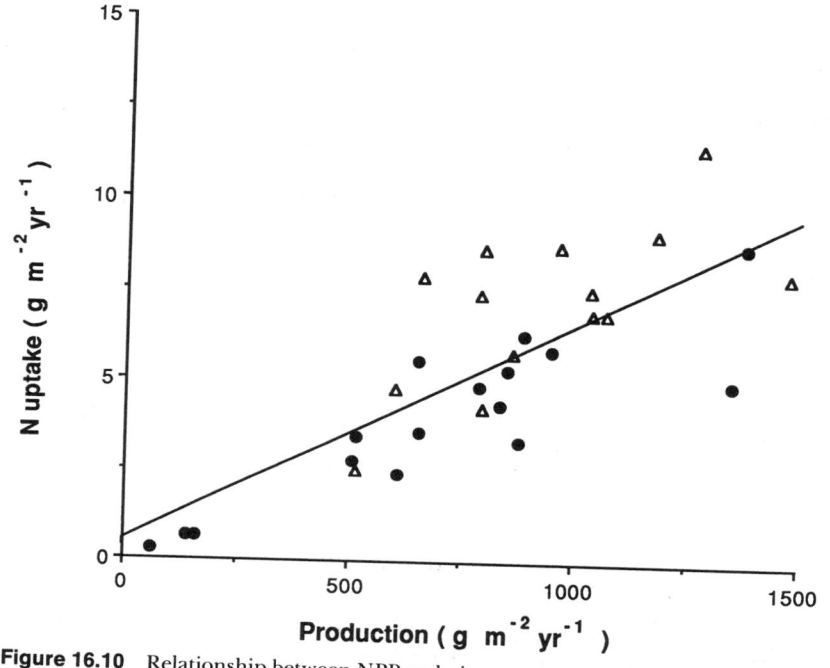

Figure 16.10 Relationship between NPP and nitrogen uptake of temperate and boreal coniferous (●) and deciduous (△) forests. [Calculated from Cole and Rapp (1981)].

the environment predictably control growth-form composition. High-resource environments support growth forms that are highly productive due to either large size or high relative growth rate, depending on time since disturbance. In contrast, low-resource environments support slowly growing plants, whose well-developed chemical defenses minimize rates of herbivory and decomposition. Growth forms have suites of interrelated physiological characteristics. Rapidly growing plants have high rates of photosynthesis, transpiration, tissue turnover, herbivory, and decomposition. Both within and among growth forms, there are important trade-offs between growth and alternative allocations such as structure, defense, storage, and reproduction. Because of the interdependencies of physiological traits, any of several interrelated traits could be used as a basis for growth-form classification. I suggest that a size-based growth-form classification be adopted, because size is one of the major determinants of plant effects on environment through patterns of energy exchange, carbon flux, nutrient flux, trophic transfers, and ecosystem responses to disturbances. A size-based growth-form classification then can be refined based on traits such as tissue chemistry, phenology, or life-history characteristics, depending on the processes to be scaled.

Growth-form characteristics not only respond to environment but feed back to reinforce natural environmental differences, largely because size and chemical differences among growth-forms strongly affect herbivory and decomposition and, therefore, nutrient availability. Remote sensing of growth forms allows mapping of ecosystem function on regional and global scales because remote sensing directly measures some ecosystem functions such as absorbed radiation and degree of environmental stress and because growth form correlates closely with many other ecosystem processes.

Acknowledgments

I thank M. Chapin, N. Chiariello, C. Field, I. Fung, S. Hobbie, and H. Reynolds for insightful comments and useful discussions. Research leading to these ideas was funded by NSF Grants DEB 8205344 and BSR 8705323 and DOE Grant DE-FG03-89ER60916.

References

Aber, J. D., and Melillo, J. M. (1982). Nitrogen immobilization in decaying hardwood leaf litter as a function of initial nitrogen and lignin content. *Can. J. Bot.* **60**, 2263–2269.

Berendse, F., Oudhof, H., and Bol, J. (1988). A comparative study on nutrient cycling in wet heathland ecosystems. I. Litter production and nutrient losses from the plant. *Oecologia* **74**, 174–184.

Bloom, A. J., Chapin, F. S., III, and Mooney, H. A. (1985). Resource limitation in plants - an economic analogy. *Annu. Rev. Ecol. Syst.* **16**, 363–392.

Bokhari, U. G., and Singh, J. S. (1975). Standing state and cycling of nitrogen in soil-vegetation components of prairie ecosystems. *Ann. Bot.* **39**, 273–285.

Bormann, F. H., and Likens, G. E. (1979). "Pattern and Process in a Forested Ecosystem." Springer-Verlag, New York.

Bryant, J. P., Chapin, F. S., III, and Klein, D. R. (1983). Carbon/nutrient balance of boreal plants in relation to vertebrate herbivory. *Oikos* **40**, 357–368.

Chabot, B. F., and Mooney, H. A. (1985). "The Physiological Ecology of North American Vegetation." Chapman and Hall, London.

Chapin, F. S., III (1980). The mineral nutrition of wild plants. *Annu. Rev. Ecol. Syst.* **11**, 233–260.

Chapin, F. S., III (1983). Nitrogen and phosphorus nutrition and nutrient cycling by evergreen and deciduous understory shrubs in an Alaskan black spruce forest. *Can. J. For. Res.* **13**, 773–781.

Chapin, F. S., III (1986). Controls over growth and nutrient use by taiga forest trees. *In* "Forest Ecosystems in the Alaskan Taiga" (K. Van Cleve, F. S. Chapin, III, P. W. Flanagan, L. A. Viereck, and C. T. Dyrness, eds.), pp. 96–111. Springer-Verlag, New York.

Chapin, F. S., III (1991a). Integrated responses of plants to stress: A centralized system of physiological responses. *BioScience* **41**, 29–36.

Chapin, F. S., III (1991b). Effects of multiple environmental stresses on nutrient availability and use by plants. *In* "Response of Plants to Multiple Stresses" (H. A. Mooney, W. E. Winner, and E. J. Pell, eds.), pp. 67–88. Academic Press, New York.

Chapin, F. S., III, Bloom, A. J., Field, C., and Waring, R. H. (1987). Interaction of environmental factors in the control of plant growth. *BioScience* **37,** 49–57.

Chapin, F. S., III, Groves, R. H., and Evans, L. T. (1989). Physiological determinants of growth rate in response to phosphorus supply in wild and cultivated *Hordeum* species. *Oecologia* **79,** 96–105.

Chapin, F. S., III, Jefferies, R. L., Reynolds, J., Shaver, G. R., and Svoboda, J. (1992). Arctic plant physiological ecology: A challenge for the future. *In* "Arctic Physiological Processes in a Changing Climate" (F. S. Chapin, III, R. L. Jefferies, J. Reynolds, G. R. Shaver, and J. Svoboda, eds.), pp. 3–8. Academic Press, New York.

Chapin, F. S., III, and Kedrowski, R. A. (1983). Seasonal changes in nitrogen and phosphorus fractions and autumn retranslocation in evergreen and deciduous taiga trees. *Ecology* **64,** 376–391.

Chapin, F. S., III, and McNaughton, S. J. (1989). Lack of compensatory growth under phosphorus deficiency in grazing-adapted grasses from the Serengeti Plains. *Oecologia* **79,** 551–557.

Chapin, F. S., III, Schulze, E.-D., and Mooney, H. A. (1990). The ecology and economics of storage in plants. *Annu. Rev. Ecol. Syst.* **21,** 423–447.

Chapin, F. S., III, and Van Cleve, K. (1981). Plant nutrient absorption and retention under differing fire regimes. *In* "Fire Regimes and Ecosystem Processes" (H. A. Mooney, J. M. Bonnicksen, N. L. Christensen, J. E. Lotan, and W. A. Reiners, eds.), pp. 301–321. U.S.D.A. Forest Service General Technical Report, WO-26, Washington.

Chiariello, N., and Roughgarden, J. (1984). Storage allocation in seasonal races of a grassland annual: Optional vs actual allocation. *Ecology* **65,** 1290–1301.

Cole, D. W., and Rapp, M. (1981). Elemental cycling in forest ecosystems. *In* "Dynamic Properties of Forest Ecosystems" (D. E. Reichle, ed.), pp. 341–409. Cambridge University Press, Cambridge.

Coley, P. D. (1986). Costs and benefits of defense by tannins in a neotropical tree. *Oecologia* **70,** 238–241.

Coley, P. D. (1988). Effects of plant growth rate and leaf lifetime on the amount and type of anti-herbivore defense. *Oecologia* **74,** 531–536.

Coley, P. D., Bryant, J. P, and Chapin, F. S., III (1985). Resource availability and plant anti-herbivore defense. *Science* **230,** 895–899.

Collatz, G. J., Ball, J. T., Grivet, C., and Berry, J. A. (1991). Physiological and environmental regulation of stomatal conductance, photosynthesis and transpiration: A model that includes a laminar boundary layer. *Agric. For. Meteorol.* **54,** 107–136.

Cook, M. G., and Evans, L. T. (1983). Some physiological aspects of the domestication and improvement of rice (*Oryza* spp.). *Field Crops Res.* **6,** 219–238.

Crocker, R. L., and Major, J. (1955). Soil development in relation to vegetation and surface age at Glacier Bay, Alaska. *J. Ecol.* **43,** 427–448.

Davidson, R. L. (1969). Effects of soil nutrients and moisture on root/shoot ratios in *Lolium perenne* L. and *Trifolium repens* L. *Ann. Bot.* **33,** 571–577.

Davis, M. B. (1981). Quaternary history and the stability of forest communities. *In* "Forest Succession" (D. C. West, H. H. Shugart, and D. B. Botkin, eds.), pp. 132–153. Springer-Verlag, New York.

Farquhar, G. D., and Sharkey, T. D. (1982). Stomatal conductance and photosynthesis. *Annu. Rev. Plant Physiol.* **33,** 317–345.

Field, C. (1991). Ecological scaling of carbon gain to stress and resource availability. *In* "Integrated Responses of Plants to Stress" (H. A. Mooney, W. E. Winner, and E. J. Pell, eds.), pp. 35–65. Academic Press, New York.

Field, C., and Mooney, H. A. (1986). The photosynthesis-nitrogen relationship in wild plants. *In* "On the Economy of Plant Form and Function" (T. J. Givnish, ed.), pp. 25–55. University Press, Cambridge.

Gates, D. M. (1980). "Biophysical Ecology." Springer-Verlag, Berlin.

Gray, J. T., and Schlesinger, W. H. (1981). Nutrient cycling in Mediterranean type ecosystems. *In* "Resource Use by Chaparral and Matorral" (P. C. Miller, ed.), pp. 259–285. Springer-Verlag, New York.

Grime, J. P. (1977). Evidence for the existence of three primary strategies in plants and its relevance to ecological and evolutionary theory. *Am. Nat.* **111,** 1169–1194.

Grime, J. P., and Hunt, R. (1975). Relative growth rate: Its range and adaptive significance in a local flora. *J. Ecol.* **63,** 393–422.

Harper, J. L. (1977). "The Population Biology of Plants." Academic Press, New York.

Horvitz, C. C., and Schemske, D. W. (1988). Demographic cost of reproduction in a neotropical herb: An experimental field study. *Ecology* **69,** 1741–1745.

Jackson, L. E., Strauss, R. B., Firestone, M. K., and Bartolome, J. W. (1988). Plant and soil nitrogen dynamics in California annual grassland. *Plant Soil* **110,** 9–17.

Matson, P. A., and Vitousek, P. M. (1990). Remote sensing and trace gas fluxes. *In* "Remote Sensing of Biosphere Functioning (R. J. Hobbs and H. A. Mooney, eds.), pp. 157–167. Springer-Verlag, New York.

Matson, P. A., Vitousek, P. M., and Schimel, D. S. (1989). Regional extrapolation of trace gas flux based on soils and ecosystems. *In* "Exchange of Trace Gases between Terrestrial Ecosystems and the Atmosphere" (M. O. Andreae and D. S. Schimel, eds.), pp. 97–108. Wiley, New York.

Matthews, E., and Fung, I. (1987). Methane emissions from natural wetlands: Global distribution, area, and environmental characteristics of sources. *Global Biogeochem. Cycling* **1,** 61–86.

McNaughton, K. G., and Jarvis, P. G. (1991). Effects of spatial scale on stomatal control of transpiration. *Agric. For. Meteorol.* **54,** 279–301.

Nambiar, E. K. S., and Fife, D. N. (1987). Growth and nutrient retranslocation in needles of radiata pine in relation to nitrogen supply. *Ann. Bot.* **60,** 147–156.

Nemani, R. R., and Running, S. W. (1989). Testing a theoretical climate-soil-leaf area hydrologic equilibrium of forests using satellite data and ecosystem simulation. *Agric. For. Meteorol.* **44,** 245–260.

Oksanen, L. (1990). Predation, herbivory, and plant strategies along gradients of primary productivity. *In* "Perspectives on Plant Competition" (J. B. Grace and D. Tilman, eds.), pp. 445–474. Academic Press, New York.

Pastor, J., and Naiman, R. (1992). Selective foraging and ecosystem processes in boreal forests. *Am. Nat.* **139,** 690–705.

Peet, R. K., and Christensen, N. L. (1987). Competition and tree death. *BioScience* **37,** 586–595.

Peterson, D. L., and Running, S. W. (1989). Applications in forest science and management. *In* "Theory and Applications of Optical Remote Sensing" (G. Asrar, ed.), pp. 429–473. Wiley, New York.

Pierce, L. L., Running, S. W., and Riggs, G. A. (1990). Remote detection of canopy water stress in coniferous forests using the NS001 Thematic Mapper Simulator and the Thermal Infrared Multispectral Scanner. *Photogram. Eng. Remote Sens.* **56,** 579–586.

Post, W.M., Pastor, J., Zinke, P. J., and Stangenberger, A. G. (1985). Global patterns of soil nitrogen storage. *Nature* **317,** 613–616.

Rappoport, H. F., and Loomis, R. S. (1985). Interaction of storage root and shoot in grafted sugarbeet and chard. *Crop Sci.* **25,** 1079–1084.

Reekie, E. G., and Bazzaz, F. A. (1987). Reproductive effort in plants. 3. Effect of reproduction on vegetative activity. *Am. Nat.* **129,** 907–919.

Reiners, W. A., Strong, L. L., Matson, P. A., Burke, I. C, and Ojima, D. S. (1989). Estimating biogeochemical fluxes across sagebrush-steppe landscapes with Thematic Mapper imagery. *Remote Sens. Environ.* **28**, 121–129.

Running, S. W. (1990). Estimating terrestrial primary productivity by combining remote sensing and ecosystem simulation. *In* "Remote Sensing of Biosphere Functioning" (R. J. Hobbs and H. A. Mooney, eds.), pp. 65–86. Springer-Verlag, New York.

Running, S. W., and Coughlan, J. C. (1988). A general model for forest ecosystem processes for regional applications. I. Hydrologic balance, canopy gas exchange, and primary production processes. *Ecol. Model.* **42**, 125–154.

Running, S. W., Nemani, R. R., Peterson, D. L., Band, L. E., Potts, D. F., Pierce, L. L., and Spanner, M. A. (1989). Mapping regional forest evapotranspiration and photosynthesis by coupling satellite data with ecosystem simulation. *Ecology* **70**, 1090–1101.

Sala, O. E., Parton, W. J., Joyce, L. A., and Lauenroth, W. K. (1988). Primary production of the central grassland region of the United States. *Ecology* **69**, 40–45.

Sato, N., Sellers, P. J., Randall, D. A., Schneider, E. K., Shukla, J., Kinter, J. L., III, Hou, Y.-T., and Albertazzi, E. (1989). Effects of implementing the simple biosphere model in a general circulation model. *J. Atmos. Sci.* **46**, 2757–2782.

Schimel, D. S., Kittel, T. G. F., and Parton, W. J. (1991). Terrestrial biogeochemical cycles: Global interactions with the atmosphere and hydrology. *Tellus* **43AB**, 188–203.

Schimper, A. F. W. (1898). "Pflanzengeographie auf Physiologische Grundlage." Verlag Gustav Fisher, Jena, Germany.

Schulze, E.-D., and Chapin, F. S., III (1987). Plant specialization to environments of different resource availability. *In* "Potentials and Limitations in Ecosystem Analysis" (E.-D. Schulze and H. Zwolfer, eds.), pp. 120–148.

Schulze, E.-D., Fuchs, M., and Fuchs, M. I. (1977). Spacial distribution of photosynthetic capacity and performance in a mountain spruce forest of northern Germany. III. The significance of the evergreen habit. *Oecologia* **30**, 239–248.

Schulze, E.-D., and Hall, A. E. (1982). Stomatal responses, water loss, and CO_2 assimilation rates of plants in contrasting environments. *In* "Physiological Plant Ecology" (O. L. Lange, P. S. Nobel, C. B. Osmond, and H. Zeigler, eds.), Vol. II, pp. 181–230. Springer-Verlag, Berlin.

Segal, M., Avissar, R., McCumber, M. C., and Pielke, R. A. (1988). Evaluation of vegetation effects on the generation and modification of mesoscale circulations. *J. Atmos. Sci.* **45**, 2268–2392.

Sellers, P. J., Mintz, Y., Sud, Y. C., and Dalcher, A. (1986). A simple biosphere (SiB) model for use within general circulation models. *J. Atmos. Sci.* **43**, 505–531.

Shukla, J., Nobre, C., and Sellers, P. (1990). Amazon deforestation and climate change. *Science* **247**, 1322–1325.

Snow, A. A., and Whigham, D. F. (1989). Costs of flower and fruit production in *Tipularia discolor* (Orchidaceae). *Ecology* **70**, 1286–1293.

Stark, N. M. (1977). Fire and nutrient cycling in a Douglas-fir/larch forest. *Ecology* **58**, 16–30.

Tilman, D. (1988). "Plant Strategies and the Dynamics and Structure of Plant Communities." Princeton University Press, Princeton, New Jersey.

Tilman, D. (1990). Mechanisms of plant competition for nutrients: The elements of a predictive theory of competition. *In* "Perspectives on Plant Competition" (J. B. Grase and D. Tilman, eds.), pp. 117–141. Academic Press, New York.

Tucker, C. J., Vanpraet, C., Boerwinkle, E., and Gaston, A. (1984). Satellite remote sensing of total dry matter accumulation in the Senegalese Sahel. *Remote Sens. Environ.* **13**, 461–474.

Tuomi, J., Niemela, P., Haukioja, E., and Neuvonen, S. (1984). Nutrient stress: An explanation for plant anti-herbivore responses to defoliation. *Oecologia* **61**, 208–210.

312 F. Stuart Chapin III

Ulaby, F., Sarabandi, K., McDonald, K., Wibitt, M., and Dobson, M. (1990). Michigan microwave canopy scattering model (MIMICS). *Int. J. Remote Sens.* **12,** 1223–1254.

Viereck, L. A., Dyrness, C. T., Van Cleve, K., and Foote, M. J. (1983). Vegetation, soils, and forest productivity in selected forest types in interior Alaska. *Can. J. For. Res.* **13,** 703–720.

Vitousek, P. M. (1990). Biological invasions and ecosystem processes: Towards an integration of population biology and ecosystem studies. *Oikos* **57,** 7–13.

Vitousek, P. M., and Reiners, W. A. (1975). Ecosystem succession and nutrient retention: A hypothesis. *BioScience* **25,** 376–381.

Vitousek, P. M., and Walker, L. R. (1987). Colonization, succession and resource availability: Ecosystem-level interactions. *In* "Colonization, Succession and Stability" (A. J. Gray, M. J. Crawley, and P. J. Edwards, eds.), pp. 207–223. Blackwell, Oxford.

Vitousek, P. M., and Walker, L. R. (1989). Biological invasion by *Myrica faya* in Hawaii: Plant demography, nitrogen fixation, ecosystem effects. *Ecol. Monogr.* **59,** 247–265.

Walker, L. R., and Chapin, F. S., III (1987). Interactions among processes controlling successional change. *Oikos* **50,** 131–135.

Wedin, D. A., and Tilman, D. (1990). Species effects on nitrogen cycling: A test with perennial grasses. *Oecologia* **84,** 433–441.

Wessman, C.A., Aber, J. D., Peterson, D. L., and Melillo, J. M. (1988). Remote sensing of canopy chemistry and nitrogen cycling in temperate forest ecosystems. *Nature* **335,** 154–156.

Whalen, S. C., and Reeburgh, R. S. (1988). A methane flux time series for tundra environments. *Global Biogeochem. Cycles* **2,** 399–409.

White, C. S. (1988). Nitrification inhibition by monoterpenoids: Theoretical mode of action based on molecular structures. *Ecology* **69,** 1631–1633.

Yarie, J. (1980). The role of understory vegetation in the nutrient cycle of forested ecosystems in the mountain hemlock biogeoclimatic zone. *Ecology* **61,** 1498–1514.

17

Grouping Plants by Their Form–Function Characteristics as an Avenue for Simplification in Scaling between Leaves and Landscapes

Todd E. Dawson and F. Stuart Chapin III

I. Introduction

The chapters in this volume offer several conceptually distinct approaches to integrating information across spatial and temporal scales. This discussion is the result of our goal to develop, expand, summarize, and integrate the general consensus among the symposium participants on which approach would be most fruitful in synthesizing functional response information for individual plants to facilitate predictions at higher levels. We assert that the cornerstone of such an effort derives from the well-known form–function relationships seen in most plants.

During the past quarter century there has been a surge of interest in and concern about the changes occurring in the atmosphere and biosphere. These changes, a result of human activities, have changed the ecological trajectory of organisms as well as the ecosystems they inhabit. Among scientists, concern about these global changes has renewed interest in how the biosphere is linked together, in the processes and mechanisms that represent these linkages, and in how the biosphere and the atmosphere interact (Botkin *et al.*, 1989). Achieving a better understanding of the important interfaces and linkages between organisms and their environment and between ecosystems and the atmosphere has raised many questions related to how various processes and functions on very different scales may or may not be linked, influence, or are influenced by processes and functions operating on scales of organization one or more levels above or below the one at which a set of observations has been made. These questions have led scientists to consider more carefully how to integrate or "scale" information in the biosphere from one level to another, and between the biosphere and the atmosphere, and whether theoretical or empirical scaling laws can be identified (see Chapter 2).

Because transfer of materials and energy is a necessary process at all levels, from individual plants to landscapes, scaling or integrating information gathered at one level of organization to another should be relatively straightforward (Chapter 10). However, no scale can be described unambiguously. Also, the precision with which we can make predictions declines as more and more complexity and information is added; for example, when we move from single leaves to whole canopies to populations to communities and, eventually, to entire landscapes. As such, information and complexity must be "grouped" (Chapter 16). Deciding on a grouping to use when scaling between different levels poses perhaps the greatest theoretical and practical challenge for investigators involved in scaling efforts. If we hope to predict changes in landscapes from our understanding of physiology, some information must be carried between levels. This information should be contained in some usable grouping that provides each higher level with the power to make meaningful predictions. Moreover, we would like these groups to be as universally applicable as possible so entirely new types of groups need not be developed for each ecosystem and ecosystem model (Chapter 8).

II. Form–Function Relationship in Plants

Vitousek (Chapter 10) has suggested that, although an understanding of ecosystem function has been based in part on physiological mechanisms, a decision about which mechanisms are relevant to the ecosystem scale is guided by the phenomena we wish to predict at that scale. As such, he suggests that scaling down (top-down) from global measurements to ecosystems and from ecosystems to the organisms that occupy them will allow our understanding of biosphere function to develop most rapidly. From this perspective we can best decide which ecosystems and processes deserve most careful scrutiny. Having given careful attention to this top-down context, bottom-up studies can be designed to investigate controls over processes that are recognized as important on the larger ecosystem or global scale.

For example, early versions of general circulation models (GCMs) treated terrestrial vegetation as a "green scum" that responded to environment according to simple physical principles. Incorporation of simple biotic controls such as canopy conductance allowed the biosphere to be treated as a "big leaf" (Sellers *et al.*, 1986). GCM models incorporating the big-leaf approach showed that differences in canopy conductance of water vapor, for example, when tropical rainforest is converted to grassland, could have profound lasting effects on climate (Shukla and Mintz, 1982; Shukla *et al.*, 1990). This realization now provides renewed interest in canopy conductance and its control. Without the top-down context, it might have been less clear which aspects of plant water relations are most important in scaling from plants to the globe; any subsequent studies on plant water relations might have had a narrower audience.

The limitation of the top-down approach is that, by definition, it treats its smallest scale empirically. In the big-leaf example, canopy conductance is considered a parameter respresentative of either forest or grassland. An understanding of the physiological basis of canopy conductance is beyond the scope of such models. However, the models provide a starting point for physiological studies that integrate the components of plant form and leaf physiology that govern canopy conductance and, perhaps other canopy exchange processes.

Given the top-down approach as a context, we feel that a bottom-up approach is most likely to provide insight into controlling mechanisms. Such approaches build on a substantial body of information collected at the level of the plant or plant part. The challenge is how to synthesize this information in a way that is useful in understanding controls over processes at higher levels. Clearly using all the plants in an ecosystem is impossible. Plants must be grouped together to simplify the task of scaling, simplification that must occur without excessive loss of predictive power. We suggest that grouping according to the relatively tight form–function relationships seen in many vascular plants (see Givnish, 1986, and Russell *et al.*, 1989, for reviews) may be the most straightforward and perhaps the most informative way to scale from leaves to landscapes because strong plant form–function realtionships are (1) well known (e.g., evergreen vs. deciduous, Chabot and Hicks, 1982; herbs vs. grasses vs. trees, Raven, 1986, and references therein) and have formed the cornerstone for many previous generalizations about plants in relation to distribution, climate (Woodward, 1987), and vegetation physiognomy (Schimper, 1898 Graetz, 1990); (2) obtained relatively easily and are the foundation of many ecophysiological investigations (Chabot and Mooney, 1985); and (3) the means by which bottom-up models predict responses at higher levels or by which top-down models are validated

(Tilman, 1988; Running, 1990; Chapters 6 and 16). For example, leaf-for canopy-level functional responses contain more detail than necessary to ascertain broad patterns of carbon flux at a landscape or regional level (Fung *et al.*, 1983; Chapter 11). However, grouping information can allow simplification, and simplification must be employed if scaling across several levels of organization is to be mathematically tractable (Chapter 2 and 15) and successful.

III. Grouping Rationale

A conventional approach to scaling, be it from the bottom up or the top down, is to use the mean and variance of a leaf-level response as an avenue to describe patterns at the landscape level (Chapter 2). Another approach would be to use a distribution of responses rather that a discrete value within the selected functional group (e.g., species, life-form, life history). This approach is much like that employed by quantitative geneticists when they use continuously varying trait values rather than discrete trait values in describing the genetic structure of a population (Falconer, 1981). Which of these, or perhaps other, strategies is most useful will depend on the scaling objective, the desired level of resolution, and the parameters being predicted. For example, if our objective is to predict water exchange between the canopy and the atmosphere for a tropical moist forest, we may require relatively detailed information on rooting depth, water relations, canopy structure, and the response of stomatal conductance to irradiance at each level of the canopy. However, to predict element cycling, it may be more important to know patterns of biomass allocation, turnover, and chemical composition. Whether the mean and variance of a particular response or a distribution of responses is used, grouping those responses by plant form and function should be done whenever the responses are relevant to the scaling question being posed (see subsequent text for exceptions).

For many ecosystems, detailed form–functional information to make landscape-level predictions already may be adequate for some variables but lacking for others. For example, if we were interested in ecosystem carbon storage, we know much more about controls over biomass allocation and carbon fixation than we do about the rules that govern allocation to lignin and subsequent effects on decomposition.

IV. Grouping Criteria

Three fundamental questions need to be asked before any form–function-oriented scaling effort takes place. (1) Where would form–functional groups be useful in our scaling efforts? (2) How all-encompassing should these groups be (e.g., all trees, or deciduous vs. evergreen) to provide meaningful information, both biologically and practically? (3) Which physiological processes within a circumscribed group are most critical to extrapolating our predictions among scales? In addressing the first question, four broad areas in ecosystem science were identified in which functional groups could be useful: biospheric–atmospheric feedbacks, carbon storage, element cycling, and surface hydrology. It is clear from the chapters in this volume that every scaling model or effort touches on or revolves around one or more of these areas. Once the ecosystem parameter to be predicted has been identified, then answering the later questions just posed should be relatively straightforward. Some examples are outlined next.

Most of the criteria that would be used to delineate form–functional groups can be fit into existing plant groupings. For example, grouping based on life history (annual, biennial, perennial) or seed dormancy characteristics long has been recognized among plant ecologists and may be not only useful but essential in deriving long-term predictions when we expect major vegetation changes to be driven by population- or community-level dynamics (Chapter 15). These same groups may be less important, however, in predicting the response of current vegetation to changing climate. Form–functional groups based on photosynthetic and water relation properties or leaf and canopy energy balance provide a basis for predicting ecosystem activity (Chapter 8) and surface hydrology (Running and Coughlan 1988). Carbon and nutrient allocation patterns and the types and quality (lignin content, litter nutrient concentration) of the carbon and nutrients provide groupings to predict carbon storage and element cycling (Chapter 16). Grouping by individual species or genus probably will not succeed because far too much detail exists for any one species, but little information has been accumulated about broad spectrum functional responses among a group of related species or genera.

A key physiological trait for most scaling questions is allocation to aboveground and belowground structures, which can be linked to broad categories of plants, principally by growth form (Chapter 16). Root:shoot ratio indicates how plants allocate photosynthate and nutrients, the relative distribution of plant organic matter in the below- and aboveground compartments, and the location where that plant litter is decomposed. This ratio influences the rate of element cycling and acts as a measure of the relative storage capacity of carbon and nutrients in each compartment. Lack of adequate information about physiological controls over belowground allocation, storage, and turnover impedes our efforts to make meaningful predictions of allocation at the landscape level and on long time scales. At present, most models employ phenomenological rules for allocation. In this regard, we must gather more information at the plant level about roots: biomass, surface area, rates of turnover, and storage properties. Another critical set of physiological processes about which even less is known involves allocation of carbon to fractions that differ in turnover within the plant (storage) and soil (decomposition).

Knowledge of leaf longevity and the leaf area index (LAI) can be useful in accounting for much of the ecosystem productivity (Chapter 8). Further, leaf longevity often is correlated with growth form and broad functional attributes (Chabot and Hicks, 1982) that would be useful in grouping plants into elements for scaling to higher levels. Our view is that life-form/growth-form groupings can provide substantial power in scaling functional information from the individual plant to the entire landscape. Grouping in this way not only provides sound rationale and criteria for characterizing broad functional categories of plants in all ecosystems, but is not an excessive oversimplification and should not affect the predictive power of higher scale models adversely. In fact, grouping by form–function relationships that are associated with particular plant growth forms are already compatible with existing ecosystem models such as the BIOME-BGC (Chapter 8) and FIFE models (Chapter 3) currently being applied to scaling questions for forest and prairie ecosystems, respectively.

V. Concluding Remarks

Scaling should carry the essential controls over processes such as the capture and retention of resources from one level to another (O'Neill *et al.*, 1986). We feel that accomplishing this goal effectively requires both top-down and bottom-up approaches. Without the top-down approach to identify critical processes and ecosystems, a synthesis of information about detailed mechanisms may be inefficient, incorporating excessive detail in some areas and ignoring other aspects that are critical to understanding processes at the higher levels. Without the bottom-up approach, top-down studies are purely descriptive and lack the basis for predictions beyond the range of circumstances that are described. The two approaches are complementary and interdependent. The challenge is to design bottom-up studies that convey mechanisms efficiently across scales. Accurate long-term predictions certainly will require incorporation of population, community, and evolutionary processes into our scaling efforts. We suggest that grouping plants by form–function relationships provides the best way to incorporate these processes in scaling from leaves to landscapes.

References

Botkin, D. B, Caswell, M., Estes, J. E., Orio, A. A. (1989). "Changing the Global Environment: Perspectives on Human Involvement." Academic Press, New York.

Chabot, B. F., and Mooney, H. A. (1985). "Physiological Ecology of North American Plant Communities." Chapman & Hall, New York.

Chabot, B. F., and Hicks, D. J. (1982). The ecology of leaf lifespans. *Annu. Rev. Ecol. Syst.* **13**, 229–259.

Chapin, F. S., III, Schulze, E.-D., and Mooney, H. A. (1990). The ecology and economics of storage in plants. *Annu. Rev. Ecol. Syst.* **21**, 423–447.

Falconer, D. S. (1981). "Introduction to Quantitative Genetics." Longman, New York.

Fung, I., Prentice, K., Matthews, E., Jerner, J., and Russell, G. (1983). Three-dimensional tracer model study of atmospheric CO_2: Response to seasonal exchanges with the terrestrial biosphere. *J. Geophys. Res.* **88**, 1281–1294.

Givnish, T. J. (ed.) (1986). "On the Economy of Plant Form and Function." Cambridge University Press, Cambridge.

Graetz, R. D. (1990). Remote sensing of terrestrial ecosystem structure: An ecologist's pragmatic view. *In* "Remote Sensing of Biosphere Functioning" (R. H. Hobbs and H. A. Mooney, eds.), pp. 5–30. Springer-Verlag, Heidelberg.

O'Neill, R. V., DeAngelis, D. L., Waide, J. B., and Allen, T. F. H. (1986). "A Hierarchical Concept of Ecosystems," Vol. 23. Princeton University Press, New Jersey.

Raven, J. A. (1986). Evolution of plant forms. *In* "On the Economy of Plant Form and Function" (T. J. Givnish, ed.), pp. 421–492. Cambridge University Press, Cambridge.

Running, S. W. (1990). Estimating terrestrial primary productivity by combining remote sensing and ecosystem simulation. *In* "Remote Sensing of Biosphere Functioning" (R. H. Hobbs and H. A. Mooney, eds.), pp. 65–86. Springer-Verlag, Heidelberg.

Running, S. W., and Coughlan, J. C. (1988). A general model for forest ecosystem processes for regional applications. I. Hydrologic balance, canopy gas exchange, and primary production processes. *Ecol. Model.* **42**, 125–154.

Russell, G., Marshall, B., and Jarvis, P. J. (1989). "Plant Canopies: Their Growth, Form and Function." Cambridge University Press, Cambridge.

Schimper, A. F. W. (1898). "Pflanzengeographie auf Physiologischer Grundlage." Fisher, Jena.

Sellers, P. J., Mintz, Y., Sud, Y. C., and Dalcher, A. (1986). A simple biosphere model (SiB) for use with general circulation models. *J. Atmos. Sci.* **43**, 505–531.

Shukla, J., and Mintz, Y. (1982). Influence of land-surface evapotranspiration on the earth's climate. *Science* **215**, 1498–1501.

Shukla, J., Nobre, C., and Sellers, P. (1990). Amazon deforestation and climate change. *Science* **247**, 1322–1325.

Tilman, D. (1988). "Plant Strategies and the Dynamics and Structure of Plant Communities," Vol. 26. Princeton University Press, New Jersey.

Woodward, F. I. (1987). "Climate and Plant Distribution." Cambridge University Press, Cambridge.

V

Integrating
Technologies for Scaling

Often progress within a discipline is limited by technology. In this last section, we focus on key technological areas that are redefining the limits of our ability to scale across levels between the leaf and the globe. Yakir and colleagues discuss the utility of stable isotopes as a tool for understanding integrated aspects of metabolism. Based on a fundamental understanding of the mechanisms causing changes in the isotopic composition of organic and inorganic compounds, it seems possible to scale across several levels of organization and gain new insights into both the temporal and spatial aspects of scaling physiological processes. In the next chapter Ustin and colleagues discuss remote sensing and the increased role that remote spectral analyses play in understanding ecosystem activity. By analyzing reflectance characteristics in different wavebands, many key aspects of metabolism and stand structure can be obtained. This developing technology is likely to play a key role in scaling between canopy and global levels in the next decade. In the last chapter Schimel discusses both the emerging roles of stable isotopes and remote sensing in integrated ecological studies. In addition, he discusses surface flux measurements and the emerging role for eddy correlation and similar approaches for understanding carbon dioxide and water vapor exchange between the atmosphere and canopy surfaces. As the new discipline of earth system science develops, each of these technologies is likely to take on greater importance, particularly as we attempt to scale physiological processes to understand the global interactions between physical and biological processes.

18

Applications of Stable Isotopes to Scaling Biospheric Photosynthetic Activities

Dan Yakir, Joseph A. Berry, Larry J. Giles,
C. Barry Osmond, and Richard B. Thomas

I. Introduction

The problem of scaling up from biological processes at the organismal level to the community or landscape is principally a matter of discarding detail, knowing something of the scale and frequency of extreme events that reset the rules for iteration, and devising appropriate means of observation and interpretation that can be sustained for long periods of time. In this chapter, we shall emphasize means of observation and show how stable isotope analyses especially are suited to large scale evaluation of photosynthetic processes in the biosphere. Some aspects of the scaling of these fundamental processes have been discussed elsewhere (Mooney and Field, 1989; Osmond, 1989; Osmond et al., 1992). Many of the potential ecological applications of stable isotopic techniques at natural abundance have been explored by Rundel et al. (1989); particular aspects of carbon isotope fractionation have been presented by Farquhar et al., 1989).

Stable isotope analysis at natural abundance was developed first in the earth sciences, especially in biogeochemistry. Biological interest in isotope analysis has been kept alive by colleagues in the earth sciences whose conclusions are sufficiently provocative to demand attention (for example, "our data indicate that suitable tree growth sites are more the rule than the exception"; Yapp and Epstein, 1982). As the principal conduits for water, carbon, and nutrient exchanges between the geosphere and the atmosphere, and as the principal respondents to soil and atmospheric quality, plant processes serve a unique role as integrators and indicators of global environmental change. The explosion of research using stable isotopes at natural abundance in plant biology can be dated from the observations of an atmospheric scientist moonlighting in archeology (Bender, 1968). Thus, unlike many other technical approaches in biological research, stable isotope methodology has developed in a ready-made big-scene context.

Although the potential for application of stable isotope analysis to key biospheric processes has been obvious for nearly half a century, only a handful of biologists have had unfettered access to sufficiently sensitive mass spectrometers in the last decade. Even fewer biologists have had access to reliable automated instrumentation for more than a few years. Undeniably, this technology is now ripe for exploitation in relation to global climatic change research, especially in the role of integration and indication of landscape-level biospheric processes. In addition, for the responsible investigator, stable isotope technologies offer prospects for validation of models, and for the adventurous investigator they access fundamental mechanisms of significance in biospheric research.

We shall focus on three categories of research applications that are relevant avenues for scaling biological processes from the leaf to the landscape:

- identification of the sources of isotopic signatures that serve as indicators of biosphere functions
- evaluation of gradients in stable isotopic signatures as a measure of sink activity
- partitioning of biological activities using mass balance calculations based on characteristic isotopic signatures of key biological processes

All biospheric activities ultimately depend on the oxidation of water by photochemical processes in the chloroplast and the subsequent biochemical reactions of photosynthesis that generate the stable carbohydrate currency of life. The biochemical processes of oxidation, reduction, oxygenation, carboxylation, and nitrification all show characteristic discrimination between the stable isotopes of carbon, hydrogen, oxygen, and nitrogen *in vitro;* this discrimination is preserved *in vivo.* These isotopic signatures show sufficient fidelity to serve the three major objectives outlined earlier.

II. Sources: The Importance of Isotopic Composition of Water in the Metabolic Compartments of Leaves

Although we know a good deal about the isotopic signatures associated with the carboxylases that ultimately stabilize this autotrophic activity

(O'Leary, 1981), we know little about the source of isotopic signatures of hydrogen and oxygen in organic matter produced during photosynthesis. Moreover, until recently we were unable to measure the isotopic composition of that fraction of leaf water that is oxidized in photosynthesis. Stevens *et al.* (1975) showed that O_2 was evolved without significant isotope fractionation during water oxidation by green and blue-green algae. Guy *et al.* (1987) confirmed these observations with isolated thylakoids and asparagus cells. In a closed microcosm, Guy *et al.* showed that photorespiration (the oxygenation of ribulose bisphosphate (RuBP) and subsequent glycolate metabolism) was responsible for an ^{18}O isotope enrichment of the gas phase that was equivalent to that found in the global atmosphere (+23.5‰ with respect to water in the oceans; the "Dole effect"). The isotopic composition of oxygen incorporated into organic matter during photosynthesis is about +27‰ relative to that of the bulk water of submerged aquatic plants, evidently as a result of carbonyl exchange reactions subsequent to synthesis (De Niro and Epstein, 1981). Estep and Hoering (1981) showed that photosynthesis discriminates against deuterium (−100 to −120‰), presumably as a result of isotope fractionation during as yet unspecified reductive events. There is evidence of further contrasting discrimination (+150 to +200‰) associated with heterotrophic metabolism (Yakir and De Niro, 1990) and with metabolic events in CAM plants (Sternberg *et al.*, 1984), also unspecified as yet.

It has been known for a long time that transpiration leads to enrichment of hydrogen and oxygen isotopes in leaf water. However, leaf water does not behave as an evaporating free water surface (Leaney *et al.*, 1985); direct evidence for isotopic heterogeneity in leaf water was obtained by Yakir *et al.* (1989). The subsequent challenge has been to identify the isotopic composition of the water fraction involved in metabolism. Two approaches have been taken.

A. Using CO_2 as a Probe

Oxygen in CO_2 and water comes to a well-defined isotopic equilibrium (Brenninkmeijer *et al.*, 1983). This reaction is extremely fast when catalyzed by the enzyme carbonic anhydrase, a ubiquitous enzyme in chloroplasts. Thus, CO_2 from photorespiration and respiration of the cells in the light should be equilibrated isotopically with chloroplastic/cytosolic water. Measurement of the oxygen isotopic composition of CO_2 respired from intact leaves under controlled steady-state conditions (Fig. 18.1A) allows us to relate the oxygen isotopic composition of the metabolic water fraction to that of total leaf water and transpiration water (Yakir, 1991). It is important to emphasize that in this approach two isotope effects must be considered: The equilibrium fractionation between CO_2 and

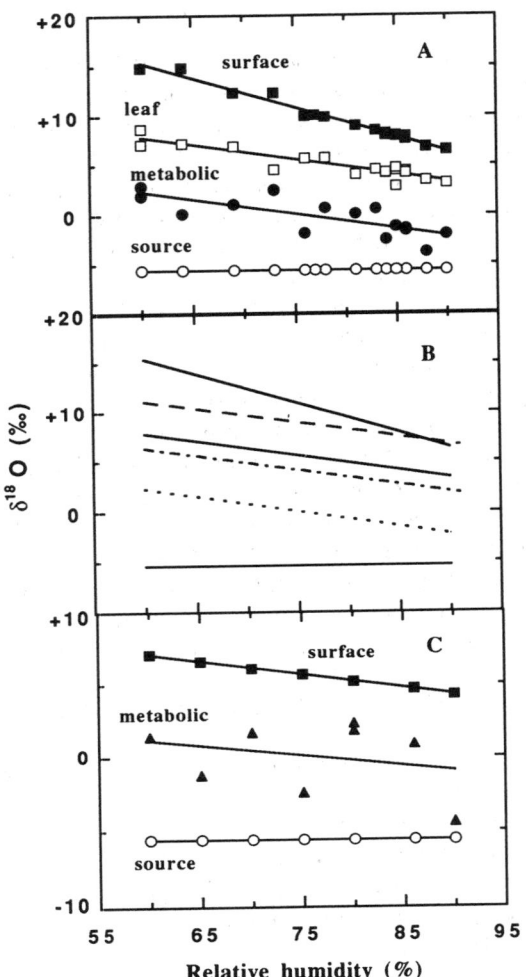

Figure 18.1 Water heterogeneity in sunflower leaf. (A) An attached sunflower leaf was sealed in an illuminated gas exchange system at each of the indicated humidities and constant conditions were maintained until an isotopic steady state was attained. The system was then flushed with CO_2-free air, the respired CO_2 was collected, its $\delta^{18}O$ value was determined, and, by correcting for equilibrium fractionation between CO_2 and water, the $\delta^{18}O$ of the metabolic water at the site of equilibrium was obtained. Leaf water and stem (source) water were then extracted to determine their $\delta^{18}O$ values. The isotopic composition of the water at the evaporating surface was calculated using the Craig model (see also Yakir, 1991). (B) Best fit lines of surface, leaf, and source water from A are compared with the best fit line of metabolic water before (dot) and after correction for possible fractionations of 4‰ (dot-dash) or the maximal 8.8‰ (dash) during diffusion of CO_2 from the leaf. (C) Leaves were treated as in A, but under a helium/CO_2 mixture (1500 μl liter^{-1} CO_2). Photosynthetic O_2 (closed triangles) evolved from water in the chloroplasts was trapped, concentrated, and fed directly into a mass spectrometer to determine $\delta^{18}O$. This isotopic composition reflected that of the intrachloroplastic water from which O_2 was evolved. Source and surface water were determined as in A. Total leaf water was extracted in only four cases (not shown) and its $\delta^{18}O$ value was always higher than that of the O_2.

water, which is well documented (Brenninkmeijer *et al.*, 1983) and must be accounted for, and kinetic fractionation due to diffusion of CO_2 from the leaf. This fractionation can be as large as 8.8‰ in diffusion into vacuum (Fig. 18.1B). It is much smaller in the experimental system in which concentration gradients were minimal, considering similar resistances by mesophyll and stomata and mixing of respired CO_2 into chamber, and as confirmed by similar results in close loop measurements when concentration gradients are eliminated.

B. Using Photosynthetic O_2 as a Probe

As mentioned earlier, water oxidation and the evolution of oxygen during photosynthesis does not involve an isotope effect. It follows that the isotopic composition of photosynthetic O_2 directly reports that of the water from which it was evolved in the chloroplast. We have developed a system that allows us to collect photosynthetic O_2 evolved from intact leaves. In this approach the product of nonfractionating, unidirectional reaction (water spliting) is quantitatively collected and therefore neither equilibrium nor kinetic isotope effects are involved.

A series of leaves of hydroponically grown sunflower plants were sealed in an illuminated gas exchange system and maintained at various humidities under a flow of 1500 μl·liter^{-1} CO_2 in helium. These conditions maximize oxygen evolution while minimizing O_2 uptake and contamination. In each case, leaf water was allowed to reach isotopic steady-state conditions as confirmed by measuring identical isotopic compositions of stem and transpiration water. The photosynthetically evolved oxygen was concentrated and fed directly into a ratio mass spectrometer for the determination of its $^{18}O/^{16}O$ ratio.

The results in Fig. 18.1C show a good agreement with those obtained previously (Yakir, 1991) using CO_2 as a probe for metabolic water and assuming only a small kinetic isotope effect during CO_2 diffusion (Fig. 18.1B). The combined results show the isotopic composition of a distinct metabolic water fraction involved in photosynthesis in sunflower leaves. This water fraction, and not total leaf water, is the source of the hydrogen and oxygen isotopic signal in plant organic matter. Similarly, the isotopic composition of this metabolic pool will influence that of molecular oxygen released by plants into the atmosphere. Moreover, because carbonic anhydrase activity is confined to the metabolic compartments of living cells, only the oxygen isotopic composition of these compartments is relevant to a consideration of biological factors influencing the $\delta^{18}O$ value of atmospheric CO_2.

It has been known for some time that seasonal trends in the $\delta^{18}O$ value of atmospheric CO_2 (Fig. 18.2) corelate with the greening of the Northern hemisphere terrestrial biosphere (Mook *et al.*, 1983). The hypothesis suggesting a role for carbonic anhydrase in facilitating the equilibration of this atmospheric CO_2 and leaf water (Francey and Tans, 1987; Friedli *et al.*, 1987) begs the question of the isotopic composition of the leaf water pool involved. Our data, which identify the $\delta^{18}O$ value of the relevant leaf water fraction directly, could help resolve the discrepancy identified by Francey and Tans (1987), who used a simple Craig model in their simulations. Most of our experiments have been done with sunflower. We have evidence already that displacement of the isotopic composition of metabolic water from that of the Craig model is species dependent and can in some cases be substantial (e.g., sunflower) while in others it is negligible. This result is not surprising given the very different morphological, anatomical, and biophysical constraints of plant water relations. With further evaluation of different species and life-forms using the newly developed techniques, it seems likely that we shall be able to use the ^{18}O signal in atmospheric CO_2 above vegetation to model landscape to global stomatal conductance and the fluxes of CO_2 exchanged by plants.

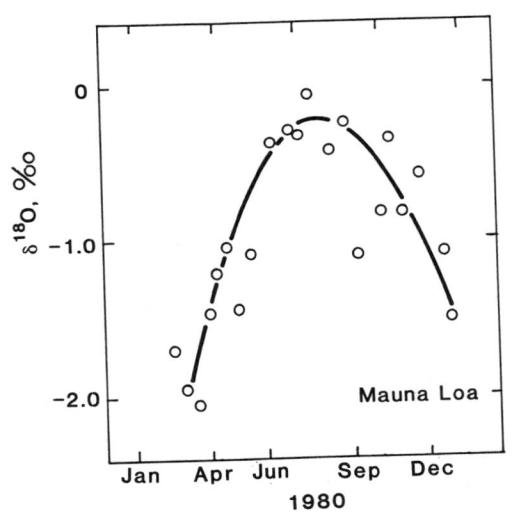

Figure 18.2   Seasonal change in $\delta^{18}O$ value of atmospheric CO_2 at Mauna Loa, Hawaii. The data show a seasonal trend and isotopic signature consistent with the equilibration of atmospheric CO_2 with leaf water in the Northern Hemisphere, facilitated by carbonic anhydrase. (Data from Mook *et al.*, 1983; interpretations proposed by Freidl *et al.*, 1987 and Francey and Tans, 1987.)

III. Gradients: The Interpretation of Gradients in Isotopic Composition and Their Value as Integrators of Photosynthetic Fluxes

The reference $\delta^{13}C$ value of atmospheric CO_2 is known with reasonable certainty. The isotope discriminations associated with RuBP carboxylase (Rubisco), PEP carboxylase, HCO_3/CO_2 interconversion, and diffusion are well known (O'Leary, 1981). Quite robust models of physical and biochemical components of carbon isotope discrimination in photosynthesis have been developed (Farquhar *et al.*, 1989) that can be applied over leaf to canopy scales, for example, to assess water use efficiency. These models provide a wealth of biological insight into plant water relations, especially if coupled with chemical separation of recent photosynthate from carbon fixed prior to stress (DeLeens *et al.*, 1989). The rapid development of this field has depended on automatic analyses of $\delta^{13}C$ values to high precision, using coupled elemental analyzer, automated-trap, or on-line mass spectrometer systems. These instruments now have the capacity, precision, and sensitivity to permit annual ring analysis from individual cores and the anatomical dissection of gradients in tissue $\delta^{13}C$ values.

Parkhurst *et al.* (1988) showed that gradients of more than 50 μbar CO_2 are common in amphistomous leaves of C_3 plants. Thick tissues of some succulent plants with CAM potentially develop even larger tissue gradients in CO_2 partial pressure as a consequence of low conductance and the high affinity of PEP carboxylase for CO_2/HCO_3^-. Such diffusional limitations should lead to substantial gradients in $\delta^{13}C$ values within photosynthetic tissues of CAM plants (Fig. 18.3). These gradients are sometimes as much as 4‰, the limit one would predict for diffusional fractionation. These large gradients imply that very low intercellular CO_2 concentrations prevail in the center of these tissues. The change in $\delta^{13}C$ value is usually greatest in the epidermal layers, which may be due in part to the more negative $\delta^{13}C$ value of epidermal fats and waxes (C. B. Osmond and S. A. Robinson, unpublished observations). Further, because the epidermis is laid down early in development, when tissues of many succulents show little CAM activity, this tissue may show much more negative $\delta^{13}C$ values (Nishida *et al.*, 1981). These factors must be evaluated further with precise anatomical observation and analysis of the C_4-carboxyl group of malic acid during dark CO_2 fixation. Although photosynthetic tissues of CAM plants may be ideal models for testing these hypotheses, increased leaf thickness of C_3 plants along environmental gradients can be correlated with changes in internal resistance and $\delta^{13}C$ values (Vitousek *et al.*, 1990).

The physical principles responsible for these gradients within tissues

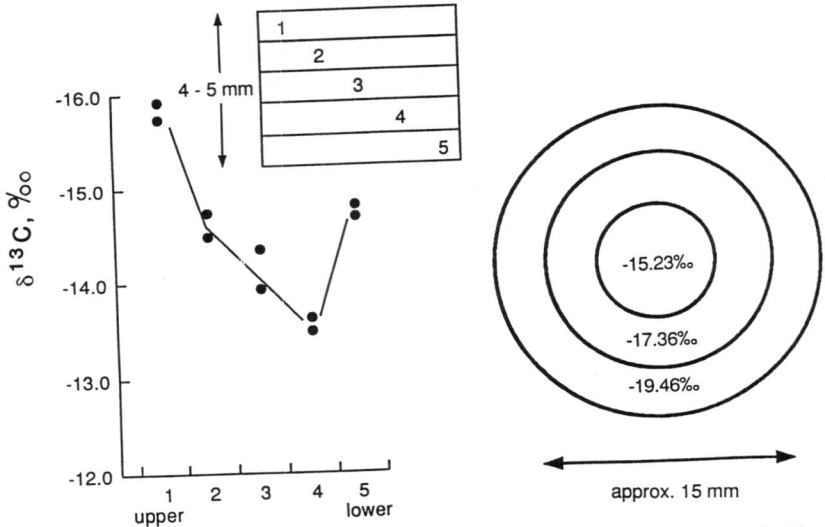

Figure 18.3 Gradients in $\delta^{13}C$ value through successive slices and cores of chlorenchyma in thick tissues of succulent plants capable of CAM. (Left) Leaf of *Crassula argentea*. (Right) Cladode of *Mammillaria longimamma*.

should be applicable to landscape scale photosynthesis but, unfortunately, not much data is yet straightforward. Evaluation of within-canopy variations in stomatal conductance in large organisms, for example, using $\delta^{13}C$ values of individual leaves, has yet to be explored in detail. In closed forests, the gradients in $\delta^{13}C$ value of source carbon due to respiratory inputs from litter are significant (Medina *et al.*, 1986). These litter gradients complicate the evaluation of gradients of within-canopy conductance from changes in $\delta^{13}C$ value of leaf carbon (Farquhar *et al.*, 1989). Sternberg (1989) exploited such gradients in source $\delta^{13}C$ value to estimate 7–8% refixation of respired carbon in a tropical moist forest. Using a double isotope technique (^{13}C, ^{18}O), he was able to evaluate humidity responses of stomata and water use efficiency at different levels in the canopy (Sternberg *et al.*, 1989). At the landscape level, there have been few attempts to examine source–sink profiles of CO_2 in the biosphere, although there have been studies with sulfur isotopes (Krouse, 1989). One study with carbon showed less negative $\delta^{13}C$ values in plants chronically exposed to pollutant gases, perhaps indicative of stomatal closure (Martin *et al.*, 1988).

Gradients in $\delta^{13}C$ value at the global scale are especially useful integrations of biospheric photosynthesis. We have argued previously that latitudinal differences in the annual variation in $\delta^{13}C$ and $\delta^{18}O$ of atmospheric

CO_2, due to photosynthetic drawdown and respiratory replenishment of CO_2, are more useful indicators of landscape level biospheric activity than the long-term trends in these values (Osmond *et al.*, 1991; see also Fig. 18.4). Pearman and Hyson (1986) developed a global-grid biomass model with three categories of carbon turnover pools. This simple model accurately simulates the annual cycle of CO_2 drawdown and input, as well as the variation in the isotopic composition of global CO_2 in the Northern hemisphere (Fig. 18.5).

Not surprisingly, their model did not work well over the Southern oceans, in which mixing between atmosphere and seawater and biological activity in the ocean remain major uncertainties. Unfortunately, when evaluating the components of marine phytoplankton autotrophy on the basis of isotopic composition, we seem to be "all at sea" (Descolas-Gros and Fontugne, 1990). These uncertainties are due in part to large variations in source carbon in different aquatic ecosystems, in part to local responses due to boundary layers, and also to diffusion in an aqueous environment. The evaluation of the aquatic "landscape" thus remains a major challenge for applications of carbon and oxygen isotopic effects. Nevertheless, we suggest that comparisons of the $\delta^{18}O$ values of atmospheric O_2 and CO_2 can help constrain estimates of global productivity and partition CO_2 exchanges between the oceanic and terrestrial biosphere.

The seemingly inexorable increase in atmospheric CO_2 concentration due to fossil fuel inputs and deforestation results in a steady trend to more negative $\delta^{13}C$ values. This gradient with time in the global atmospheric CO_2 concentration is well preserved in the bristlecone pines of the White Mountains. These trees record a change of about -2%₀ over the past 300 years. Corn cobs may be even more reliable indicators over a longer time frame (Marino and McElroy, 1991).

IV. Partitioning: Evaluating Photosynthetic Pathways within Ecosystsms, Carbon Allocation below Ground, and Integration with Nitrogen Fixation

The $\delta^{13}C$ values of phytoplankton and their responses to environment, which presently defy evaluation, illustrate our limited abilities to identify sources and to define gradients of dissolved inorganic carbon in seawater, before we can use mass balance arithmetic over the short life of these simple organisms. Fortunately, we have much more information when it comes to application of this arithmetic to the inaccessible processes of below-ground performance of land plants, which are very important in landscape-level considerations. Although a great deal more "ground work" needs to be done, the potential for developing predictive

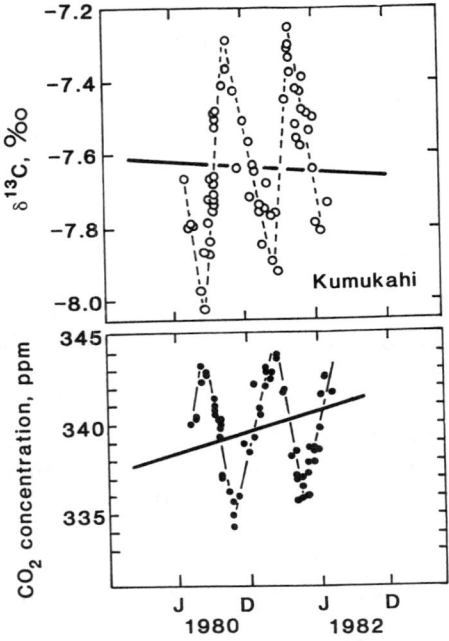

Figure 18.4 Seasonal changes in atmospheric CO_2 concentration and $\delta^{13}C$ value of atmospheric CO_2 at Kumukahi, Hawaii. The data show the isotopic signature that links the summer drawdown of CO_2 to isotope discrimination by Rubisco. (Data from Mook *et al.*, 1983.)

capabilities and monitoring long-term vegetational processes, using the $\delta^{13}C$ and $\delta^{15}N$ values of plant root systems, is already evident.

Naturally, attention has been focused on situations containing a mixture of C_3 and C_4 plants, or a mixture of legumes and non legumes, in which distinct isotopic signatures can be evaluated. Drawing on the success of agronomists and entomologists in applying $\delta^{13}C$ values to assess dietary preferences of cows and grasshoppers (Ludlow *et al.*, 1976; Boutton *et al.*, 1978), it became evident that mass balance calcualtions based on total root zone $\delta^{13}C$ values were good indicators of C_3 and C_4 competition underground (Svejcar and Boutton, 1985). Svejcar *et al.* (1988) used these methods to establish that defoliation of a C_4 grass conserved water for growth and root development of a C_3 legume. Several landscape-level applications have been developed in ecological (Dzurec *et al.*, 1985; Tieszen and Archer, 1990) and agricultural (Balesdent *et al.*, 1988) systems.

Competitive interactions may be explored best using the replacement series methods of de Wit in combination with stable isotope analysis.

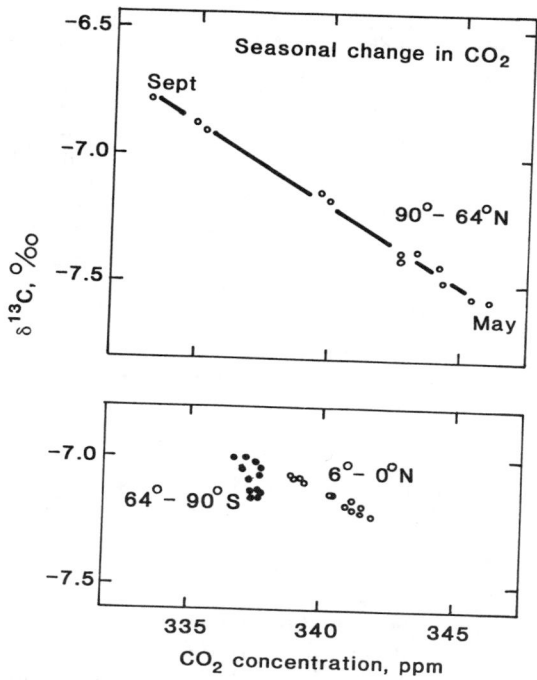

Figure 18.5 Simulation of the coordinated changes in CO_2 concentration and $\delta^{13}C$ value of CO_2 in the atmosphere at different latitudes, using a three-reservoir model. (Redrawn from Pearman and Hyson, 1986.) The model conforms to observations in northern latitudes (cf. Fig. 18.4) and shows damping at equatorial and southern latitudes, as expected from oceanic interactions and the paucity of biospheric activity, respectively.

Wong and Osmond (1991) have shown that few of the conventional wisdoms about competitive advantages between C_3 and C_4 plants can be taken for granted with respect to differences in water use efficiency and nitrogen use efficiency. For example, underground biomass of the C_3 plant was insensitive to the presence of C_4 plants at high nitrogen and low light, but susceptible at low nitrogen and high light. C_4 plants displayed the same responses. A stimulation of C_3 root biomass by elevated CO_2 was evident in high light and low nitrogen but, surprisingly, underground biomass of the C_4 plant also was increased by elevated CO_2.

Although the signals are much smaller, the high precision of modern isotope analysis systems also makes it possible to use $\delta^{15}N$ value mass balance calculations to evaluate nitrogen fixation and inorganic nitrogen uptake in natural systems. Pioneering work of Delwiche and Steyn (1970) and Virginia and Delwiche (1982) suggests that nodule-to-landscape scal-

Table 18.1 Interactions between Nitrogen Supply and Elevated Atmospheric CO_2 Concentration with Nodulation and N_2 Fixation in *Gliricidia*, as Deduced from $\delta^{15}N$ Values

	Treatment			
	350 μbar CO_2		650 μbar CO_2	
Parameter	14 mM N	No N	14 mM N	No N
Nodule number	29.6 ± 12.9	79.6 ± 15.4	40.8 ± 13.6	73.4 ± 20.5
Mass per nodule (mg)	6.5 ± 1.3	6.6 ± 0.9	26.5 ± 7.3	7.8 ± 1.4
$\delta^{15}N$ of leaf (‰)	−2.24 ± 0.1	+ 1.45 ± 0.02	+ 1.86 ± 0.09	+ 1.25 ± 0.09
Percentage leaf N from N_2 fixation[a]	34	100	56	100

[a] Calculated from

$$(\delta^{15}N_F - \delta^{15}N_S)/(\delta^{15}N_F - \delta^{15}N_A),$$

where $\delta^{15}N_F = \delta^{15}N$ of fertilizer NH_4NO_3 (−2.64 ‰); $\delta^{15}N_S = \delta^{15}N$ of leaf from plants supplied with NH_4NO_3; and $\delta^{15}N_A = \delta^{15}N$ of leaf from plants without added nitrogen fertilizer.

ing is possible. Experiments by Thomas *et al.* (1991) with the nitrogen-fixing tree *Gliricidia* confirm the potential of this method in elevated CO_2 research (Table 18.1). Although the response of nodule number to saturating nitrogen nutrition was uninfluenced by elevated CO_2, the $\delta^{15}N$ value of plant matter indicates that elevated CO_2 substantially increased dependence on nitrogen fixation in the presence of saturating nitrogen (From 34 to 56% of nitrogen fixed). This result is consistent with the vastly increased size of nodules, and the presumed greater supply of carbohydrate for nitrogen fixation under elevated CO_2.

V. Summary

This brief summary highlights progress in stable isotope research that facilitates leaf-to-landscape integration of significant photosynthetic processes and their ecological consequences. Although fundamental mechanisms of isotopic fractionations of oxygen and hydrogen remain to be evaluated, the potential applications of this knowledge are great. Already we see that understanding the $\delta^{18}O$ value of leaf water, which transmits a signal to atmospheric CO_2 via leaf carbonic anhydrase, may help integrate stomatal conductance and transpiration on the landscape to global scale. The gradients of $\delta^{13}C$ value within leaves and cladodes of succulent plants present interesting new approaches to the related problem of internal resistance, but understanding at the cellular level must be expanded. Sources and gradients of isotopic signatures in canopies have barely been explored yet, and multiple isotope analyses hold great promise for leaf-to-landscape integration. In all these facets of ecological research, stable isotopes provide methods for testing models of system function at the leaf and landscape levels. Stable isotope analyses also provide unique access to key nonphotosynthetic processes underground, which underlie competitive relationships and ecological change. Most investigators would agree that communities and ecosystems with C_3 and C_4 species are good indicators and integrators of the biospheric response to elevated CO_2 and increasing temperature, changes that still depend on soil nutrition and water availability. The scope for creative application of $\delta^{13}C$ and $\delta^{15}N$ to understanding population dynamics in these systems is potentially vast.

Taking a broader view of biospheric processes, and speculating on responses of heterotrophs to climatic change, we should remember the lessons of the past. Stable isotopic composition of paleosols in the Siwalik Sequence on the Potowar Plateau of Pakistan reveal a marked change from a C_3 to C_4 dominated landscape 6–7 million years ago (Quade *et al.*, 1989). This change evidently coincided with the disappearance of *Sivapithecus* (a relative of tree-living orangutans) from the fossil record, the disappearance of trees, and the onset of Asian monsoons. One hopes that the selective pressures of research funding during the late 20th century may send the dominant mammal descendants of *Sivapithecus* back to the trees, to sample tissues for stable isotopic analyses, and to integrate the leaf-to-landscape processes needed to understand global climatic change.

Acknowledgments

The SIRA mass spectrometer used in our research was funded by grants from Duke University and the North Carolina Biotechnology Center. This research was supported in part by Grants DE-FG05-89 ER 14005 (DOE) and DCB-9006830 (NSF) to Barry Osmond.

References

Balesdent, J., Wagner, G. H., and Mariotti, A. (1988). Soil organic matter turnover in long-term field experiments as revealed by the ^{13}C natural abundance tracer technique. *Soil Sci. Soc. Am. J.* **52,** 118–124.

Bender, M. M. (1968). Mass spectrometric studies of carbon-13 variation in corn and other grasses. *Radiocarbon* **10,** 468–472.

Boutton, T. W., Cameron, G. N., and Smith, B. N. (1978). Insect herbivory on C_3 and C_4 grasses. *Oecologia* **36,** 21–32.

Brenninkmeijer, C. A. M., Kraft, P., and Mook, W. G. (1983). Oxygen isotope fractionation between CO_2 and H_2O. *Isotope Geosci.* **1,** 181–190.

Deleens, E., Marcotte, L., Schwebel-Duge, N., and Vartanian, N. (1989). Stable isotope study: Long-term partitioning during progressive drought stress in *Brassica napus* var *oleifera*. *Plant Cell Environ.* **12,** 615–620.

Delwiche, C. C., and Steyn, P. L. (1970). Nitrogen isotope fractionation in soils and microbial reactions. *Environ. Sci. Technol.* **4,** 929–935.

De Niro, M. J., and Epstein, S. (1981). Isotopic composition of cellulose from aquatic organisms. *Geochim. Cosmochem. Acta* **42,** 496–506.

Descolas-Gros, C., and Fontugne, M. (1990). Stable carbon isotope fractionation by marine phytoplankton during photosynthesis. *Plant Cell Environ.* **13,** 207–218.

Dzurec, R. S., Boutton, T. W., Caldwell, M. M., and Smith, B. N. (1985). Carbon istope ratios of soil organic matter and their use in assessing community composition changes in Curlew Valley, Utah. *Oecologia* **66,** 17–24.

Estep, M. F., and Hoering, T. C. (1981). Stable hydrogen isotope fractionation during autotrophic and mixotrophic growth of microalgae. *Plant Physiol.* **67,** 474–477.

Farquhar, G. D., Ehleringer, J. R., and Hubick, K. T. (1989). Carbon isotope discrimination in photosynthesis. *Annu. Rev. Plant Physiol. Plant Mol. Biol.* **40,** 503–537.

Francey, R. J. and Tans, P. (1987). Latitudinal variation in oxygen-18 in atmospheric CO_2. *Nature* **327,** 495–497.

Friedli, H., Sigenthaler, V., Rauber, D., and Oeschger, H. (1987). Measurements of concentration, $^{13}C/^{12}C$ and $^{18}O/^{16}O$ ratios of tropospheric carbon dioxide over Switzerland. *Tellus* **39B,** 80–88.

Guy, R. D., Fogel, M. F., Berry, J. A., and Hoering, T. C. (1987). Isotope fractionation during oxygen production and consumption in plants. *In* "Progress in Photosynthesis Research" (J. Biggins, ed.), Vol. III, pp. 597–600. Martinus Nijhoff, The Hague.

Krouse, H. R. (1989). Sulfur isotope studies of the pedosphere and biosphere. *In* "Stable Isotopes in Ecological Research" (P. W. Rundel, J. R. Ehleringer, and K. A. Nagy, eds.), Vol. 68, pp. 424–444. Springer-Verlag, New York.

Leaney, F. W., Osmond, C. B., Allison, G. B., and Ziegler, H. (1985). Hydrogen isotope composition of leaf water in C_3 and C_4 plants: Its relationship to the hydrogen-isotope-composition of dry matter. *Planta* **164,** 215–220.

Ludlow, M. M., Troughton, J. H., and Jones, R. J. (1976). A technique for determining the proportion of C_3 and C_4 species in plant samples using stable natural isotopes of carbon. *J. Agric. Sci. (Cambridge)* **87,** 625–632.

Marino, B. D., and McElroy, M. B. (1991). Isotopic composition of atmospheric CO_2 inferred from carbon in C_4 plant cellulose. *Nature* **349,** 127–131.

Martin, B., Bytnerowicz, A., and Thorstenson, Y. R. (1988). Effects of our pollutants on the composition of stable carbon isotopes, $\delta^{13}C$, of leaves and wood, and on leaf injury. *Plant Physiol.* **88,** 218–224.

Medina, E., Montes, G., Cuevas, E., and Roksandic, Z. (1986). Profiles of CO_2 concentration and $\delta^{13}C$ values in tropical rainforests of the Upper Rio Negro Basin, Venezuela. *J. Trop. Ecol.* **2,** 207–217.

Mook, M. G., Koopmans, M., Carter, A. F., and Keeling, C. D. (1983). Seasonal latitudinal and secular variations in the abundance and isotopic ratios of atmospheric carbon dioxide I. Results from land stations. *J. Geophys. Res.* **88,** 10,915–10,933.

Mooney, H., and Field, C. B. (1989). Photosynthesis and plant productivity: Scaling to the biosphere. *in* "Photosynthesis" (W. R. Briggs, ed.), pp. 19–44. A. R. Liss, New York.

Nishida, K., Roksandic, Z., and Osmond, B. (1981). Carbon isotope ratios of epidermal and mesophyll tissues from leaves of C_3 and CAM plants. *Plant Cell Physiol.* **22**, 923–926.

O'Leary, M. H. (1981). Carbon isotope fractionations in plants. Phytochemistry **20**, 553–557.

Osmond, C. B. (1989). Photosynthesis from the molecule to the biosphere: A challenge for integration. *in* "Photosynthesis" (W. R. Briggs, ed.), pp. 5–17. A. R. Liss, New York.

Osmond, C. B., Yakir, D., Giles, L., and Morrison, J. (1992). From corn shucks to global greenhouse. *In* "Photosynthetic Carbon Metabolism and Regulation of Atmospheric CO_2 and O_2" (N. E. Tolbert, ed.), American Society of Plant Physiologists, Rockville Maryland.

Parkhurst, D. F., Wong, S. C., Farquhar, G. D., and Cowan, I. R. (1988). Gradients of intercellular CO_2 levels across leaf mesophyll. *Plant Physiol.* **86**, 1032–1037.

Pearman, G. I., and Hyson, P. (1986). Global transport and inter-reservoir exchange of carbon dioxide with particular reference to stable isotope distributions. *J. Atmos. Chem.* **4**, 81–124.

Quade, J., Cerling, T. E., and Bowman, J. R. (1989). Development of Asian monsoon revealed by marked ecological shift during the latest Miocene in northern Pakistan. *Nature* **342**, 163–166.

Rundel, P. W., Ehleringer, J. R., and Nagy, K. A. (eds) (1989). "Stable Isotopes in Ecological Research," Vol. 68, Springer-Verlag, New York.

Sternberg, L. D. S. L. O. (1989). A model to estimate carbon dioxide recycling in forests using $^{13}C/^{12}C$ ratios and concentrations of ambient carbon dioxide. *Agric. For. Meteorol.* **48**, 163–173.

Sternberg, L. D. S. L. O., Mulkey, S. S., and Wright, S. J. (1989). Oxygen isotope ratio stratification in a tropical moist forest. *Oecologia* **81**, 51–56.

Sternberg, L. O. De Niro, M. J., and Ting, I. P. (1984). Carbon, hydrogen and oxygen isotope ratios of cellulose from plants having intermediate photosynthetic modes. *Plant Physiol.* **74**, 104–107.

Stevens, C. L. R., Schultz, D., Van Baalen, C., and Parker, P. L. (1975). Oxygen isotope fractionation during photosynthesis in a blue-green and a green alga. *Plant Physiol.* **56**, 126–129.

Svejcar, T. J., and Boutton, T. W. (1985). The use of stable carbon istope analysis in rooting studies. *Oecologia* **67**, 205–208.

Svejcar, T. J., Boutton, T. W., and Christiansen, S. (1988). Rooting dynamics of *Medicago sativa* seedlings growing in association with *Bothriochloa eaucasia*. *Oecologia* **77**, 453–476.

Thomas, R. B., Richter, D. D., Ye, H., Heine, P. R., and Strain, B. R. (1991). Nitrogen dynamics and growth of seedlings of an N-fixing tree (*Glircidia sepium* (Jacq.) Walp.) exposed to elevated atmospheric carbon dioxide. *Oecologia* **88**, 415–421.

Tieszen, L. L. and Archer, S. (1990). Isotopic assessment of vegetation changes in grassland and woodland systems. *In* "Plant Biology of the Basin and Range" (C. B. Osmond, L. F. Pitelka, and G. M. Hidy, eds.), Vol. 80, pp. 293–321. Springer-Verlag, New York.

Virginia, R. A., and Delwiche, C. C. (1982). Natural ^{15}N abundance of presumed N_2-fixing and non-N_2-fixing plants from selected ecosystems. *Oecologia* **54**, 317–325.

Vitousek, P. M., Field, C. B., and Matson, P. A. (1990). Variation in foliar $\delta^{13}C$ in Hawaiian *Metrosideros polymorpha*: A case of internal resistance? *Oecologia* **84**, 362–370.

Wong, S. C., and Osmond, C. B. (1991). Elevated atmospheric partial pressure of CO_2 and plant growth. III. Interactions between *Triticum aestivim* (C_3) and *Echinochloa frumentacea* (C_4) during growth in mixed culture under different CO_2, N nutrition and irradiance treatments, with emphasis on below-ground responses estimated using the $\delta^{13}C$ value of root biomass. *Aust. J. Plant Physiol.* **18**, 137–152.

Yakir, D. (1991). Water compartmentation in plant tissues: isotopic evidence. *In* "Water and Life" (G. N. Somero, C. B. Osmond, and L. S. Bolis, eds.), pp. 205–222. Springer-Verlag, Heidelberg.

Yakir, D., and De Niro, M. J. (1990). Oxygen and hydrogen isotope fractionation during cellulose metabolism in *Lemna gibba* L. *Plant Physiol.* **93**, 325–332.

Yakir, D., De Niro, M. J., and Rundel, P. W. (1989). Isotopic inhomogeneity of leaf water: Evidence and implications for the use of isotopc signals transduced by plants. *Geochim. Cosmochim. Acta* **53**, 2769–2773.

Yapp, C. J., and Epstein, S. (1982). Climatic significance of the hydrogen isotope ratios in tree cellulose. *Nature* **297**, 636–639.

19

Remote Sensing of Ecological Processes: A Strategy for Developing and Testing Ecological Models Using Spectral Mixture Analysis

Susan L. Ustin, Milton O. Smith, and John B. Adams

I. Introduction

Regional and global environmental issues require that ecologists address the applicability of ecological models across diverse scales. How well physical processes (e.g., photosynthesis, respiration, evapotranspiration), measured and understood on leaf-to-canopy scales, can be extrapolated to large regions is uncertain. In many cases, it is impossible to test models extended to larger scales because the relevant measurements are lacking at the appropriate resolutions. Remote sensing is one of the emerging technologies that has the potential to extend measurements over spatial scales ranging from the microscopic at shorter wavelengths to the global over a broad range of wavelengths. Remotely sensed images can be obtained from a variety of sensors, from portable CCD cameras to satellites. Further, remote sensing offers tools to formulate and test ecological hypotheses at larger scales. Figure 19.1 illustrates the spatial scales of most direct methods of environmental measurements and observations in comparison with scales of current satellites. The most useful contributions of remote sensing technology to ecology are likely to be based on frequently repeated multispectral measurements covering very large areas.

The new satellite systems provide not merely photographic representations of the surface of the Earth but also physical measurements of the absorptance, reflectance, and emittance properties of landscapes,

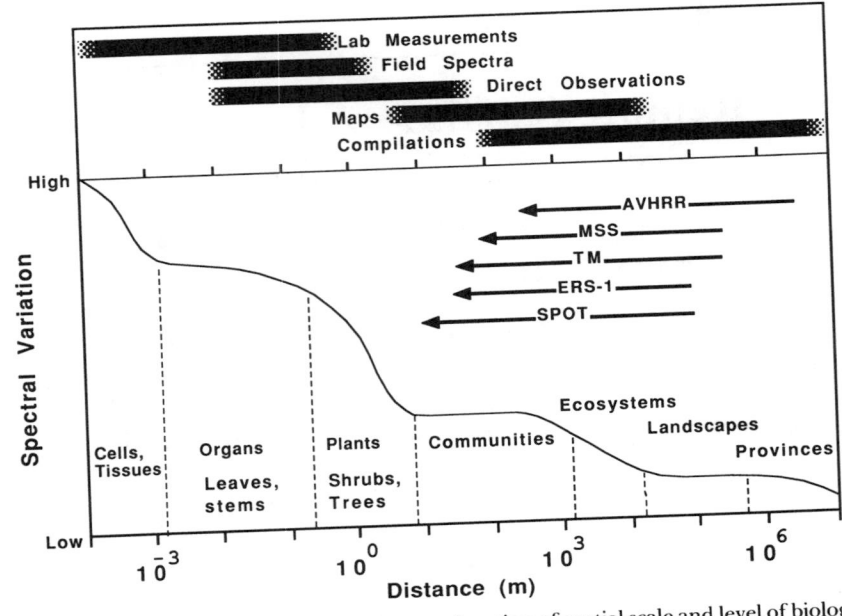

Figure 19.1 Relative spectral variation as a function of spatial scale and level of biological organization, from biochemicals to global biosphere. The spatial scales typical of direct measurements and satellite sensors are indicated. Current satellites are the Advanced Very High Resolution Radiometer (AVHRR), the Landsat series Multispectral Scanner (MSS), the Thematic Mapper (TM), the European Systeme Probatoire d'Observation de la Terre (SPOT), and the European Radar Satellite (ERS-1). The minimum pixel resolution is shown at the tip of the arrows and the field-of-view of the sensor is shown by the line length.

obtained in a spatial matrix and repeated at frequent temporal intervals. Remote sensing measurements convert photons received by a sensor from pixels (smallest resolvable surface areas) arrayed in a spatial context into voltages that are digitized. Information about the surface is derived from the spectral characteristics and their spatial and temporal patterns. These data can be used to explore ecological properties and processes in ecological models only after the data are converted from the raw digital numbers at each wavelength interval to ecological properties, a process that usually requires three models: remote sensing models that calibrate and convert the raw data into more usable forms, connecting models that translate the data into ecological variables, and ecological models that use measured variables to predict states and processes.

A fundamental property of matter is that it absorbs and emits energy at specific wavelengths; the absorption spectrum is determined by the chemical composition and structure of the compounds present. Many plant compounds now are identified routinely using spectroscopic assays in the laboratory (e.g., Weyer, 1985; Marten *et al.*, 1989), suggesting that spectral characteristics could be used to measure biogeochemical properties at other scales. Vegetation spectra are most varied at the level of biochemical constituents and cell structure (Fig. 19.1). As spatial scales increase, spectral variation decreases nonlinearly, with a trajectory that is understood incompletely. The changing variance results largely from the averaging of some components, including vegetation, soils, and other extraneous effects, for example, atmospheric conditions, as spatial scales increase. There are fewer spectrally unique components at coarser scales because most materials are mixtures of the "pure" materials from finer scales. One of the key issues in relating remote sensing to ecological models is the identification of the factors that define the spectral variance across scales. Ultimately, the overlapping levels of measurement should make it possible to construct testable ecological models that cross all scales.

Approaching problems at new scales often requires conceptualizing the component issues with fresh perspectives. Current ecological models generally interpret multispectral images in terms of preexisting paradigms [e.g., estimates of leaf area index (LAI), identification of species or community types] rather than develop new measurement definitions that take full advantage of the qualities and scales of remote sensing images. The analytical tools used to interpret images also have limited the development of new organizational definitions because most of the tools are not designed to test hypotheses, an essential step in synthesizing new paradigms. None of the current remote sensing analysis procedures, including spectral mixture analysis (SMA), appears to function correctly under all environmental conditions. Finally, most approaches use only a few spectral bands and are unable to make use of the array of data available. SMA is, however, one of the few image-processing approaches that can use the spectroscopic information from a wide range of sensors. It is flexible for a variety of applications and provides a constant frame of reference from which to make quantitative interpretations of biophysical changes over space and time. These features can be used to develop a robust strategy for testing and validating ecological models at large scales.

A. Current and Future Earth Observing Satellites

Many airborne and spaceborne remote sensing systems are available for research, presenting a diverse array of optical, radar, and thermal sensors. The current satellites sensors are listed in Fig. 19.1. Although no single sensor was designed to provide measures of all the vegetation properties of ecological interest, all satellites have the potential to provide at least some ecological information. The traditional approach to vegetation measurements using satellite data involves contrasting signals from red and near-infrared channels, chosen because vegetation and soils display large differences in reflectance because of the strong chlorophyll absorption feature of green plants. Determining ratios of red to near-infrared reflectance is the most commonly applied procedure for detecting vegetation (Jackson, 1983; Sellers, 1985,1987; Tucker and Sellers, 1986); thus, most ecologically oriented remote sensing studies have relied on this type of information (Roughgarden *et al.*, 1991).

We are beginning a new era in remote sensing with instruments of far greater capability than current satellites. Procedures must be developed that can make full use of the information these instruments will produce. Over the next decade, the Earth Observing System (EOS) with its many new sensors will become available; aircraft prototypes are available now for research. Collectively, the EOS sensors will measure most regions of the solar spectrum that are transmitted through the atmosphere (Rasool, 1987; Covault, 1989; Wickland, 1991). If ecologists continue to treat data from these new sensors as surrogates for traditional ecological measurements (e.g., biomass, leaf area index, and species-based community descriptions) without reconsidering ecological paradigms, it is unlikely that remote sensing will alter present concepts or realize its potential contributions in ecology. The challenge of reformulating concepts and developing new paradigms may have contributed to the, thus far, limited application of remote sensing in ecology.

II. Relevant Ecological Measurements

At the simplest level, remotely sensed images visually describe spatial landscape patterns: the location, areal extent, and changes over time of communities and ecosystems. Applications to deeper ecological questions require firm linkages between environmental properties and electromagnetic fluxes at different wavelengths. The first step in developing the connections is to consider the variables required by ecological models and the data provided by present and future satellites.

All landscapes are complex. They are composed of heterogeneous mosaics of varying community composition and structure, perhaps impossible to characterize fully at the landscape scale. As an example, consider a temperate forest ecosystem and the changes in age distribution, density, gap fraction, and species composition that occur as the forest matures. Figure 19.2 depicts a chronosequence, from early second growth forest to climax forest, and illustrates how some forest properties vary with time (Peet, 1981; Peet and Christensen, 1987). The relatively stable LAI reveals few of the significant ecosystem changes. Most remote

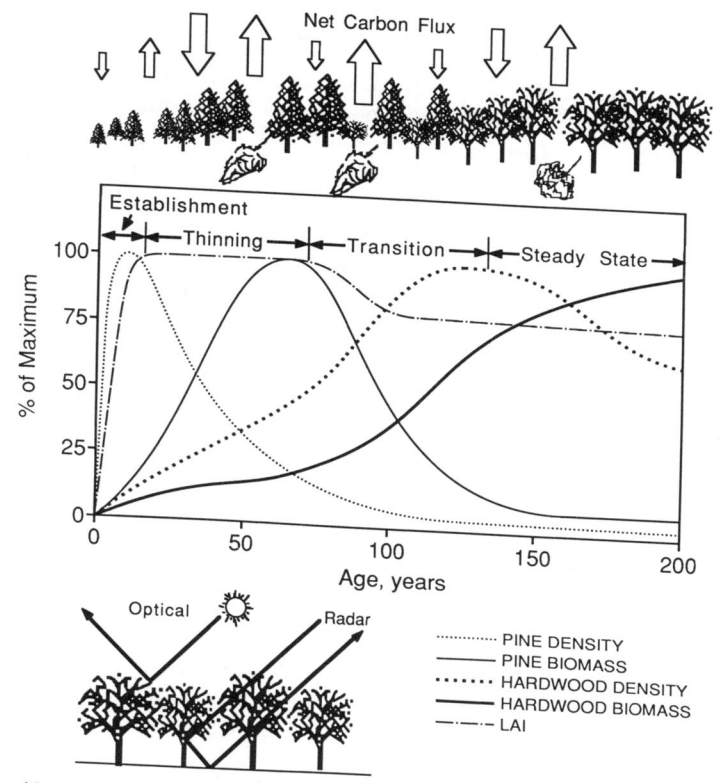

Figure 19.2 Typical chronosequence of secondary pine-hardwood forests succession in the eastern United States. The upper tree profile depicts community dynamics and canopy properties expressed as percentage of maximum values, illustrating changes in ecosystem structure, canopy gaps, LAI, and biomass. The typical range of maximum LAI is 6–8 (Nehmeth, 1971), of pine density is 1,000–30,000 tress/ha, of hardwood density is 5000–6000 trees/ha, of above ground pine biomass is 20–40 kg/m² , and of above ground hardwood biomass is 18–30 kg/m² (Peet and Christensen, 1987). The lower tree profile illustrates differences in the canopy location of maximum signal derived from optical and radar sensors.

sensing models applied to ecological problems have focused on estimating LAI, although many other canopy properties could be estimated. Detailed information about the surface structure can be obtained by using multiple viewing angles, a range of wavelengths, and by combining optical and radar signals. Geometric models have been used to estimate stand structure and gap fractions from image texture (e.g., optical models, Li and Strahler, 1985, 1986, 1988; Smith *et al.*, 1990a,b; microwave models, Sun and Simonett, 1988; McDonald *et al.*, 1990; Ulaby *et al.*,

1990) but these have received little ecological application thus far. Geo-metric remote sensing models assess both the large-scale geometry of the landscape (topography, texture, community distributions) and the fine-scale architecture of the canopy. Therefore, they should be applicable to a wide range of environmental problems.

The magnitude of net CO_2 flux, as indicated by the size and direction of arrows in Fig. 19.2, varies with structural properties and successional stage. Water, other atmospheric gases, and nutrients also vary in magnitude with the structural properties of the system. Because the rates of exchange of gases and nutrients, and the energy flow, change with the structure of the ecosystem, knowledge of the location and distribution of physiognomic types and their density, size, and biomass will provide key parameters for assessing biophysical processes and ecosystem dynamics. Multispectral measurements do not correspond to these entities directly, but it is increasingly practical to infer these and other canopy properties from the spectral measurements.

III. Current Approaches to Remote Sensing

Several approaches to developing physically interpretable remote sensing models have been proposed, each with some strong features and some limitations (e.g., see Asrar, 1989; Ulaby and Elachi, 1990; Wickland, 1991). Because a complete review of these models is beyond the scope of this chapter, we present a few examples using SMA as a model analysis tool to illustrate the types of ecological properties under investigation in remote sensing research and some of the implications for understanding ecological processes over a range of spatial scales.

The path from satellite data to ecological interpretation involves several steps. Here, we concentrate on the step that converts the remote sensing data into some parameter or parameters relevant to an ecological analysis. Developing ecological interpretations from relevant parameters may involve a range of additional steps. For example, if spatial patterns in chlorophyll or canopy water can be detected spectrally (Tucker, 1977, 1979, 1980; Jackson, 1986; Gao and Goetz, 1990), how can such information be incorporated into gas exchange or biometeorological process models? If other biochemicals can be detected, as suggested by reports of spatial patterns in canopy lignin and cellulose (Wessman *et al.*, 1988; Elvidge, 1990; Roberts *et al.*, 1990; Wessman, 1990), how can this information be incorporated into a better understanding of community relationships and function? Lignin may serve as a surrogate for the biophysical properties that control microbial decomposition rates (Meentemeyer, 1978; Melillo *et al.*, 1982). Lignin gradients in the upper canopy surface, as measured by spaceborne sensors, may, however, be related more closely to the architectural distribution of foliar and stem biomass than to soil nutrient conditions. Considerable ecological research remains to be done before such parameters can be used to interpret the biophysical state of the ecosystem.

A. Spectral Mixture Analysis of Images

Spectral mixture analysis has a complex history in remote sensing and analytical spectroscopy (Adams *et al.*, 1991), but we restrict our discussion to efforts to interpret multispectral images as mixtures of surface materials (e.g., vegetation and soils) and processes (e.g., illumination, atmospheric effects, and instrument calibration) within a single analytical framework (e.g., Adams *et al.*, 1986, 1989, 1992; Gillespie *et al.*, 1990a; Roberts *et al.*, 1990; Sabol *et al.*, 1990; Smith *et al.*, 1990a,b). This analysis assumes that the pixel spectra that make up an image are composed of mixtures of the spectra of several dominant scene components (Fig. 19.3). SMA transforms the pixel-to-pixel spectral variability of images into concentrations of reference "end members." The reference end members are the reflectance spectra of materials (e.g., plant organs, litter, soils, rocks collected from the site or from similar ecosystems) measured under specified conditions. The methods used to define the number and types of end members vary with the application, but include statistical procedures (e.g., factor analysis) to identify the intrinsic dimensionality of the data, or specification of materials of known interest. The selected end members must be distinct and have spectra that are not reproduced by mixtures of the other end members in the scene. The criteria for selecting the set of end members are (1) that the set accounts for the image spectral variability and (2) that the set produces end member concentrations within physically realistic limits (i.e., between 0 and 100%).

B. Ecological Measurements from Remote Sensing Data

It is useful to define two levels of components in an image, namely, dominant and subordinate spectral factors. Dominant spectral factors correspond to scene components that affect the overall shape of the spectrum, whereas subordinate factors typically correspond to subtle absorption features that usually are localized over a few bands. The dominant spectral variation in images is caused by mixtures of a few surface materials distributed over the landscape in varying proportions. Typically, a given scene contains four to eight identifiable components. The number of identifiable components is relatively insensitive to the number of spectral bands. Even the 224-band Airborne Visible Infrared Imaging Spectrometer (AVIRIS), which acquires a continuous spectrum over the 400- to 2500-nm wavelength range, does not increase this num-

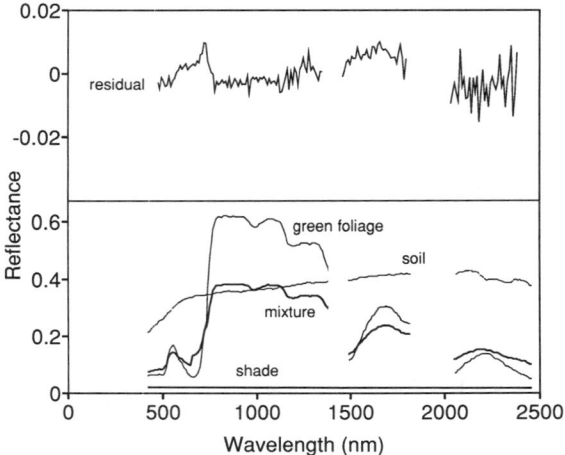

Figure 19.3 Three reflectance end members used to model spectral variation from an AVIRIS scene covering part of Owens Valley, California, a typical mixed pixel spectrum, and a residual spectrum. Mixtures of these end members—49% vegetation (foliage from semiarid shrub species), 19% shade, 30% granitic (gray) soil, and 0% weathered (tan) soil (not shown)—provide a best fit to the measured pixel. Image mixtures are calculated from the multispectral variation on a pixel-by-pixel basis, using a simple linear calibration (Smith *et al.,* 1990a,b). The residual spectrum represents the remaining pixel variation unaccounted for by the model.

ber. Relatively few scene components have been identified over a broad range of scales (from meters to globe); these include a few types of foliage, wood, litter, a few contrasting types of soils, and shade or shadow. However, not all components are resolvable in a given image because of the particular mixtures and their spectral contrasts.

To illustrate how these properties may be used to interpret the ecology of a region, we applied SMA to a small segment of an AVIRIS image from Owens Valley, California, near the town of Independence. Mixtures of only four spectral materials—shade, foliage, and two soil types (tan and gray)—account for 98% of the image variation. The spectra for these materials are shown in Fig. 19.3. The dominant surface factors (Plate 3) were very similar using image data of Owens Valley acquired in three different seasons (spring, fall, and winter) and from two instruments [Landsat Thematic Mapper (TM) and AVIRIS]. The same end members emerged from each analysis. Seasonal differences appeared as varying proportions of shade, soil, and vegetation. Despite differences in spectral and spatial resolutions, seasons, time-of-day, atmospheric depth, and other factors, SMA predictions were consistent.

The suite of identifiable end members varies somewhat between studies

and sites. In the Amazon Basin near Manaus, Brazil, Adams *et al.* (1990) identified shade and only one soil type, but were able to separate the vegetation end member into foliar (green) and wood-litter components. TM studies of temperate conifer forests and arid regions (e.g., the Grand Desertio) produce similar suites of scene components. The fact that a similar suite of end members is obtained from several different optical sensors over a wide range of ecosystems and seasons suggests that the end members should be considered spectral building blocks for constructing landscape models. Differences in landscape processes are likely to depend on how the units are assembled and how, or which, components interact.

Analytical flexibility is important for cases in which only some of the ecologically important surface components can be identified. Generally, fine-grained studies for small regions will define more end members than coarse-grained studies of larger regions. Thus, repeated analyses on spatial or spectral subunits of the image may increase the number of identified components, or additional components may be identified from the residual or unmodeled spectral variance, as described next. For the AVIRIS example, we find that the vegetation abundances are dependent on the spectral region included in the analysis (Plate 4). This wavelength-specific sensitivity results because photosynthetically active radiation is absorbed efficiently in the visible region, whereas little energy is absorbed in the near-infrared range. These differences result in less sensitivity in the visible region than in either of the infrared regions. The fact that the spatial patterns do not coincide fully indicates that the information content from different wavelength regions is not identical. The different levels of end member recognition as the spectral range is varied produce patterns analogous to fractals.

C. Quantifying Scene Components

The fractional abundances of the end members and residual must sum to one for each pixel. End member concentrations can be displayed (Plate 5) and analyzed hierarchically using geostatistical or geometric approaches (Mousset-Jones, 1980; Woodcock and Strahler, 1987; Davis, 1989). Soil types are determined by the parent geological material, surface deposition, and weathering patterns (Plate 5A,C). Here, the gray soils are less weathered than the older tan soils, although both are derived from Sierra Nevada granitic sources (Smith *et al.*, 1990a). The vegetation fractions are proportional to the areal abundance of projected canopy cover (Plate 5D) and are independent of community composition. In Owens Valley, the vegetation end member is a composite spectrum that includes both photosynthetic and nonphotosynthetic canopy biomass. It identifies not a specific species but a canopy type typical of the semiarid vegetation of the region. Repeated observations provide a mechanism to evaluate changes in the abundance, distribution, or types of end members, as might occur if land use, climate, or other factors (e.g., herbivory or pathogens) change.

The shade end member (Plate 5B) accounts for illumination differences at all spatial scales in the images. It is less directly interpretable in ecological terms than soil and vegetation maps since shade is dependent on sun angle and topography and varies with time. However, the related properties of incident radiance and net radiation have ecological significance in the context of surface energy exchange (Smith *et al.*, 1990b). The effects of topographic shading may be removed using a terrain correction model; the residual surface roughness then includes patterns caused by canopy architecture (Smith *et al.*, 1990a,b). These patterns may be used to infer community physiognomy (e.g., grasses, shrubs, and trees) and surface texture, which may be processed further to estimate canopy closure and gap structure at appropriate scales. Further development of canopy radiation transfer models should improve estimates of additional canopy architectural properties (see reviews Ross, 1981; Goel, 1988,1989; Ross and Myenni, 1990). These models will provide a strong theoretical basis for linkage to canopy gas exchange and energy balance.

There is potential for further development of texture information to improve interpretations of landscapes. Identification of community or ecosystem boundaries (e.g., to describe habitat fragmentation and loss and the distribution of corridors) provides information about the composition and physical integrity of the ecosystems in the landscape. Texture and gradient analysis may be used to examine relationships between

---

Plate 3 (A) AVIRIS vegetation end member from the Sierra Nevada bajada (left edge of image) and the floor of Owens Valley, California, including the city of Independence (right edge of image), at junction of Independence Creek (upper center, extending from the left to right) and the toe of the alluvial fan, obtained July, 1989. (B) Thematic Mapper (TM) vegetation end member from May 1985, and (C) TM, December 1982. Note that A is a higher resolution image and covers only the central area within the black frame in B. The Sierra Nevada bajada is missing from the left side and the Owens River is missing from the right side of the images. Images are color-density sliced into low (0–20%, gray), intermediate (21–30%, yellow), and high (>30% green) vegetation cover classes. The vegetation end member image overlays the shade end member (shown as gray tones), which adds some information about topography to the display.

Plate 4 Vegetation end member concentrations calculated from three spectral regions. The visible wavelengths (473–643 nm, A) have limited spectral contrast, making mixtures of vegetation, soils, and shade difficult to separate. The regions in the near-infrared (783–877 nm, B) and the shortwave infrared (1286–2375 nm, C) have greater contrast between vegetation and background materials, and increased spectral resolution of vegetation. The relative abundance of each end member corresponds directly to the image brightness. Note that these images only include a small area around Independence, California, seen in the upper right in Plate 3A.

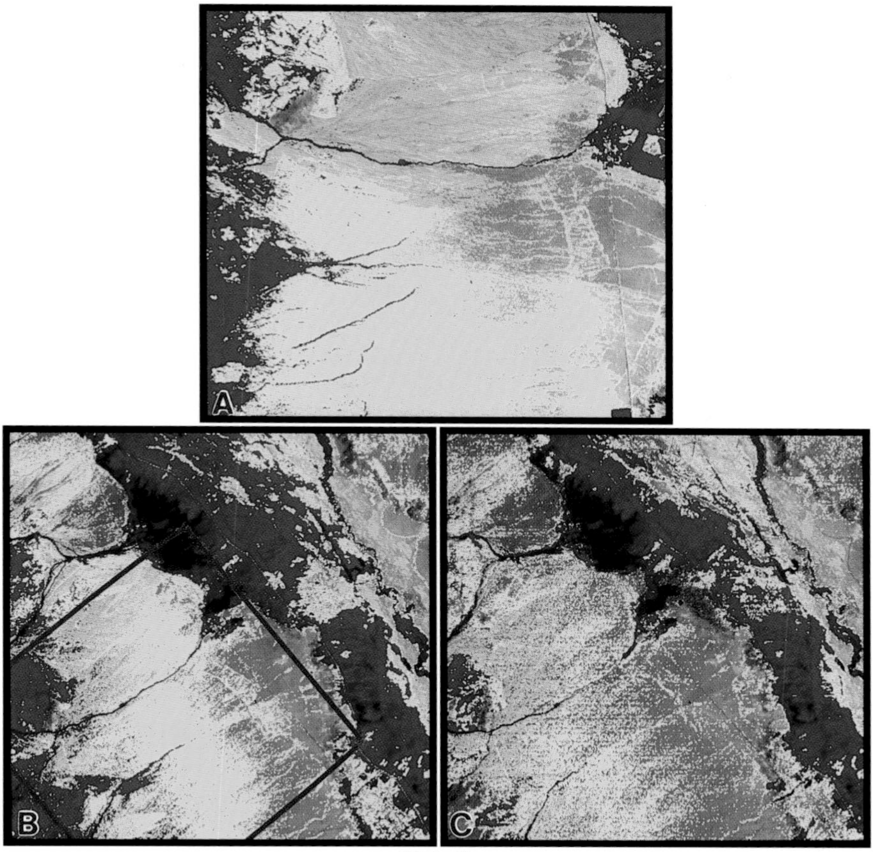

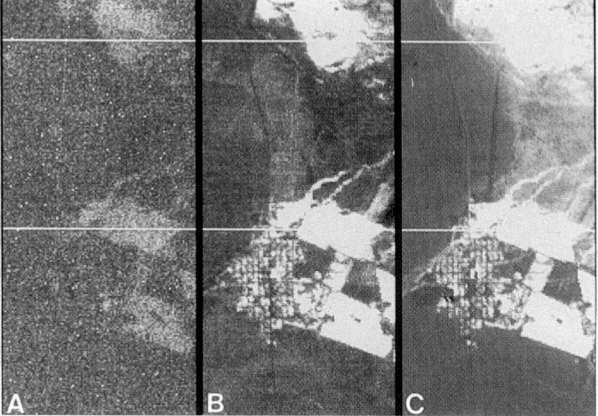

Plate 3

Plate 4

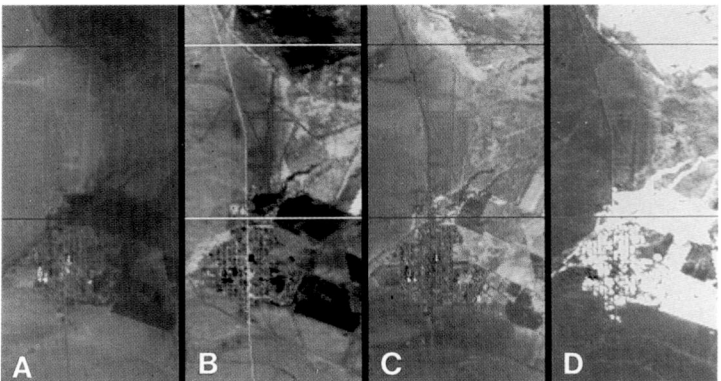

Plate 5

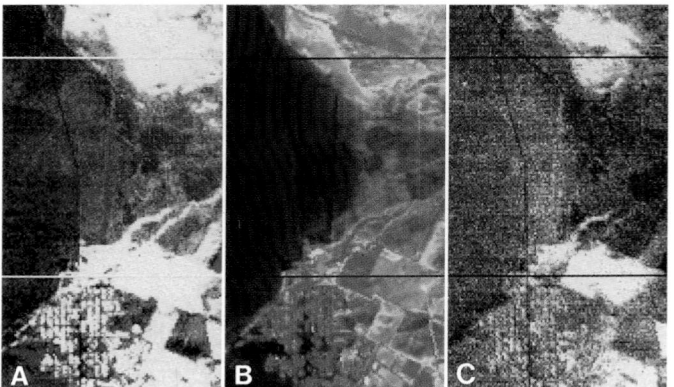

Plate 6

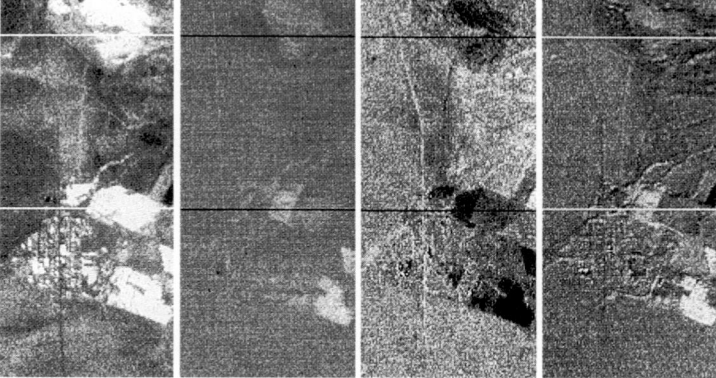

Plate 7

physiognomic units and the proportions of photosynthetic and nonphotosynthetic biomass. It may be practical to develop canopy models based on remote sensing of functional characteristics derived from canopy geometries or physiogomic types (e.g., Horn, 1971; Tilman, 1988).

The interactions and association of end members can be evaluated for edaphic and climate-related patterns. In Owens Valley, communities are distributed edaphically. Spectral classification was possible because of the combined effects of soil and vegetation gradients (Ustin *et al.*, 1986). Temporal differences also provide an opportunity for separating community types since phenological patterns vary widely among communities in a landscape. Improved methods of visualizing complex data relationships are needed. The association of end members can be examined using false-color images (Plate 6A,B). Areas predominantly of one end member are red, blue, or green; mixtures are intermediate. The spatial interactions among end members that are apparent in Plate 6 were not obvious in the raw image data or in the gray-scale end member images (Plates 4 and 5). For example, Plate 6 demonstrates that the vegetation patterns differ depending on the wavelength interval used. In the visible region, there is little spectral contrast among surface types, while there is maximum contrast in the near-infrared. These differences result in less sensitivity for detecting vegetation in the visible region than in either of the infrared regions. It may be possible to use wavelength-specific sensitivity to characterize vegetation properties under different conditions. More detailed examination of scene components, as illustrated in

---

Plate 5 Fraction images of the four reflectance end members derived from 171 AVIRIS bands. Atmospheric water vapor bands and bands of low signal/noise were excluded from the analysis. Images from left to right show the gray soil (A), shade (B), tan soil (C), and vegetation (D) end members. The relative abundance of each end member corresponds directly to the image brightness; fractions sum to unity.

Plate 6 Composite false-color images formed from three end members or residuals. Images derived from analyses of satellite data may be recombined into new composite images displaying properties not visualized directly in the original data. (A) Composite image showing interactions among the visible (blue), near-infrared (green), and shortwave infrared (red) vegetation fractions. The spatially distinct areas differentially contribute to the composite vegetation end member from the three spectral regions. (B) Composite image of the three end members: tan soil (red), vegetation (green), and gray soil (blue). The hue varies with the magnitude of the numerical value; colors depend on the relative proportions of the end members in each pixel. (C) Composite image of residuals at 525-nm (blue), 809-nm (green), and 1100-nm (red) regions. The high residuals are not random but show clear wavelength-specific spatial associations suggesting biogeochemical differences in surface conditions.

Plate 7 Residual images (difference between calibrated reflectance and estimated mixture spectrum) show areas where mixtures of the four end members do not fit the measured spectral variation at specific wavelengths (shown from left to right are 574, 986, 1254, and 1333 nm).

Plate 6B, provides a mechanism to probe ecosystem interactions among edaphic, topographic, and community structure factors.

D. Identifying Major Scene Components

Determining the identity and the characteristics or state of vegetation and soils is one of the most frequent uses of remotely sensed data. However, achieving this goal has been elusive; vegetation and soil characteristics have been hindcast more frequently than predicted. Using SMA, unknown components in images may be identified by comparing them with reference spectra using spectral matching or other statistical distance procedures (e.g., Clark *et al.,* 1990; Goetz *et al.,* 1990; Smith *et al.,* 1990a,b). The basic similarity in physiology and biochemistry of plants restricts the range of spectral variability, compared with that in soils and geologic minerals. Pigments and water produce the most important absorption features in plant canopies. The spectral features of plants differ mainly in the magnitude of band depths, widths, and shapes, but not in wavelength position. Cumulative responses to a variety of stress agents produce a similar suite of spectral changes as chlorophyll and water are lost from the canopy. Thus, direct searching for specific features is less helpful than in geological applications.

The presence of unique absorption bands can be used to infer the chemical and physical state of the surface. Although spectral matching can be used directly on the image data (e.g., Goetz *et al.,* 1985; Wessman *et al.,* 1988; Wessman, 1990), many absorption bands of biological interest are weak, for example, lignin or cellulose, or the compounds may be in low concentration in the environment, for example, unusual soils. In these cases, it may not be possible to detect the characteristic absorption bands directly. Sometimes a spectrum of interest resembles a mixture of other materials. For example, the spectrum of dry grass resembles mixtures of shade, green foliage, and soil. In the Owens Valley example (Fig. 19.3 and Plate 5), the green vegetation end member itself is a spectral mixture of foliage and stems. It is impractical to have a reference set of spectra for all possible environmental materials and their mixtures. Thus, stratifying the images using SMA before applying more specific interpretative procedures may improve identifications of specific compounds.

E. Error Analysis

Validation is, perhaps, the most overlooked aspect of image analysis. Field data are usually insufficient to validate the results of image analyses directly because of differences in scales and the inability to sample the full range of spectral variability. Efforts have focused on integrated analyses requiring multidisciplinary simultaneous constraints (e.g., Smith *et al.,* 1990b) as a way to provide intrinsic validation. Instrument and atmospheric calibrations are integrated into the system of equations used in the SMA; potential solutions that require calibrations outside of instrument operation range and atmospheric conditions (available from external sources) provide a check on the model.

The residual spectral variation, remaining after SMA, can be displayed as an image in which both the spatial and spectral patterns can be used to evaluate sources of analysis error. The residual fraction is usually near the magnitude of instrument noise (<3% maximum brightness). Weather fronts, air pollution, moisture gradients, and pathogens have very different spatial and temporal patterns; these provide some basis of assigning casual factors for observed patterns. For example, mismodeling of atmospheric conditions produces different error patterns than does incorrect instrument calibration or improper end member selection.

F. Indentifying Minor Scene Components

Areas in the scene that show poor fit with the mixture model provide a diagnostic tool for developing a physical explanation for departures. The four end members from the AVIRIS example (Plate 5) left only a mean 2% residual spectrum over all bands. However, the residuals at specific bands may be higher or a few pixels in the image may have high mean residuals. The residual spectrum may be used to identify minor scene components, such as specific biochemicals or minerals, present either in low concentration over the scene or having spatially restricted distributions. The spatial distribution of the residual variance at three wavelengths of the 171 used in the SMA illustrated in Plate 6C. Spatially distinct patches on the valley floor correspond to irrigated and mesic plant communities that are modeled poorly by the semiarid shrub end member. Examinations of partial correspondences between vegetation and other end members, as illustrated in Plate 6, provide a mechanism by which to examine ecological interpretations carefully. Because of the difficulty in validating remotely sensed images, or even visualizing the information in such a complex data set, careful cross-evaluation of many alternative descriptions of the data provides confidence for conclusions. These figures demonstrate how our understanding of the image data varies with our perspective and how an approach such as SMA can be used to test and validate ecological interpretations. The interactions of residuals from different wavelength regions may be examined visually through color, intensity, and spatial patterns (Plate 6C) or statistically (e.g., with clustering routines). The distinct color patterns show that many of the patches have unique spectral assemblages. These spatial patterns may provide clues for identifying new or additional end members.

The varying patterns provide clues for identifying surface conditions and biogeochemistry. Most departures from the SMA model represent irrigated patches on the valley floor and occur on the near-infrared plateau (986 nm), the trailing edge of the infrared plateau (1254 nm), the visible bands (574 nm), and in the vicinity of bands where liquid water has absorptions (e.g., 1333 nm), all of which are shown in Plate 7. Although the vegetation end member provides the best overall fit for the entire Independence Creek watershed, the area shown in Plate 7 was chosen to illustrate spatial and intensity differences in residuals at different wavelengths. A typical residual spectrum of a pixel is shown in Fig. 19.3. The residual spectra of pixels selected from the image may be examined to identify the wavelength regions showing the most significant departures from the SMA model. Figure 19.4, selected from 12 different vegetation patches on the valley floor, shows high residuals occurring at the long wavelength edge of chlorophyll absorption band (650–750 nm), on the near-infrared plateau (750–950 nm), and near the water bands in the shortwave infrared region (1400–1500 nm). Compared with the vegetation end member, the residuals show that vegetation from these sites has higher reflectance (positive residual) in the chlorophyll region (671 nm), about equal reflectance in the 945-nm region, and lower reflectance (negative residual) in the 1648-nm region. These spectral patterns indicate that the selected vegetation end member was less green and drier than vegetation in these patches, consistent with their more mesic condition. Changes in the proportions of photosynthetic and non-photosynthetic canopy can be followed over time or space, either through changes in the choice of end members or by changes in the residual spectrum.

Although the residuals shown in Plates 6C and 7 and Fig. 19.4 do not correspond to absorptions of specific biochemicals, such identifications are feasible (Gillespie *et al.*, 1990a; Roberts *et al.*, 1990). In most cases, the residuals show deviations from the modeled spectra by only a few bands in width (10–50 nm) and by a small percentage of maximum reflectance, consistent with many biogeochemical absorption features. Many of the 12 areas selected from the valley floor show similar positive and negative trends, but if areas with more diverse surface conditions were examined, the patterns could be distinctly different.

IV. Conclusions

Remote sensing has significant potential for providing the synoptic landscape-scale data needed to develop models at new levels of ecological organization. Realizing the full potential of this technology will require

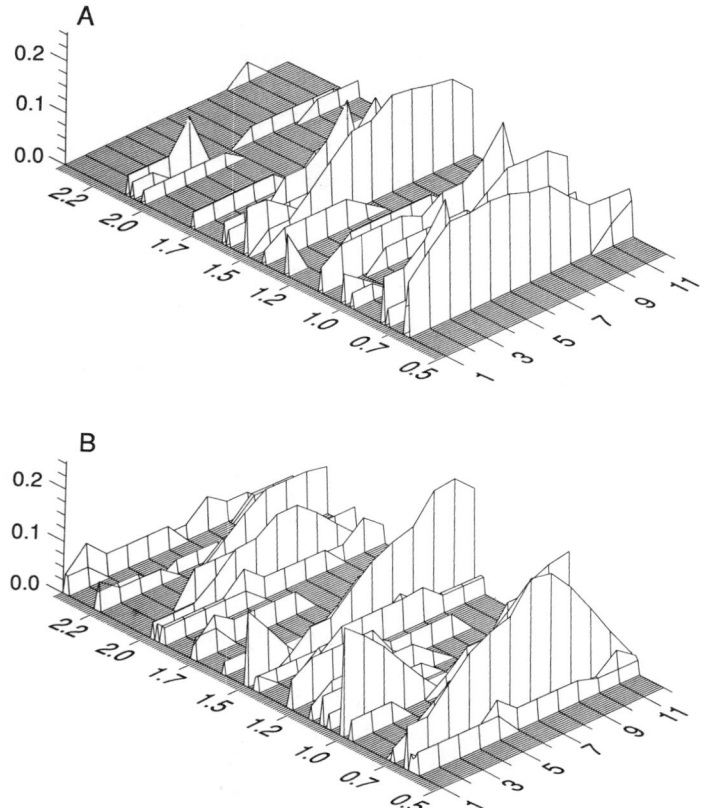

Figure 19.4 Contour plot of (A) positive (higher reflectance) and (B) negative (lower reflectance) residuals across the 400- to 2500-nm spectrum, extracted from 12 separate locations (means of 9 pixels) on the valley floor. The maximum deviation from the model was $<\pm0.3$ and the mean $<\pm0.2$. Biogeochemical properties represented by the residual spectra may be identified by comparison with spectra of known materials through spectral matching routines.

developing new paradigms in ecology and remote sensing analysis. We used SMA as an example of an organizational strategy to transform raw images into variables more directly related to ecological models. SMA can be applied to many types of remotely sensed data; it has been used to analyze Viking Lander data from Mars (Adams *et al.*, 1986), multiband thermal infrared data (Gillespie, 1990b), Thematic Mapper and other satellite data (Smith *et al.*, 1990a,b), and AVIRIS data (Roberts *et al.*, 1990). The reference spectra provide a constant frame of reference from which to interpret spectral variability in images, allowing evaluation of spatial and temporal characteristics of the terrestrial landscape.

Both major and minor sources of spectral variance, including biogeochemical conditions, can be derived and analyzed using SMA. By characterizing the sources of image variance, SMA can function in series with other analyses to classify the land surface, characterize the structure of the landscape, and estimate some properties related to physiological or biogeochemical states. The SMA procedure is well adapted to provide the first step in such a hierarchical series by providing a mechanism for testing alternative hypotheses.

The technology of imaging spectrometry is new, and its potential remains relatively untested, largely as a result of the inability of most remote sensing procedures to use the full range of spectral information and the lack of ecological models capable of using the information from this new technology. Several steps are crucial to developing remote sensing models generally useful for ecological applications: (1) identification of the spectral components; (2) explicit tests of assumptions linking the spectral components to ecological characteristics; and (3) ecological models formulated to use the spatial, temporal, and spectral information from spaceborne sensors.

V. Summary

If ecologists wish to develop models that use remote sensing data to validate our emerging conceptual views of Earth ecosystems, it will be necessary to create a new ecological paradigm consistent with the spectral data from satellite systems. An overall strategy to incorporate remotely sensed images into ecological models requires an examination of conceptual frameworks within ecology and the image processing tools used to relate remote sensing data to ecological processes. We discussed applications of remotely sensed images and links to variables needed for ecological models. SMA is one approach to processing digital image data over a broad range of spatial scales and spectral wavelength regions. It provides a method for hypothesis testing and is particularly useful for evaluating imaging spectrometer or other multiband data. SMA produces measurements with a constant frame of reference, a necessary condition for interpreting the structural and biochemical conditions of ecosystems and landscapes.

Acknowledgment

Research reported here was supported by the Land Processes Branch of NASA.

References

Adams, J. B., Smith, M. O., and Johnson, P. E. (1986). Spectral mixture modeling: A new analysis of rock and soil types at the Viking Lander 1 site. *J. Geophys. Res.* **91,** 8098–8112.

Adams, J. B., Smith, M. O., and Gillespie, A. R. (1989). Simple models for complex natural surfaces: A strategy for the hyperspectral era of remote sensing. *Proc. IGARSS '89.* 16–21.

Adams, J. B., Kapos, V., Smith, M. O., Filho, R. A., Gillespie, A. R., and Roberts, D. A. (1990). A new Landsat view of land use in Amazonia. *Int. Soc. Photogramm. Remote Sens. Manaus, Brazil.* **28,** 177–185.

Adams, J. B., Smith, M. O., and Gillespie, A. R. (1992). Imaging Spectroscopy: Data analysis and interpretation based on spectral mixture analysis. *In* "Remote Geochemical Analysis: Elemental and Mineralogical Composition" (C. M. Pieters and P. Englert, eds.). LPI and Cambridge University Press, Cambridge (in press).

Asrar, G. (ed). (1989). "Theory and Applications of Optical Remote Sensing." Wiley, New York.

Clark, R. N., Middlebrook, B., Livo, E., and Gallagher, A. (1990). Laboratory spectral and radiometric calibration of AVIRIS. *In* "Proc., Airborne Science Workshop: AVIRIS." Jet Propulsion Laboratory, Pasadena, California.

Covault, C. (1989). Mission to planet earth: Major space effort mobilized to blunt environmental threat. *Aviation Week Space Tech.* March 13, 34–43.

Davis, J. C. (1989). "Statistics and Data Analysis in Geology." Wiley, New York.

Elvidge, C. D. (1990). Visible and near infrared reflectance characteristics of dry plant materials. *Int. J. Remote Sens.* **11,** 1775–1795.

Gao, B.-C., and Goetz, A. F. H. (1990). Column atmospheric water vapor and vegetation liquid water retrievals from airborne imaging spectrometer data. *J. Geosphys. Res.* **95**(D4), 3549–3564.

Gillespie, A. R., Smith, M. O., Adams, J. B., Willis, S. C., Fischer, A. F., III, and Sabol, D. E. (1990a). Interpretation of residual images: Spectral mixture analysis of AVIRIS images, Owens Valley, California. *In* "Proc., Airborne Sci. Workshop: AVIRIS." Jet Propulsion Laboratory, Pasadena California.

Gillespie, A. R., Smith, M. O., Adams, J. B., and Willis, S. C. (1990b). Spectral mixture analysis of multispectral thermal infrared images. *In* "Proc., Airborne Sci. Workshop: TIMS." Jet Propulsion Laboratory, Pasadena, California.

Goel, N. S. (1988). Models of vegetation canopy reflectance and their use in estimation of biophysical parameters from reflectance data. *Remote Sens. Rev.* **4,** 1–212.

Goel, N. S. (1989). Inversion of canopy reflectance models for estimation of biophysical parameters from reflectance data. *In* "Theory and Applications of Optical Remote Sensing" (G. Asrar, ed.), pp. 205–251. Wiley, New York.

Goetz, A. F. H., Vane, G., Solomon, J. E., and Rock, B. N. (1985). Imaging spectrometry for earth remote sensing. *Science* **228,** 1147–1153.

Goetz, A. F. H., Gao, B.-C., Wessman, C. A., and Bowman, W. D. (1990). Estimation of biochemical constituents from fresh, green leaves by spectrum matching techniques. *Proc. Int. Geosci. Remote Sens. Symp.* **2,** 971–974.

Horn, H. H. (1971). "The Adaptive Geometry of Trees." Princeton University Press, Princeton, New Jersey.

Jackson, R. D. (1983). Spectral indices in n-space. *Remote Sens. Environ.* **13,** 409–421.

Jackson, R. D. (1986). Remote sensing of biotic and abiotic plant stress. *Ann. Rev. Phytopathol.* **24,** 265–287.

Li, X., and Strahler, A. H. (1985). Geometric-optical modeling of a conifer forest canopy. *IEEE Trans. Geosci. Remote Sens.* **GE-23,** 508–514.

Li, X., and Strahler, A. H. (1986). Geometric-optical bidirectional reflectance modeling of a conifer forest canopy. *IEEE Trans. Geosci. Remote Sens.* **GE-24,** 906–919.

Li, X., and Strahler, A. H. (1988). Modeling the gap probability of a discontinuous vegetation canopy. *IEEE Trans. Geosci. Remote Sens.* **26,** 161–170.

Marten, G. C., Shenk, J. S., Barton, F. E., II (eds.) (1989). Near infrared reflectance spectroscopy (NIRS): Analysis of forage quality. USDA Res. Ser. Handbook #643.

McDonald, K. C., Dobson, M. C., and Ulaby, F. T. (1990). Using MIMICS to model L-band multiangle and multitemporal backscatter from a walnut orchard. *IEEE Trans. Geosci. Remote Sens.* **28,** 477–491.

Meetenmeyer, V. (1978). Macroclimate and lignin control of litter decomposition rates. *Ecology* **59,** 465–472.

Melillo, J. M., Aber, J. D., and Muratore, J. F. (1982). Nitrogen and lignin control of hardwood leaf litter decomposition dynamics. *Ecology* **63,** 621–626.

Mousset-Jones, P. F. (ed.) (1980). "Geostatistics." McGraw-Hill, New York.

Nehmeth, J. C. (1971). "Dry Matter Production in Young Loblolly (*Pinus tueda* L.) and Slash (*Pinus eliotti* Englm.) plantations. Ph.D. Thesis, North Carolina State University, Raleigh, North Carolina.

Peet, P. K. (1981). Changes in biomass and production during secondary forest succession. *In* "Forest Succession: Concepts and Application" (D. C. West, H. H. Shugart, and D. B. Botkin, eds.), pp. 324–338. Springer-Verlag, New York.

Peet, P. K., and Christensen, N. L. (1987). Competition and tree death. *BioScience* **37,** 586–595.

Rasool, S. I. (1987). Potential of remote sensing for the study of global change. *Adv. Space Res.* **7,** 1–97.

Roberts, D. A., Smith, M. O., Adams, J. B., Sabol, D. E., Gillespie, A. R., and Willis, S. C. (1990). Isolating woody plant material and senescent vegetation from green vegetation in AVIRIS data. *In* "Proc., Airborne Sci. Workshop," pp, 42–57. Jet Propulsion Laboratory, Pasadena, California.

Ross, J. K. (1981). "The Radiation Regime and Architecture of Plant Stands." W. Junk, The Hague.

Ross, J. K., and Myneni, R. B. (eds.) (1990). "Photon-Vegetation Interactions: Applications in Optical Remote Sensing and Plant Ecology." Springer-Verlag, New York.

Roughgarden, J., Running, S., and Matson, P. (1991). What does remote sensing do for ecology. *Ecology* **72,** 1918–1922.

Sabins, F. F., Jr. (1987). "Remote Sensing: Principles and Interpretation." Freeman, New York.

Sabol, D. E., Jr., Adams, J. B., and Smith, M. O. (1990). Predicting the spectral detectability of surface materials using spectral mixture analysis. *Proc. IGARSS* **90,** 967–970.

Sellers, P. J. (1985). Canopy reflectance, photosynthesis and transpiration. *Int. J. Remote Sens.* **6,** 1335–1372.

Sellers, P. J. (1987). Canopy reflectance, photosynthesis, and transpiration. II. The role of biophysics in the linearity of their interdependence. *Remote Sens. Environ.* **21,** 143–183.

Smith, M. O., Ustin, S. L., Adams, J. B., and Gillespie, A. R. (1990a). Vegetation in deserts. I. A regional measure of abundances from multispectral images. *Remote Sens. Environ.* **29,** 1–26.

Smith, M. O., Ustin, S. L., Adams, J. B., and Gillespie, A. R. (1990b). Vegetation in deserts. II. Environmental influences on regional abundance. *Remote Sens. Environ.* **29,** 27–52.

Sun, G. Q., and Simonett, D. S. (1988). Simulation of L-band HH Microwave Backscattering from Coniferous Forest Stands: A Comparison with SIR-B Data. *Photogramm. Eng. Remote Sens.* **54,** 1195–1201.

Tilman, D. (1988). "Plant Strategies and the Dynamics and Structure of Plant Communities." Princeton University Press, Princeton, New Jersey.

Tucker, C. J. (1977). Spectral estimation of grass canopy variables. *Remote Sens. Environ.* **6,** 11–26.

Tucker, C. J. (1979). Red and photographic infrared linear combinations for monitoring vegetation. *Remote Sens. Environ.* **8,** 127–150.

Tucker, C. J. (1980). Remote sensing of leaf water content in the near-infrared. *Remote Sens. Environ.* **10,** 23–32.

Tucker, C. J., and Sellers, P. J. (1986). Satellite remote sensing of primary production. *Int. J. Remote Sens.* **7,** 1395–1416.

Ulaby, F. T., and Elachi, C. (eds.) (1990). "Radar Polarimetry for Geoscience Applications." Artech House, Norwood, Pennsylvania.

Ulaby, F. T., Sarabandi, K., McDonald, K., Whitt, M., and Dobson, M. C. (1990). Michigan Microwave Canopy Scattering Model (MIMICS). *Int. J. Remote Sens.* **11,** 1223–1253.

Ustin, S. L., Adams, J. B., Elvidge, C. D., Rejmanek, M., Rock, B. N., Smith, M. O., Thomas, R. W., and Woodward, R. A (1986). Thematic mapper studies of semiarid shrub communities. *BioScience* **36,** 446–452.

Wessman, C. A., Aber, J. D., Peterson, D. L., and Melillo, J. M. (1988). Remote sensing of canopy chemistry and nitrogen cycling in temperate forest ecosystems. *Nature* **335,** 154–156.

Wessman, C. A. (1990). Evaluation of canopy biochemistry. *In* "Remote Sensing of Biosphere Functioning" (R. J. Hobbs and H. A. Mooney, eds.), pp. 135–156. Springer-Verlag, New York.

Weyer, L. G. (1985). Near-infrared spectroscopy of organic substances. *Appl. Spect. Rev.* **21,** 1–43.

Wickland, D. E. (1991). Mission to planet earth and global ecology. *Ecology* **72,** 1923–1933.

Woodcock, C. E., and Strahler, A. H. (1987). The factor of scale in remote sensing. *Remote Sens. Environ.* **21,** 311–332.

20

New Technologies for
Physiological Ecology

David S. Schimel

I. Introduction

The role of theory and technology in science is often synergistic, that is, new theory may require new types of measurements and new measurement technologies may allow the development of new or the rejection of old theory. In the study of physiological ecology on larger scales (or over longer intervals), theoretical understanding and new technologies are progressing hand-in-hand with the result of significant progress in areas that, in some cases, were not even prominent on the physiological research agenda until a few years ago. A key example is the role of vegetation in modifying regional and global climate through the surface energy budget. In this discussion, I will try to emphasize aspects of new technologies as they fit into emerging new understanding.

II. Discussion

A. Flux Measurements

The measurement of fluxes of water, momentum, heat, CO_2, and other gases using eddy correlation and other micrometeorological techniques is an important new methodology in a number of disciplines, notably atmospheric chemistry, micrometeorology, and physiological ecology. Several reviews describe the technique in detail (Fowler and Duyzer, 1989). In short, simultaneous measurements are made of vertical wind (a vector quantity) and concentration of the relevant scalar quantity

([CO_2], [H_2O], temperature); the flux is computed using the covariance of the two quantities. Eddy correlation techniques and instrumentation have developed sufficiently over the past two decades to be in common use now.

Eddy correlation measurements give physiological ecologists access to a direct measurement of canopy gas exchange, a perennial issue in the field. However, a number of issues arise in the application of flux measurement technology. First, the technology is still difficult and a major investment of effort is required to obtain, install, and operate the equipment. Considerable expertise and experience are required to ensure data quality. The cost of the equipment, although dropping, is high. A full installation capable of measuring heat, momentun, and water fluxes cost as much as $50,000 in 1990. Few off-the-shelf instruments exist for gases such as CO_2, O_3, or CH_4, although all these species may be measured with eddy correlation.

The physiological ecologist interested in whole-canopy exchange of CO_2, may have gotten a technique capable of more than he or she desired. Eddy correlation for CO_2 measures the whole flux: photosynthesis and plant and soil respiration. The measurement is sometimes referred to as "instantaneous net ecosystem production." Therefore, although eddy correlation seems to be the solution to the perennial leaf-to-canopy problem (Jarvis and McNaughton, 1986), it in fact creates many new problems since soil respiration and photosynthesis are controlled by different processes, often studied by different research communities. New and more complex models are needed to explain eddy correlation CO_2 fluxes, incorporating not only canopy photosynthesis but also root dynamics and soil carbon turnover! The use of eddy correlation measurements in studies of gas exchange forces plant physiologists to work with biogeochemistry, in which case new technology has led to new intellectual linkage.

Several issues persist in the application of eddy correlation measurements. Terrain is the first. Although it is said that micrometeorologists are like Dorothy (they always want to be back in Kansas), studies show that earlier assumptions about slope and fetch may be relaxed. Slopes up to 20% are acceptable, if lee slopes are avoided. "Footprints," the area sampled by the eddy correlation system, may be substantially smaller then early estimates, allowing fairly short upwind fetches, especially when short towers are employed or when measurements are made over rough canopies. Models for footprint analysis are improving and aid in choosing experimental sites and interpreting results. The upshot of the relaxation of terrain and canopy requirements for flux measurements is that the technique can be applied over a wide range of natural systems to obtain critical data.

Next, sampling with eddy correlation remains a problem. Although eddy correlation integrates fluxes over the troublesome 1- to 100-m scale of chamber-to-chamber variability, the high cost and difficult logistics involved in making flux measurements using eddy correlation imply that the number of environments in which measurements can be made in a single study is typically small, often only one. In contrast, large numbers of chambers can be deployed over environmental gradients with relative ease to analyze variability associated with soils, hydrology, chemistry, or vegetation (despite difficulties of fine-grained variability). As a demonstration of this problem, the FIFE experiment (Chapter 3) deployed 22 flux measurement systems of the eddy correlation and Bowen ratio types in a 15-km area, yet convincing evidence exists that the area remained undersampled (Chapter 3).

Eddy correlation and other micrometeorological methods are ideal for several types of intensive investigations. For example, once a site is selected that is representative of an area of interest, as established by preliminary chamber sampling or through similarity of vegetation or soils, the eddy correlation fluxes are generally better estimates of the true flux from that area because the eddy correlation fluxes do integrate over the difficult scale of core-to-core variability (Schimel *et al.*, 1988) and because eddy correlation eliminates artifacts and bias resulting from use of chambers (Mosier, 1989). Eddy correlation and other micrometeorological techniques are also of great value in a modeling context, especially for fluxes controlled directly or indirectly by photosynthetically active radiation or the surface energy balance. Emission (H_2O, NH_3) and deposition (O_3 CO_2) fluxes of many species are controlled by the whole plant canopy responding to incoming radiation or soil moisture and modulated by the biological state of the canopy. A number of models have been developed based on physical theory and leaf physiology to describe whole-canopy response; these models are best validated with canopy flux measurements. Eddy correlation measurements fit well into a sequence of modeling, measurement, and validation or modification of the initial model. These models, describing whole-canopy or whole-system response to seasonal and diurnal forcing, are a critical link between biological and atmospheric science.

B. Isotope Techniques

Although flux measurements provide a new view of instantaneous fluxes between ecosystems and the atmosphere, the natural fractionation of isotopes provides completely different insights into biology and biophysics. Isotopes provide an integrated view of the long-term behavior of plants and other organisms, ranging from the seasonal integration of water relations derived from $\delta^{13}C$ in leaves to the millenial view provided by carbon, oxygen, and hydrogen isotopes in ice cores (Oescher and Langway, 1989). Organic carbon in soils and paleosols records important information about its carbon source (plants) over time, as do soil carbonates.

Isotopes can integrate in both time and space, as is demonstrated by the use of carbon isotopes in the identification of sources and source regions in the global carbon budget (Chapter 11). Although the individual sources and sinks involved in exchange of CO_2 with the atmosphere are too small and too numerous to enumerate directly with precision using conventional techniques, the analysis of ^{13}C and ^{14}C in CO_2 provides direct information. Ambiguity remains after isotope analysis due to inadequate knowledge of source or sink signatures, lack of unique signatures, or inadequate sampling, but the isotope budget and its distribution in space provides a crucial additional constraint on source and sink amounts.

Carbon isotopes have been valuable in reducing uncertainty in the global budget of CO_2. However, they have been much less successful for CH_4 for a number of reasons. First, methane is destroyed in the atmosphere by photochemical reaction with associated fractionation. This term must be estimated and is imperfectly known (Cantrell *et al.*, 1990). Second, biological oxidation of methane in soils, waters, and sediments prior to and after emission from the atmosphere causes further fractionation (S. Tyler, personal communication). Third, in part because of oxidation, but also because of sparse data and other uncertainties, the source signatures for emitted methane are more variable and less well characterized than equivalent values for CO_2. The upshot of all this information is that it is very unclear whether the carbon isotopes in methane have contributed much to reduce uncertainty. Several developments will be helpful: (1) the value of the fractionation coefficient for the photochemical destruction of CH_4 is becoming better known, (2) many more measurements of source terms are now available, (3) the biology and isotope geochemistry of biological methane consumption is becoming better known, and (4) the physiology of production, transport (in plants with aerenchyma), and consumption of methane is becoming better known. If CO_2 is any guide, the better the understanding of the biology, the more powerful the applicaiton of isotope techniques will become.

A tremendous number of biological and geological materials contain isotopic records. These materials include the atmosphere itself, preserved atmosphere in ice cores, surface and groundwater, leaf water, soil organic and inorganic carbon, carbonates, coral, corn kernels preserved in archeological settings and by contemporary agricultural archives, tree rings, live plant tissue, soil gases, and live and preserved consumer organisms. These materials record information over a wide range of time and space scales, from the near microscopic to the global, and from daily or less to millennial. Sampling strategy is an essential but often neglected aspect of collecting material for isotopic analysis. Two examples illustrate this problem. First, macroclimate is modulated by local landscape factors—topography, soils, and vegetation. Comparative analysis of isotopic records to deduce macroclimate must take into account the modulating local effects on large-scale patterns. Any individual sample of plant, soil, or other material will reflect large-scale influences through the filter of the microenvironment. Landscape variability and microhabitat should be considered when designing sampling strategies for isotopic materials. The local filter is even a factor in very long records such as ice cores where both the regional climate (affecting isotopic temperature records) and the firnification (snow to ice) process affect the isotopic signature preserved. A second example is variability in the vertical zone. Gradients of humidity within the canopy will affect records of water relationships, depending on sampling height. The isotopic composition of CO_2 within a canopy can vary also, because of fractionation during photosynthesis and the presence of the soil source at the bottom of the canopy. Again, a proper sampling strategy is a prerequisite to rigorous interpretation. There are still too few mass spectrometers, and sample preparation is too lengthy, for the number of samples in a typical isotope investigation to approach the sample sizes frequently encountered in studies using other technologies. More instruments and more automated or streamlined sample preparation methods are required.

C. Remote Sensing

Remote sensing is a tool that has become of interest to physiological ecologists over the past decade or so. Remote sensing has the ability to measure biophysical attributes of the plant canopy, or surrogates directly. In addition, some work indicates that some biochemical attributes of the plant canopy may be measured remotely.

Remote sensing occupies an interesting position in the time and space scales it addresses. Remote observations are snapshots in time, often available only during limited times of the day; they, however, cover very large areas of land. Traditionally, ecological studies have addressed instantaneous behavior of small areas; large-area studies have tended to emphasize the distribution of slowly changing attributes of ecosystems. Certainly, remote sensing studies emphasizing vegetation classification are consistent with this traditional ecological division of the time–space domain. However, direct biophysical retrievals have been emphasized in remote sensing studies and their time series in flux studies (Sellers, 1985). These more process-oriented remote sensing studies have emphasized thermal information and the normalized difference vegetation index (NDVI), now thought to be an indicator of photosynthetic light absorption (Sellers, 1985). Numerous empirical studies show strong relationships between the NDVI and gas exchange (Desjardins *et al.*, 1990; Sellers *et al.*, 1990). The time integral of NDVI is thought to be a reasonable predictor of net primary productivity.

The use of satellite time series data is very helpful in applying remote sensing to physiological process studies on large scales, but raises some difficult technical issues as well. The first difficulty is atmospheric correction. The effects of differential absorption of light in the atmosphere must be accounted for to ensure that changes in radiances measured by a spaceborne sensor over time are due to changes in the surface of the Earth and not in the air column. The second problem is biodirectional reflectance. The spectral response of a plant canopy is a complex function of the angle from which the sensor views the scene and the angle of the sun. These factors can be accounted for but require sophisticated models. The third difficulty is phenology, which influences those properties of the canopy that influence its reflectance, for example, leaf angle distribution, layering, optical depth function, and leaf optical properties. Therefore, simple instantaneous relationships between canopy attributes and spectral reflectance may not hold up over a time series.

Most studies of physiological or biochemical properties of canopies have begun with empirical observations and correlations, as with the vegetation index, or were extrapolated from laboratory studies (Wessman *et al.*, 1988). Models that begin with physical and chemical properties of vegetation and use radiative transfer calculations to predict top-of-the-canopy radiances would be a valuable tool for sensitivity analysis prior to initiating field studies. Currently, no such models exist, although the theory of radiative transfer in canopies is approaching the level of sophistication at which production of such models is possible.

Finally, several problems still hamper the widespread use of remote sensing in ecology. First, adequate computing power is now within reach of most investigators but extant software addresses few of the analysis problems encountered in physiological applications of remote sensing. It is rare that adequate data are available for rigorous atmospheric correction of time series data. The analytical and technological aspects of remote sensing are specialized sufficiently that most first-rate remote sensing-based research is conducted by teams, often working in a group situation. This imposes obvious constraints that are financial, logistical, and often cultural in nature. There is some truth to the notion that much very unexciting groundwork must be laid before remote sensing can be used to support really exciting science. This situation is one of positive feedback because people will not enter the field until the payoff is evident, yet the field will not mature to that point without effort. Only in the past few years, with NASA support of projects such as FIFE and the Earth Observing System, has the number of ecologists using remote sensing approached such critical mass.

References

Cantrell, C. A., Shetter, R. E., McDaniel, A. J., Calvert, J. G., Davidson, J. A., Lowe, D. C., Tyler, S. C. Cicerone, R. J., and Greenberg, J. P. (1991). Carbon kinetic isotope effect in the oxidation of methane by the hydroxyl radical. *J. Geophys. Res.* **95** 22,455–22,462.

Desjardins, R. L., Schuepp, P. H., and MacPherson, J. J. (1990). Saptial and temporal variations of CO_2 sensible and latent heat fluxes over the FIFE site. *In* "Symposium on the First ISLSCP Field Experiment" (F. G. Hall and P. J. Sellers, convenors), pp. 46–48. American Meterological Society, Boston.

Fowler, D., and Duyzer, J. H. (1989). Micrometerological techniques for the measurement of trace gas exchange. *In* "Exchange of Trace Gases between Terrestrial Ecosystems and the Atmosphere" (M. O. Andreae and D. S. Schimel, eds.), pp. 189–208. Wiley, Berlin.

Jarvis, P. G., and McNaughton, K. G. (1986). Stomatal control of transpiration: Scaling from leaf to region. *Adv. Ecol. Res.* **15**, 1–49.

Mosier, A. R. (1989). Chamber and isotope techniques. *In* "Exchange of Trace Gases between Terrestrial Ecosystems and the Atmosphere" (M. O. Andreae and D. S. Schimel, eds.), pp. 175–188. Wiley, Berlin.

Oescher, H., and Langway C. C. (eds.) (1989). "The Environmental Record in Glaciers and Ice Sheets." Wiley, Berlin.

Sellers, P. J. (1985). Canopy reflectance, photosynthesis and transpiration. *In. J. Remote Sensing* **6**, 1335–1372.

Sellers, P. J., Heiser, M., and Walthall, C. W. (1990). A comparision of surface biophysical properties and remotely sensed variables from FIFE. *In* "Symposium on the First ISLSCP Field Experiment" (F. G. Hall and P. J. Sellers, convenors), pp. 117–120. American Meterological Society, Boston.

Schimel, D. S., Simkins, S., Rosswall, T. H., Mosier, A. R., and Parton, W. J. (1988). Scale and the measurement of nitrogen trace gas fluxes from terrestrial ecosystems. *In* "Scales and Global Change" (T. H. Rosswall, R. G. Woodmansee, and P. G. Risser, eds.), pp. 179–193. Wiley, New York.

Wessman, C. A., Aber, J. D., Peterson, D. L., and Melillo, J. M. (1988). Remote sensing of canopy chemistry and nitrogen cycling in temperate forest ecosystems. *Nature* **335**, 154–156.

Subject Index

grouping plants and, 314
growth forms and, 304
leaf-to-canopy scaling and, 45, 48–49, 51
  photosynthesis, 63–73
new technologies and, 359, 361
population structure and, 264
prospects for scaling and, 227
remote sensing and, 340, 344
water vapor and carbon dioxide exchange
  and, 80, 87, 90
Energy balance
canopy, spatial information and, 22
water vapor and carbon dioxide exchange
  and, 80, 90
Energy exchange
growth forms and, 296–300, 307
remote sensing and, 348
Environment
biological systems and, 233–234
  individual plants, 234–236
  models, 241
  natural ecosystems, 243–244
  patterns, 238, 240–241
  simplicity of models, 245–248, 250
bottom-up models and, 115–117
ecophysiologists and, 159–161, 163
ecophysiology and, 1–2, 128–129
forest ecosystem model and, 151
functional units in ecology and, 231
global carbon balance and, 199, 208, 211,
  214–215, 217
global carbon cycle and, 179
global dynamics and, 170–172
grouping plants and, 314
growth forms and, 287, 306–308
  ecological controls, 292–294
  ecosystem, 295–296, 300–301
  feedback, 303–304
  physiology, 288–289, 291
leaf-to-canopy scaling and, 41, 43–44,
  47–48
  photosynthesis, 53, 63–73
leaf to ecosystem level integration and,
  39–40
local level concepts of scale and, 8,
  12–13, 17
population structure and, 256–257,
  259–260, 272
prospects for scaling and, 224
remote sensing and, 341–342, 344
stable isotopes and, 324, 329, 331

water vapor and carbon dioxide exchange
  and, 78, 81, 84, 93, 106
Environmental heterogeneity, biological
  systems and, 246–247
Enzymes
ecophysiologists and, 163
stable isotopes and, 325
water vapor and carbon dioxide exchange
  and, 81
Equatorial zone, global carbon balance and,
  208–209, 212–213
Erodium, biological systems and, 238
Establishment, local level concepts of scale
  and, 18
Eulerian models
local level concepts of scale and, 16
water vapor and carbon dioxide exchange
  and, 82–84
Evaporation
canopy, spatial information and, 32
forest ecosystem model and, 147, 150
growth forms and, 296
stable isotopes and, 325
water vapor and carbon dioxide exchange
  and, 78, 80, 100, 106–107
  information, 91
  leaf-to canopy scaling, 82
Evapotranspiration
bottom-up models and, 121
canopy, spatial information and, 34
ecophysiology and, 138
forest ecosystem model and, 141, 145
growth forms and, 296–298
leaf-to-canopy scaling and, 49
prospects for scaling and, 223, 225, 227
Evolution
biological systems and, 246, 251
bottom-up models and, 117, 123
canopy, spatial information and, 34
ecophysiology and, 2
grouping plants and, 318
local level concepts of scale and, 7–8
prospects for scaling and, 226
water vapor and carbon dioxide exchange
  and, 78, 91
Extinction coefficient, leaf-to-canopy
  scaling and, 62–65

Feedback
bottom-up models and, 118, 123–124
ecophysiology and, 129, 137–138

Physiological Ecology
A Series of Monographs, Texts, and Treatises

Continued from page ii

DATE DUE

2198836			

DEMCO 38-297